4th International Seminar on Ocean and Coastal Engineering, Environmental and Natural Disaster Management (ISOCEEN 2016)

IOP Conference Series: Earth and Environmental Science
Volume 135

Surabaya, Indonesia
13 December 2016

ISBN: 978-1-5108-6059-9
ISSN: 1755-1307

Printed from e-media with permission by:

Curran Associates, Inc.
57 Morehouse Lane
Red Hook, NY 12571

Some format issues inherent in the e-media version may also appear in this print version.

Printed by Curran Associates, Inc. (2018)

For permission requests, please contact the Institute of Physics
at the address below.

Institute of Physics
Dirac House, Temple Back
Bristol BS1 6BE UK

Phone: 44 1 17 929 7481
Fax: 44 1 17 920 0979

techtracking@iop.org

Additional copies of this publication are available from:

Curran Associates, Inc.
57 Morehouse Lane
Red Hook, NY 12571 USA
Phone: 845-758-0400
Fax: 845-758-2633
Email: curran@proceedings.com
Web: www.proceedings.com

Table of contents

Volume 135

4th International Seminar on Ocean and Coastal Engineering, Environmental and Natural Disaster Management (ISOCEEN)

13 December 2016, Surabaya, Indonesia

Accepted papers received: 9 March 2018
Published online: 28 March 2018

Preface

4th International Seminar on Ocean and Coastal Engineering, Environmental and Natural Disaster Management (ISOCEEN)

Peer review statement

Papers

Indonesian Sea Accident Analysis (Case Study From 2003 – 2013)

Y Arya Dewanto and D Faturachman.....1

Efficacy of Water Chamber Type Seawall to Dissipate Incident Wave and Its Performance to Extract Wave Power

Firman Husain, M Rusdi Alwi and Ashury.....9

Effects of stern-foil submerged elevation on the lift and drag of a hydrofoil craft

K Suastika and Apriansyah.....18

The Assessment of the Ultimate Hull Girder Strength of RO-RO Ship after Damages

Muhammad Zubair Muis Alie, Ganding Sitepu and Samuel Izaak Latumahin.....26

The Influences of Stirring and Cow Manure Added on Biogas Production From Vegetable Waste Using Anaerobic Digester

N O Abdullah and E S Pandebesie.....34

The Evaluation of Cone Capsule as an Alternative Hull form for Portable Tsunami Lifeboat to Support Evacuation System in the Coastal Regions and Small Islands

Ahmad Fauzan Zakki, Suharto and Aulia Windyandari.....41

The Study on Stability and Seakeeping Characteristics of the Glass Bottom Boat Trimaran in Karimunjawa Island

Berlian Arswendo Adietya, Aulia Windyandari and Ahmad Fauzan Zakki.....49

Analysis Of Resistance And Effective Wake Friction Due To Addition Of Stern Tunnels On Passenger Ship Using Cfd

D Chrismianto, Tuswan and P Manik.....57

Study of Water Quality Changes due to Offshore Dike Development Plan at Semarang Bay

M Wibowo and B A Hakim.....65

Evaluating Inundation in Urban Drainage Systems in Tamalanrea District Makassar Based Ecodrainase

Rizky Alfidhdha and Nieke Karnaningroem.....75

The Discrepancy between The Programs and Disaster Management Policy in Klapanunggal District, Bogor, West Java

D A Puspito Sari, I Listiyowati, T Nefianto and Lasmono.....82

Determining Domestic Container Shipping as an Enforcement of Indonesian International Hub Port

H I Nur, S D Lazuardi, F Hadi and M Hapis.....90

Freight Calculation Model: A Case Study of Coal Distribution

I T Yunianto, S D Lazuardi and F Hadi.....98

Fast Missile Boat Project Planning using CPM and What If Analysis Method

Silvianita, R Firmansyah, D M. Rosyid, Suntoyo and D M Chamelia.....104

Analysis Of Inland Waterway Transport For Container Shipping: Cikarang To Port Of TanjungPriok

T Achmadi, H I Nur and L R Rahmadhon.....110

Seasonal-Temporal Variability of Sea Surface Temperature Derived from Satellite and In Situ Data: Case Study of Indian Ocean Western Part of Sumatera Island

A Wahyudi, S B Susilo and J P Panjaitan.....117

Application Side Casing on Open Deck RoRo to Improve Ship Stability

Hasanudin, I K.A.P Utama and Jeng-Horng Chen.....124

Physicochemical Characteristic of Municipal Wastewater in Tropical Area: Case Study of Surabaya City, Indonesia

I M W Wijaya and E S Soedjono.....132

Boundaries Delineation of Marine Management Sharing According to Local Government Law No. 23/2014 (Case Study: Surabaya, Sidoarjo, Bangkalan and Sampang)

Khomsin and S Intan Ary Prayogi.....138

Potential of Fishing Port Development in the East Java

N Rosana and V D Prasita.....144

Longitudinal Pipeline Scour Propagation Induced by Wave-Current Interaction For the South Sumatra-West Java Submarine Pipeline

Suntoyo, B. Perkasa, T.J. Atikasari and A. Wisudawan.....149

Mooring Design Selection of Aquaculture Cage for Indonesian Ocean

Y Mulyadi, N Syahroni, K Sambodho, M Zikra, Wahyudi and H B P Adia.....156

The Usage of Geographical Information System in the Selection of Floating Cages Location for Aquaculture at Prigi Bay, Trenggalek Regency, East Java

H D Armono, B G Mahaputra and M Zikra.....168

Hydrodynamic and Sediment Transport Modelling of Suralaya Coastal Area, Cilegon, Indonesia

A. H. Fattah, Suntoyo, H. A. Damerianne and Wahyudi.....179

Experimental Study on Interference Effects of Two Tandem Cylinders Wrapped Around by Triple Helical Rods with Gap on Induced Drag

R.W. Prastianto, K. H. Dwipayana, N. Syahroni and B. Pumbarino.....188

Project Scheduling Based on Risk of Gas Transmission Pipe

Silvianita, A Nurbaity, Y Mulyadi, Suntoyo and D M Chamelia.....194

Numerical Analysis of Residual Stress and Distortion Use Finite Element Method on Inner Bottom Construction of Geomarin IV Survey Ship with Welding Sequence Variations

N Syahroni, A B W Hartono and M Murtedjo.....203

IOP Conf. Series: Earth and Environmental Science **135** (2018) 011001 doi:10.1088/1755-1315/135/1/011001

The 4th International Seminar on Ocean and Coastal Engineering, Environmental and Natural Disaster Management (**ISOCEEN) 2016**

Preface

International Seminar on Ocean and Coastal Engineering, Environmental and Natural Disaster Management (ISOCEEN) is an annual internasional conference event held by cooperation among Institut Teknologi Sepuluh Nopember (ITS) especially Department of Ocean Engineering, Tohoku University, Japan, HZ University of Applied Sciences, Netherlands and The Coral Triangle Initiative on Coral Reefs, Fisheries and Food Security (CTI-CFF). This forum was more focused on creating and expanding professional network to foster the relationship between the University, industry, business and communities across the country. The Organizing Committee worked many years for preparing of this seminar and their efforts resulted to full success. Based on the cooperation, there are some conference that we already held before, as follows :

- Japan – Indonesia Workshop on Estuary and Climate Change (JIWECC), August 8th -10th Surabaya, Indonesia
- The 1st International Seminar on Ocean and Coastal Engineering, Environmental and Natural Disaster Management (ISOCEEN), November 11, 2013,. Surabaya, Indonesia.
- The 2nd International Seminar on Ocean and Coastal Engineering, Environmental and Natural Disaster Management (ISOCEEN), December 2, 2014, Surabaya, Indonesia.
- The 3rd International Seminar on Ocean and Coastal Engineering, Environmental and Natural Disaster Management (ISOCEEN), December 10, 2015, Surabaya, Indonesia.
- The 4th International Seminar on Ocean and Coastal Engineering, Environmental and Natural Disaster Management (ISOCEEN), December 13, 2016, Surabaya, Indonesia.

ISOCEEN 2016 was aimed on providing a discussion, for exchanging of knowledge, researches and of recent solutions for many researchers and experts in the field of Ocean, Offshore, Coastal engineering, Environmental and Disaster Management. The results of discussions are presented in this book. We hope this book will be interesting and usefull for many engineers and scientists whose activity related with researches of ocean and development of coastal areas.

Finally, on behalf of the organizing committee of the seminar, sincere appreciation is expressed to all authors who contribute to our seminar. Special thanks are also due to all keynote speakers, invited talks and chairpersons for the efforts in preparing the manuscripts and managing the sessions, respectively.

Suntoyo, PhD
Chairman of Organizing Committee

IOP Conf. Series: Earth and Environmental Science **135** (2018) 011001 doi:10.1088/1755-1315/135/1/011001

Committees

Advisory Committee:

- Rector of ITS Surabaya, Indonesia
- Dean of Marine Technology Faculty - ITS, Indonesia
- Head of Ocean Engineering Department, Faculty of Marine Technology - ITS, Indonesia

Organizing Committee:

- Suntoyo, PhD (Chairman)
- Silvianita, PhD (Secretary)
- Yoyok Setyo Hadiwidodo, PhD
- Dr. Hasan Ikhwani, MSc
- Dr. Wahyudi
- Dr. Eng. Rudi Walujo Prastianto
- Dr. Eng. Muhammad Zikra
- Nur Syahroni, PhD
- Herman Pratikno, PhD
- Agro Wisudawan, MT
- Dirta Marina Chamelia, MT

International Scientific Committee:

- Prof. Hitoshi Tanaka, Dr. Eng. (Tohoku University, Japan)
- Prof. Widi Agoes Pratikto, PhD (ITS, Indonesia)
- Prof. Daniel M. Rosyid, PhD (ITS, Indonesia)
- Prof. Eko Budi Djatmiko, PhD (ITS, Indonesia)
- Prof. Mukhtasor, PhD (ITS, Indonesia)
- Suntoyo, PhD (ITS, Indonesia)
- Makoto Umeda, Dr. Eng. (Tohoku University, Japan)
- Dr. Joost Stronkhorst (HZ University of Applied Sciences, Netherlands)
- Ir. Liliane Geerling R.T.D. (HZ Universityof Applied Sciences, Netherlands)
- Ahmad Sana, PhD (Sultan Qaboos University, Oman)
- Priyantha Ranjan, PhD (Curtin University, Australia)

IOP Conf. Series: Earth and Environmental Science **135** (2018) 011001 doi:10.1088/1755-1315/135/1/011001

- Dr. Eng. Totok Suprijo (ITB, Indonesia)
- Dr. Eng. Purwanto Bekti Santoso (Unsoed, Indonesia)
- Dr. Eng. Sitang Pilailar (Kasetsart University, Thailand)
- Dr. Nguyen Trung Viet (Water Resource University, Vietnam)
- Dr. Taufiqur Rachman (Hasanuddin University, Indonesia

IOP Conf. Series: Earth and Environmental Science **135** (2018) 011002 doi:10.1088/1755-1315/135/1/011002

Peer review statement

All papers published in this volume of *IOP Conference Series: Earth and Environmental Science* have been peer reviewed through processes administered by the proceedings Editors. Reviews were conducted by expert referees to the professional and scientific standards expected of a proceedings journal published by IOP Publishing.

IOP Conf. Series: Earth and Environmental Science **135** (2018) 012001 doi:10.1088/1755-1315/135/1/012001

Indonesian Sea Accident Analysis (Case Study From 2003 – 2013)

Y Arya Dewanto[1] ,D Faturachman[2]

[1]Marine Engineering Department, Faculty of Ocean Technology, Darma Persada University, Indonesia
[2]Marine Engineering Department, Faculty of Ocean Technology, Darma Persada University, Indonesia

Email :aryadewanto@yahoo.com

Abstract. There are so many accidents in sea transportation in Indonesia. Most of the accidents happen because of low concern aspects of the safety and security of the crew. In sailing, a man as transport users to interact with the ship and the surrounding environment (including other ships, cruise lines, ports, and the situation of local conditions). These interactions are sometimes very complex and related to various aspects of. Aware of the multiplicity of aspects related to the third of these factors, seeking the safety of cruise through a reduction in the number of accidents and the risk of death and serious injuries due to accidents and goods transported is certainly not enough attempted through mono-sector approach, but rather takes a multi-sector approach to the efforts. In this paper, we described the Indonesian Sea Transportation accident analysis for eleven years divided into four items: total of ship accident type, ship accident factor, total of casualties, region of ship accidents. All data founded from Marine Court (*Mahkamah Pelayaran*). From that 4 items we can find Indonesia Sea Accident Analysis from 2003-2013.

1. Introduction

Majority (80 to 85%) of all recorded maritime accidents are generally attributed to human error or associated with human error. Contribution of human error to maritime accidents has increased over a ten-year period 1991 to 2001 [2]. Most of the accidents are the result of senseless and avoidable human errors. The concern about human factors is growing as human error is significantly implicated in so many marine accidents. Many of the failures are actually the result of errors (i.e. latent failures) that have been designed and constructed into highly complex systems especially system integration and interfacing. Indonesia as an archipelago country consisting of thousands of Islands and has a vast sea area so indispensable transportation mode is ocean freight as a means of mobility and driving force of economic development nationwide [15]. Even in parts of Indonesia, the ship is one of the means of transport used to get in touch with the outside world. A great potential this is very useful when coupled with the assurance of the safety and security of the marine mode of transportation.

Sea transport holds a very important role in maritime countries, such as in Indonesia whose territory is an archipelago. Associated with sea transport, there are three aspects that are interlinked with each other, i.e. traffic and sea transportation, port and also safety and security covering their cruise [1].

Ship accident that occurred recently in the Indonesian waters can occur at anytime and anywhere. Due to unforeseen circumstances, the various parties associated with industry transportation create a variety of sea conditions are very strict with regard to the safety and security of the ship in accordance with the condition of the waters of Indonesia. International institutions specializing in the maritime areas of the IMO (International Maritime Organization) to make the International Convention SOLAS (Safety of Live at Sea) as well as various derivatives of its implementation in the form of codes such as the ISM (International Safety Management) Code, ISPS (International Port Facilities Security) Code, the IMDG (International Maritime Dangerous Goods) Code [7]. SOLASis a fixed rule that is mandatory for all countries to implement it, it is no exception that come into Indonesia ratified it. Therefore, the Government of Indonesia implement it by making the ACT No. 17 of 2008, who then makes a wide range of Government Regulation (PP) as a technical guide its implementation, as the consummation of legislation and other regulations.

2. Literature Review

The maritime transportation affects most of goods and passenger's movement among islands in an archipelago country, Indonesia. As the result, the reliability of maritime transportation to link among islands will impact to the development of country especially in the frame of economic sector. The maritime transportation could be divided into two categories which are sea transportation and crossing transportation. The crossing transportation bridges between places as the continuation of road or railway by crossing the straits. The ship operation system contains complex interrelation among technical factor, environment factor as well as human factor that control the ship handling process. One important idea for careful consideration is ship handling. Ship handling is defined as the practice of guiding a ship which controls ship movement through ship controllability means, visual monitoring means and instrumental monitoring. Therefore, ship navigation operation and ship manoeuvring are the integrated part of the ship handling. The successful of ship handling will reduce accident from happening. In other word, if there is a problem with ship handling process such as the difficulty of controlling ship behaviour, it might lead to accident. It could be generalized that ship handling difficulty is potentially directed toward move ship accident [8].

Accidents occurring in sea, rivers, lakes, and crossing that reached Marine Court in 2005-2010 was mostly due to human error (65%), and only a few accidents in the waters caused by natural factors [4]. Given the reasons mentioned above, all accidents can be minimized if prevention efforts are seriously performed by all parties so as not to stumble on the same stone. Water transport accidents occur mainly due to overcrowding and navigation system, which is characterized by a large number of passengers and goods compared to the draft Commission. For passengers who do not have the expertise and skills in emergency situations, it is important to note that users of the waterway in the category of vulnerable population groups. Efforts to ensure the safety of passengers and crew must be considered as a serious issue, including this trivial security equipment such as buoys. Current conditions, many ships that do not have safety equipment should be able to buoy passengers and crew when the vessel having accident. Most of the accidents occur due to the low awareness on the aspects of security and safety of the crew. The figures differ from the manifest of passengers and number of passengers on the ground become common place. Transportation is the lifeblood of society and the economy in Indonesia. Transportation development activities in Indonesia are out of various dimensions (marine transport and others) and increasing. This is an impacts of economic activity and socio-cultural activities and community. In addition, the process of regulatory reform in the field of national transportation deregulation has also triggered an increase in transport activity. To understand fully that human consciousness towards the preservation of the environment are increasingly high, so that sea transportation accidents which can cause damage to the environment (pollution) should be a significant consideration. In order to further integrate transport infrastructure and facilities that meet the requirements of security and safety of transport, it is necessary to make a standardization of regulation system and procedures, as well as human resource professionals to realize the service organization of the transport and works in order to hold everything intact. Then it is necessary to have a system of good governance, where Governments have function in the transportation services which include coaching in the aspects of setting up, monitoring and controlling the system [14].

3. Problem Identification

3.1 Characteristic of Sea Transportation Accident

The ship is major means of sea transport, where many people rely on it for sustenance. Each time the safety of human life at sea is threatened, both the sailors and the people on board are attached. From the facts and the data obtained, the sea had swallowed many accident victims and property which is not small in size. Ship accidents can happen anywhere, anytime and to anyone. For that reason, the crew and passengers need to know about ways to escape if there is an accident happening including on board, first aid and fire safety responsibilities. There is urgent need for training of the crew, especially in the areas of safety for the crew members, as well as rescue techniques, as required by the Convention of the IMO (International Maritime Organization) and the State Governments concerned. Many of the accidents victim at sea are caused by a lack of basic security knowledge and protection of the environment, according to the IMO, its numerous deaths which occurred in the sea caused by the human factor.

Characteristics of accidents in general are:
a. an accident as a rare occurrence;
b. as an event that accidents do not know when to expect;
c. accidents as those events Multiple Factors.

3.2 Regulation and Law about Ship Accident

Principles of safety transport is bought to the Government attention for long time, such as the policy, established in 1999 Presidential Decree number 105 in year of 1999 on the establishment of the Komite Nasional Keselamatan Transportasi (KNKT) or National Transportation Safety Committee/NTSC [9].

Government Regulation No. 1 year of 1998 concerning the examination of the ship accident split into five examination categories, namely:

a. Sinking ship;
b. Ship fired;
c. Ship collision;
d. Ship accidents which cause the soul of man and wildlife losses of property and
e. Ship was run a ground or crashed out.

Examination of the ship accident consisted of a preliminary examination by Syahbandar (Harbor Master) and advanced examination by The Marine Court (Mahkamah Pelayaran). Whereas the Legislation of the Republic of Indonesia number 17 year of 2008 about the shipping/ cruise Article 245 states that: Accidents aboard the events experienced by the vessel that may threaten the safety of the ship and/or the human spirit in the form of: sinking ship, ship fired, ship collision, and ship was run aground or crashed out.

Later, in the Article 256 about the investigation of the ship accident stated that:

(1) The Accident Investigation Board conducted by the National Transportation Safety Committee to search for facts in order to prevent the occurrence of the ship accident with the same causes;
(2) Investigation as referred to subsection (1) made against each ship accident;
(3) Investigations conducted by the National Transportation Safety Committee referred to subsection (1) is not to determine which errors or omissions on the occurrence of the ship accident.

To minimize the occurrence of sea accidents, top priority is to rescue the soul in order to satisfy all the rules by the standards and even more to ensure the safety at sea, which requires a support from around the world. There are three organizations that govern the safety of the ship. The IMO (International Maritime Organization), ILO (International Labor Organization) and ITU (International Telecommunication Union), Indonesia is one of the three members of the organization and has ratified the convention. To ensure safety at sea, it is required for a standard (the rules) which applies nationally and internationally, which are:

IOP Conf. Series: Earth and Environmental Science **135** (2018) 012001 doi:10.1088/1755-1315/135/1/012001

a. National Standard includes:
1. The Legislation of the Republic of Indonesia Act No. 17 year of 2008 about a shipping/ cruise that spelled out in the regulation of the Government and the Minister's decision and the regulation;
2. The Legislation of the Republic of Indonesia Act No. 3 year of 1988 replacement Act No. 5 year of 1964 about the Telecommunications that comes with PP No. 10 of 1974 on Public Telecommunications;
3. Government Regulation No. 7 year of 2000 about seamanship that arrange regulates about competence, expertise and skills qualifications for crew and the captain/Master on all ships except sailing ship, sail boat with motor, motor boat with a size less than GT 35, private yacht that was used to trade and specialized ships.

b. International Standard includes:
1. SOLAS (Safety of Life at Sea) 1974 and the amendment [16];
2. MARPOL (Marine Pollution) 73/78 and the protocols [11];
3. Load Line Convention 1966 [10];
4. COLLREG (Collision Regulations) 1972 [3];
5. Tonnage Measurement 1966;
6. STCW (Standards of Training Certificate and Watch keeping) 1978 Amendment 1995;
7. ILO No. 147 Year of 1976 about the Minimum Working Standards for Crew Commerce [6];
8. ILO Convention No. 185 year of 2008 about SID (Seafarers Identification Document) which has been ratified by The Legislation of the Republic of Indonesia law No. 1 year of 2009 [6];
9. SAR (Search & Rescue) Convention [16];
10. GMDSS (Global Maritime Distress Safety System) [5];
11. ISM (International Safety Management) Code [12];
12. ISPS (International Ship and Ports Security) Code [13].

In addition to the conventions mentioned above there is one rule that could not be released from the safety of the voyage which set about Radio Communication close relation to the Radio Regulations (RR), Telegraph and Telephone Regulation under the Convention on the International Telecommunication Union (ITU).

4. Result and Discussion

4.1. Indonesian Ship Accident Data

The accident happened on a river, Lake, and river crossings that are up to the Marine Court over voyage caused by human error, and just a little accident in the waters caused by natural factors. Tracing the reason mentioned above should all the events of the accident can be minimized while there are preventive efforts from all parties so as not to stumble on the same stone. The Ship Accidents Data shown on Table and Figure 1 – 4.

Table 1. Total of Ship Accident According to Marine Court Decision 2003-2013

No	Type of Ship Accident	2003	2004	2005	2006	2007	2008	2009	2010	2011	2012	2013
1	Sunk	16	10	8	12	9	13	11	4	7	16	3
2	Collision	10	11	10	9	4	15	9	2	5	7	8
3	Grounded	5	9	4	5	5	2	3	6	3	3	14
4	Fired	6	3	5	6	9	4	5	5	4	6	5
5	Others	1	3	3	6	5	1	5	2	2	2	3
	Total	38	36	30	38	32	35	33	19	21	34	33

Source: Mahkamah Pelayaran, Setjen, Kemenhub

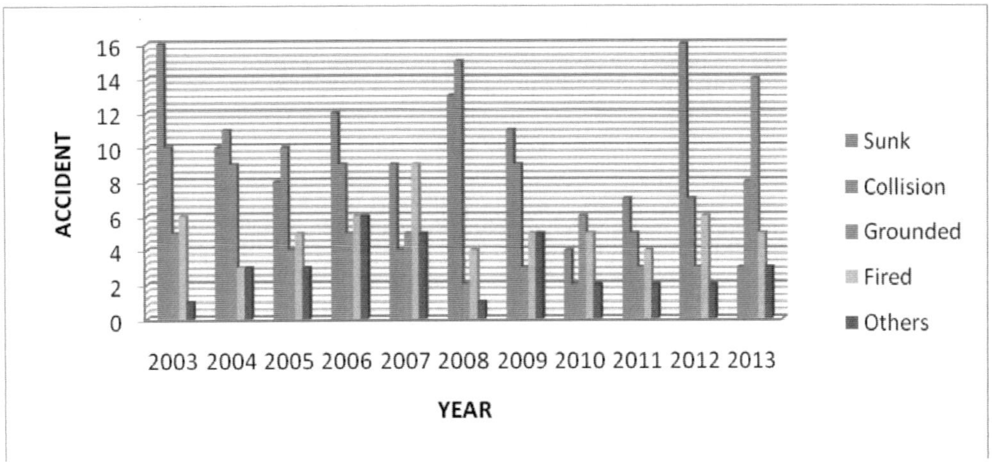

Figure 1. Total of Ship Accident TypesAccording to Marine CourtDecision 2003-2013

Table 2. Total of Marine Court Decision by Factor Ship's Accident 2003-2013

No	Ship's Accident Factor	2003	2004	2005	2006	2007	2008	2009	2010	2011	2012	2013
1	Human Error Factor	16	19	16	20	19	22	20	13	19	17	18
2	Force Majeure Factor	8	11	7	12	8	10	7	1	0	11	7
3	Others Factor	14	6	7	6	5	3	6	5	2	6	8
	Total	38	36	30	38	32	35	33	19	21	34	33

Source: Mahkamah Pelayaran, Setjen, Kemenhub

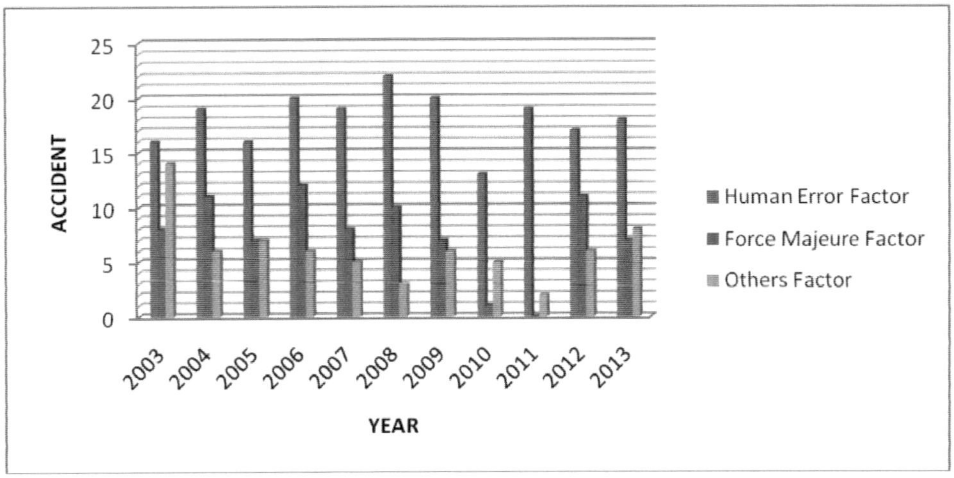

Figure 2. Total of Marine Court Decision by Factor Ship's Accident 2003-2013

IOP Conf. Series: Earth and Environmental Science **135** (2018) 012001 doi:10.1088/1755-1315/135/1/012001

Table 3. Total of Victims According to Marine Court Decision 2003-2013

No	Ship's Accident Victims	2003	2004	2005	2006	2007	2008	2009	2010	2011	2012	2013
1	Injured	0	19	9	0	1	0	15	1	0	14	0
2	Missing	8	15	145	146	363	16	223	0	12	35	1
3	Death	7	59	32	43	136	43	59	44	32	68	8
	Total	15	93	186	189	500	59	297	45	44	117	9

Source: Mahkamah Pelayaran, Setjen, Kemenhub

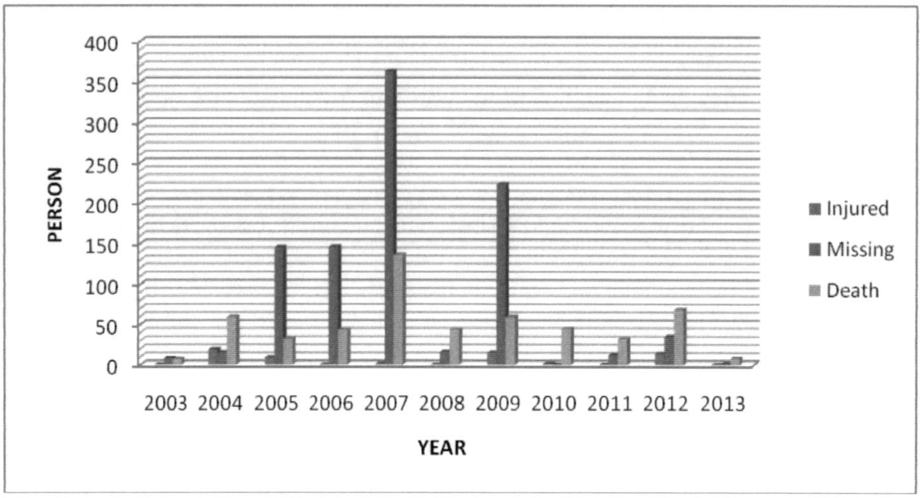

Figure 3. Total of Victims According to Marine Court Decision 2003-2013

Table 4. Total of Marine Court Decision by Ship Area of Accident 2003-2013

No	Ship Area of Accident	2003	2004	2005	2006	2007	2008	2009	2010	2011	2012	2013
1	Western Part Indonesia	26	24	16	20	23	22	22	9	8	20	21
2	Centre Part Indonesia	11	12	11	6	7	13	11	8	10	12	12
3	Easter Part Indonesia	1	0	3	12	2	0	0	2	3	2	0
	Total	38	36	30	38	32	35	33	19	21	34	33

Source: Mahkamah Pelayaran, Setjen, Kemenhub

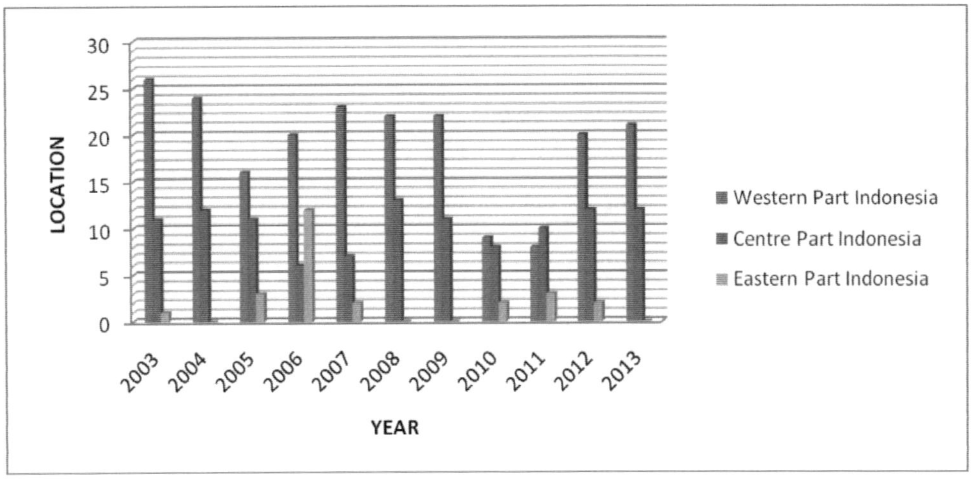

Figure 4. Total of Marine Court Decision by Ship Area of Accident 2003-2013

4.2 Problem Solving

According to research before [4], if it is restricted to the scope of the company (in terms of the micro), it appears that the occurrence of accidents owing to the discrepancy between the three major elements of production (sub human systems, environment and management physic) resulting in the occurrence of an action and the circumstances are not safe.

But directly of the accident can be grouped into two outlinescauses, namely:

a. Unsafe actions of humans (Unsafe Acts) for example:work without any authorize-failed to give a warning,working with the wrong speed and so on.
b. A state of insecurity (Unsafe Condition) for example: thesafety equipment on board which are damaged or are not usable, environmental and weather on the waters is not goodformaliciousitemsthatcanshipexploding/burning.

5. Conclusion

1. From the Marine Court Decision from 2003-2013, after processing the data, by percentage:
   a) Total of Ship Accident: sunk 31%, collision 26%, grounded 17%, fired 17% and others 9%.
   b) Total of Factor Ship's Accident: human error 57%, force majeure 24%, others 19%.
   c) Total of Victims: missing 62%, death 34%, injured 4%.
   d) Total of Ship Area of Accident: western part Indonesia 61%, centre part Indonesia 32%, eastern part Indonesia 7%.
2. The safety of the ship is affected by the ship's supplies, ships, load functions charge and driver skills. In order for the safety of passengers and crew aboard stay awake, then supplies the ship had to be adapted to the standard of safety, the use of the ship as its main function, the burden of the charge does not exceed the limit load required, the driver of a ship actually sailed the ship accomplished and master cruise lines took place.
3. The high number of accidents sea in Indonesia shouldbe paid attention all sides, not only government, but also the owner of a ship related institutions and the citizens who must be more active in providing information. Of the results of observations, the main cause of the accident the sea isbecause the excess of the capacity of transport set, whether itis transportation of goods and people. Do not even rarelyusers shipping services exert oneself aboard a ship thoughthe ship was filled with determination can be placed on board.

Acknowledgments

Authors wishing to acknowledge assistance or encouragement from Darma Persada University, special work by technical staff and students. Also for financial support from Kemenristekdikti(Ministry of Research and Higher Education) Republic of Indonesia special for the Hibah Dosen Pemula (Grant Lecturer Novice) in year of 2016.

References

[1] Abrahamson, B.J. 1980. International Ocean Shipping: Current Concepts and Principles, West View Press, Inc. Boulder, Colorado.
[2] Baker, Clifford C. and Seah, Ah Kuan. 2004. Maritime Accidents and Human Performance: The Statistical Trail, ABS Technical Papers 2004, 225-240.
[3] Collision Regulation, 1972.
[4] Faturachman, Danny, Mustafa Shariman. 2012. Sea Transportation Accident Analysis in Indonesia, Procedia Vol. 40 (2012) 616-621.
[5] Global Maritime Distress Safety System.year 1992.
[6] ILO (International Labor Organization) Convention No. 147 year 1976 & No. 185 year 2008.
[7] IMO, IMO/MSC Circular. 1997. Interim Guidelines for the Application of Formal Safety Assessment to the IMO Rule-making Process. London.
[8] Inoue, K. 1998. Evaluation Method on Ship-handling Difficulty for Navigation Restricted and Congested Waterways, Kobe University of Mercantile Marine, Kobe.
[9] Komisi Nasional Keselamatan Transportasi (National Transportation Safety Committee) 2009, Number of Marine Court Accident by Factor Ship Accident, Jakarta.
[10] LoadLine Convention. year 1966.
[11] Marine Pollution. year 1973/1978.
[12] International Safety Management Code. Year 1993.
[13] International Ship and Port facility Security Code. Year 2003.
[14] P.T. Trans Asia Konsultan. 2009. Laporan Analisis Tren Kecelakaan Laut, Jakarta.
[15] Pomeroy, R.V. and Tomlinson, C. M. 2000. A System Approach to Integrating the Human Element into Marine Engineering Systems, Proceeding of the Conference on Human Factors in Ship Design and Operation,Royal Institution of Naval Architects, September.
[16] Search and Rescue on Maritime Convention. year 1979.
[17] The Safety of Live At Sea (SOLAS) Convention. 1974.

IOP Conf. Series: Earth and Environmental Science **135** (2018) 012002 doi:10.1088/1755-1315/135/1/012002

Efficacy of Water Chamber Type Seawall to Dissipate Incident Wave and Its Performance to Extract Wave Power

Firman Husain[1], M Rusdi Alwi[2] and Ashury[1]

[1] Ocean Engineering Department, Engineering Faculty of Hasanuddin University, Indonesia
[2] Marine Engineering Department, Engineering Faculty of Hasanuddin University, Indonesia
Email :firman.husain@unhas.ac.id

Abstract. This study examined performance a water chamber type of seawall to dissipate incident wave and also extract wave power. The model of water chamber is consists of two water chamber type. The low reflection wave resulted by flow separation in lower end of vertical wall and it occurs after wave resonant condition. This structure is appropriate for short and long wave condition. A horizontal plate was setup in the end of vertical wall in order to improve performance of water chamber type seawall. This trial test taken to dissipate much longer waves condition and it's related to utilize the wave energy inside the water chamber. Additional trial of the cross section, such as an adoption of a bevel edge instead of a right angle edge was also examined. It was seen that the seawall of a water chamber type with a horizontal plate is very effective to reduce reflected waves for both short and long waves. Also it was confirmed that large amplifications of pumping mode wave motions in the two water chambers appear for comparatively wide range of wave frequency

1. Introduction

The climate change effect was influence increase frequency of wave attack to the coastal area. This condition is sometimes leads to fatal damage along coastal shore. It condition was also appearing many problem of human life especially that occupied around coastal area. To minimize the risk of wave attack, set up of a wave protection for example seawall to withstand incoming waves is needed to ensure many activities in coastal area keep it running.

In recent time, various type of seawall has been constructed in many places around the world in order to defend coastal area from the wave brunt. While the other seawall models is examine stage in the laboratory scale in order to investigate performance of seawall model to reduce wave reflection and wave overtopping. Application of conventional vertical seawall to protect coastal area resulting the highly reflected wave and it sometimes cause damage in top area of seawall. Tanimoto and Yoshimoto [1] were proposed a type wave barrier called vertically slotted caisson type. This structure has much rectangular hole as serve as the wave dissipater. This type of dissipater seems a good performance to reduce reflected incoming waves and overtopping waves in comparatively deep seas areas. However the structure is suitable for limited conditions only. Typically a water chamber width should be about a quarter length of wave. An improved model of the slotted wall type presented by Cox et al [2]. He has introduced a double wave barriers composed of a perforated wall in the seaward and a solid wall in the shoreward. The effectiveness of this type for reducing wave reflections was examined extensively and it was reported that the performance was still limited to the comparatively short waves. Furthermore the construction cost may become expensive, especially to build perforated walls and supporting structures.

In the present paper, a type of seawall of highly dissipative has been proposed. This type of water chamber is consists of two chambers and called the double water chamber type of seawall. This model is similar to the double curtain wall proposed by Nakamura et al [3]. To expand the effective range of wave frequency of reflected wave dissipations and also wave amplifications in the chambers, a horizontal plate below the first (front) water chamber was installed. So the first and second chambers separated by a horizontal plate. Comparison test between the double water chamber types of seawall with and without horizontal plate has been carried out. Another type separator called bevel edge shapes was tested to explore the magnification of wave amplification in the second water chamber.

2. Experiment

In order to examine various model seawalls, a series of experiments have been conducted in the wave flume to investigate performance of the model. All experiment activity has been running in the regular wave condition only.

2.1. Experimental apparatus

The model is tested in the flume tank with 30m long, 1m wide and 1.25m high as seen in the figure 1. The wave maker of wave flume using the piston-type put in one side and a wave absorber was installed in the other end side. The experimental model scale is assumed to be about 1/25 of the prototype. Five wave gauges of a capacitance type were used in the experiment and arranged in the wave flume. The first gauge was set in front of the wave maker; the second and third placed in front of the model to measure the incident and reflected wave; the fourth and five gauge put inside the first and second chamber respectively.

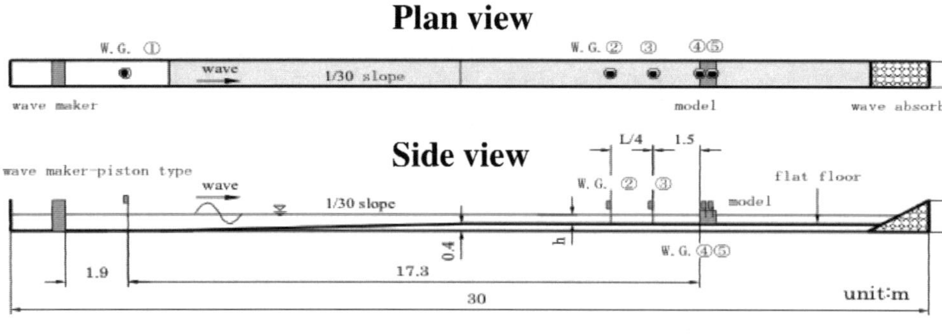

Figure 1. Wave flume and experimental installation

All typical model bodies of the double water chamber type seawall made from the transparent acrylic plates with 10 mm thickness. It aimed to observe the effect of an opening height of the bottom channel. The seawall model was installed at about 19.2m apart from the wave maker. Breadths of the two chambers are fixed. Hence the first chamber width located offshore side is 25 cm (B1) and the second one nearshore side is 23cm (B2). In the experiment, comparison tests between double water chamber type seawall with and without a horizontal plate (see figure 2 and figure 3) have been carried out. The two different heights, say e2=5cm and 10cm, were adopted in the experiment. The dimension of e2 is varied by moved vertically of horizontal plate include the second vertical wall for giving the necessary opening height e2. In an additional trial, the bevel edge model (see figure 4) as separated in second chamber instead of right angle shaped model has been examined.

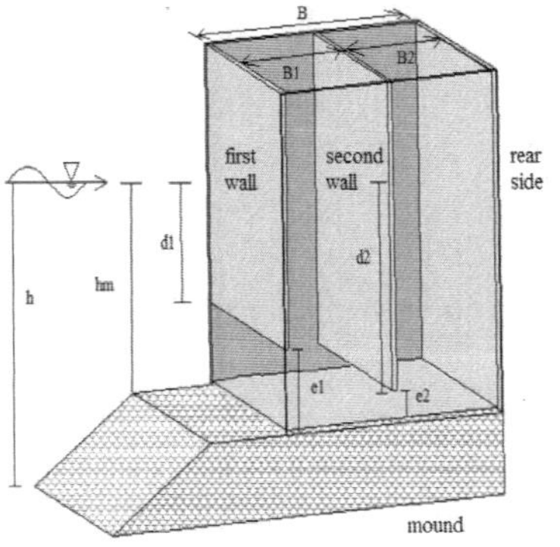

Figure 2. Model bodies of double water chamber without a horizontal plate

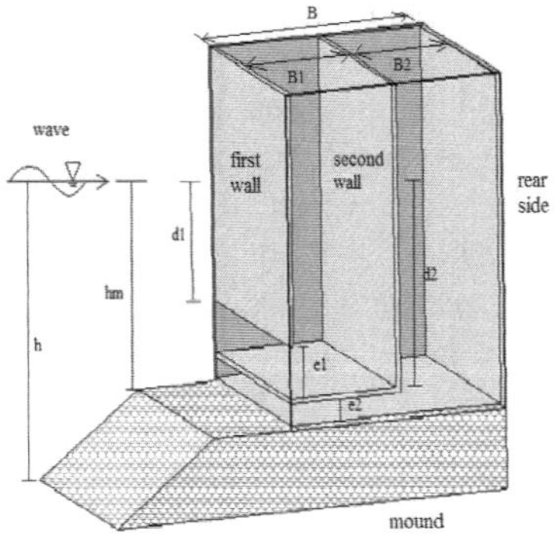

Figure 3. Model bodies of double water chamber with a horizontal plate (right angel)

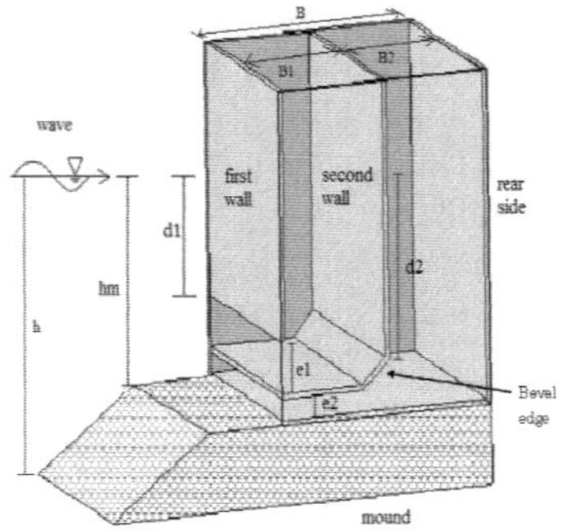

Figure 4. Model bodies of double water chamber with a bevel edge

The dimensions of these three types of a double water chamber seawall models are summarized in Table 1. The type 2 consist of two kinds namely type 2a and type 2b and difference lies in height of second water channel (*e2*).

Table 1. Principal dimensions of the model bodies

Model bodies	Type 1	Type 2 (a)	Type 2 (b)	Type 3
Total breadth (*B*)	50cm	50cm	50cm	50cm
Breadth of first chamber (*B1*)	25cm	25cm	25cm	25cm
Breadth of second chamber (*B2*)	23cm	23cm	23cm	23cm
Draft depth of first chamber (*d1*)	18.5cm	18.5cm	13.5cm	18.5cm
Draft depth of second chamber (*d2*)	29.5cm	29.5cm	24.5cm	29.5cm
Height of first water channel (*e1*)	15cm	10cm	10cm	10cm
Height of second water channel (*e2*)	5cm	5cm	10cm	5cm
Mound water height (*hm*)	34cm	34cm	34cm	34cm
Water depth (*h*)	50cm	50cm	50cm	50cm

2.2. Experimental condition

In the experiment, the range of wave period was from 0.7 to 3.5 s. For each wave period condition, the two different wave height conditions, H= 8 cm and 16 cm, were used. For comparatively short wave period conditions, only the wave condition of H=8cm was adopted because of wave breaking. Height of water depths from the bottom flat is 50 cm (hm) and from the mound is 34.5 cm (d1). The heights of both are kept constant through the experiment.

3. Numerical analysis

The numerical analysis used in this study is based on the wave source distribution method with a linear damping proportional to a fluid velocity, developed by Nakamura and Ide [4]. In the numerical analysis, at first, the fluid domain around the seawall is divided into two regions, hence non-damping and damping wave region. Some distance from the front of body to the seaward of semi-infinite region is assumed to be non-damping wave region, where the potential flow theory without non-damping can be applied. The surrounding wave region about the seawall structure is assumed to be a damping wave region, where wave energy dissipation takes place that is proportional to a fluid velocity

According to the previous study [3], the damping wave region was set to be equal to the region from 1/60 of wavelength in front of the seawall face to the shore side of it. The same criterion was adopted in this study. In the damping wave region, it was assumed that the vortices are uniformly distributed in the fluid domain. The resultant equation of motion is similar to the one derived by Sollitt and Cross [5] for the homogeneous porous media except for an inertial term of porous materials,

$$\frac{\partial \bar{q}}{\partial t} = -\frac{1}{\rho}\nabla(p+\gamma z) - f_c \omega \bar{q} \tag{1}$$

Where, $\bar{q}$ is the fluid velocity vector around the body, ρ is fluid mass density, ∇ is the gradient operator, p is the fluid pressure, γ is the fluid weigh density, z is the vertical coordinate, ω is the angular frequency of wave, f_c is the equivalent linear damping coefficient.
The fluid motion described by equation (1) is irrotational, therefore the velocity potential can be defined in the idealized damping wave region. To represent the velocity potential in each region, the wave source distribution method was firstly applied. Hereinafter is considering the matching boundary conditions between two wave regions, i.e. the non-damping and damping regions. The following equation of the Green's functions was used in the analysis for non-damping region.

$$G\,(x,z;X,Z) = -\frac{i}{k}\frac{(k^2-k_0^2)}{(k^2-k_0^2)h+k_0}\cosh k(h+z)\times\cosh k(h+\bar{Z})\exp(ik_x|x-X|)$$

$$-\sum_{n=1}^{\infty}\frac{1}{k_n}\frac{(k_n^2+k_0^2)}{(k_n^2+k_0^2)h-k_0}\cos k_n(h+z)\times\cos k_n(h+Z)\exp(-k_n|x-X|) \tag{2}$$

Where, (x,z) is a field point in the wave filed, (X,Z) is a source point, h is a water depth, k is a progressive mode wave number, k_n is an evanescent mode wave number and k_o is a wave number corresponding to deep water wave condition. k and k_n are obtained by solving the following dispersion relation equations,

$$\omega^2 = gk\tanh(kh), \omega^2 = -gk_n\tan(k_nh)\ \ (n = 1,2,3,\cdots) \tag{3}$$

For a damping wave field, the corresponding Green's function G_D derived by Nakamura and Ide [4]. It is shown as,

$$G_D(x,z;X,Z = -\sum_{n=0}^{\infty}\frac{i}{\mu_n}\frac{(\mu_n^2-\lambda^2)}{(\mu_n^2-\lambda^2)h+\lambda}\cos\mu_n(h+z)\times\cos\mu_n(h+z)\exp(i\mu_n|x-X|) \tag{4}$$

Where, λ is defined by eq. (5), in which μ_n is a complex wave number, which is given by solving the following dispersion relation in a damping wave field,

$$\lambda = \frac{\omega^2}{g}(1+if_c) = \mu_n\tanh(\mu_nh) \tag{5}$$

IOP Conf. Series: Earth and Environmental Science **135** (2018) 012002 doi:10.1088/1755-1315/135/1/012002

4. Result of Experiment

4.1. Wave reflection

Figure 5 shows the result of wave reflection coefficient Cr as a function of L/B. Here, L is a length of wave corresponding to the water depth at the toe of a sloped mound, and B is a total breadth of the double water chamber type seawall. In the figure, the reflection coefficients of the two different cases are shown, typically with and without a horizontal plate below the first water chamber. Also, numerical computation results are shown. The equivalent linear damping coefficient fc=0.3 was used after some trial and error approach.

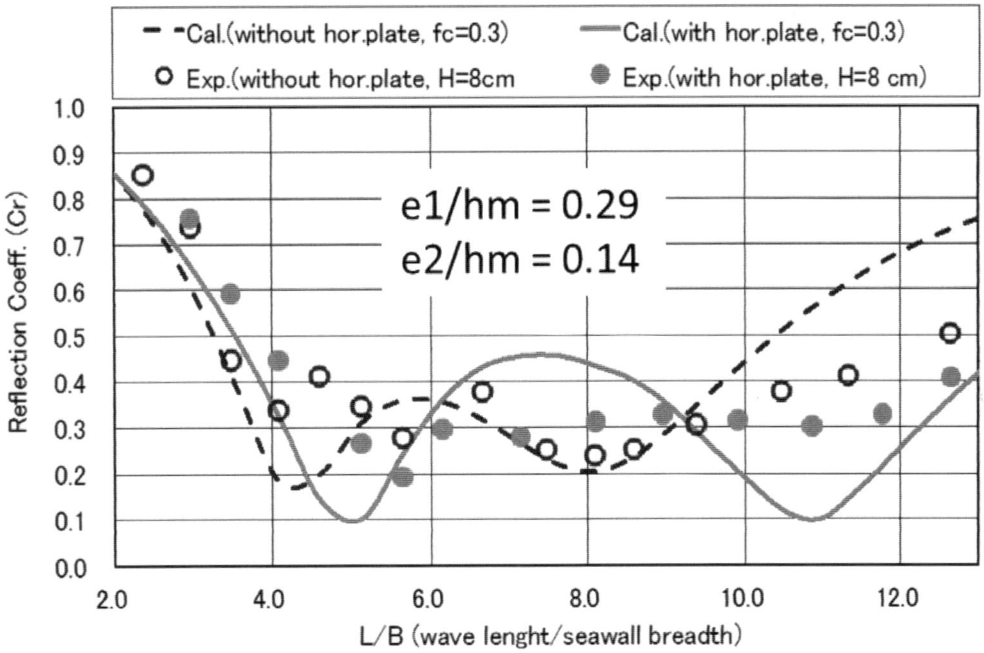

Figure 5. Comparison Cr of the water chamber with and without horizontal plate.

It can be seen that there are two minimum reflection points for different values of L/B regardless of with or without a horizontal plate. The tendency is very clear for the computation results. It is caused by the excitations of pumping mode wave resonance in the first and second water chambers. Under the resonant condition, flow separations and resultant vortex formations around the lower edge of the curtain walls become evident and predominant. Therefore, the wave energy dissipation increases under these resonant conditions.

By addition of the horizontal plate, the minimum reflection point corresponding to larger value of L/B shifts to much longer wave condition, i.e. from L/B=8 to 11. It may be resulted from increased amount of an oscillating water mass for the second water chamber by the horizontal plate. Therefore, the new type of double water chamber seawall is effective to dissipate reflected waves for much wider range of wave frequency.

Figure 6 shows the result of reflection coefficient for the case with a horizontal plate, but for different bottom channel height e2, say 5cm and 10cm. It can be seen that the effective range of wave frequency becomes narrow with increasing the bottom channel height e2. However, the absolute value of Cr for the case of e2=10cm becomes smaller than the case of e2=5cm for intermediate range of wave frequency.

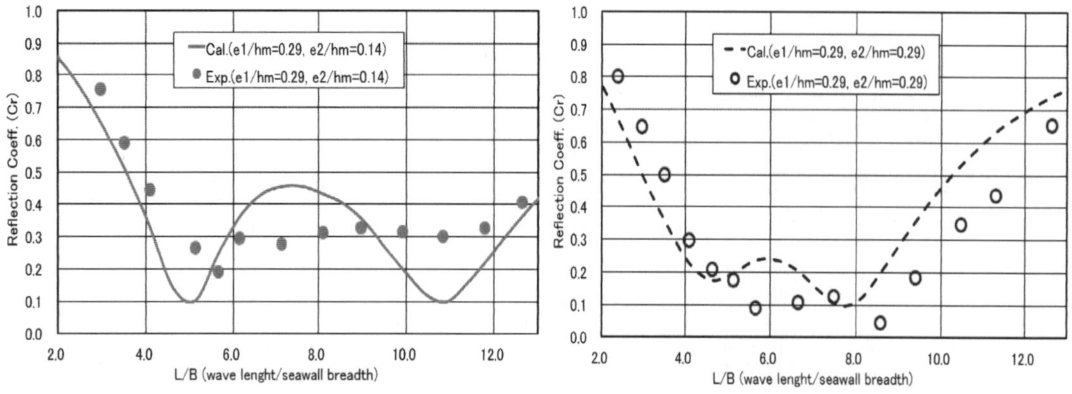

Figure 6. Cr of different channel heights below a horizontal plate (e2=5cm and 10cm. H=8cm)

Figure 7 shows the computation result of reflection coefficient for various values of e2, i.e., e2=3cm to 15cm in the model body. From this figure, we can see the similar tendency to figure 6. It becomes clear that the minimum reflection point corresponding to the longer wave condition shifts gradually with increasing the channel height e2. However the reflection coefficient between the two minimum reflection points becomes larger with decreasing e2. It is a kind of secondary effect for expanding the effective range of wave frequency.

From the practical point of view, it may be important to estimate the reasonable value of e2 by considering both the effective range of wave frequency and also the absolute value of Cr between the two minimum reflection points. Under the limited condition such as given in figure 7, the channel height of e2=5cm is a kind of optimum condition

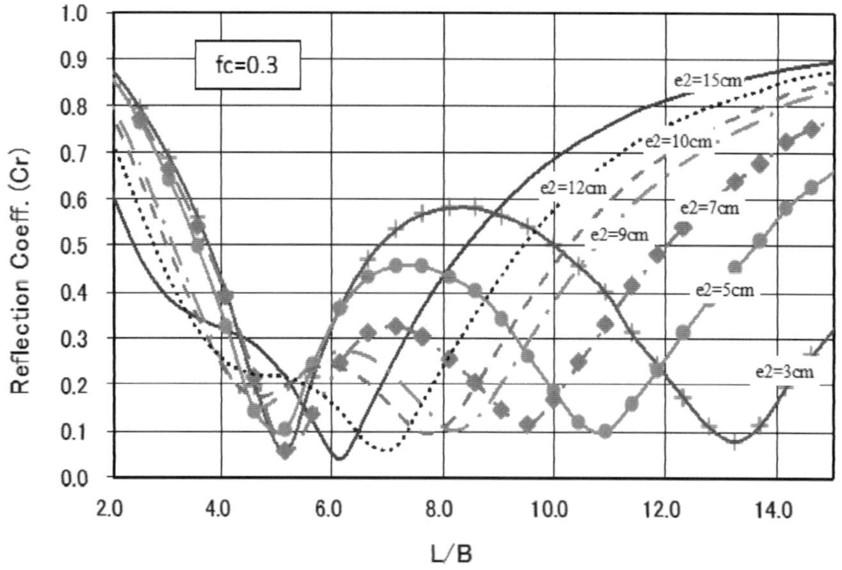

Figure 7. Computational result of Cr from various heights of e2

4.2. Wave amplification

For effective wave energy extractions, much larger wave amplifications in the water chamber are usually required for wider range of wave frequency. A wave power extraction device based on the well-known oscillating water column type is typical. Figure 8 shows typical results of the wave amplifications in the two water chambers. The wave amplification is defined as the ratio of a wave height in each water chamber (Hc) to an incident wave height (H). In the experiment, we have confirmed that almost the same wave height appears in the water chamber because of the excitation of pumping mode wave motions in the chambers. In figure 8, the results corresponding to the case of e2=5cm are shown. However, it includes the two different results corresponding to the case of a right angle shaped plate and that of a bevel edged shaped plate. The latter cross section was adopted to magnify the wave amplification in the second wave chamber.

From this figure, it can be seen that there is no or very little difference on the wave amplifications between the two different cross sections. For both the models, the wave amplification reaches peak at first in the first water chamber for shorter wave conditions. Then, it does in the second water chamber for longer wave conditions. The consistency between experimental and computational results on the wave amplification becomes worse near the peaks of wave amplification on the two water chambers. It may be considered that the wave amplification near the peak conditions is strongly dissipated by the existence of sharp edges.

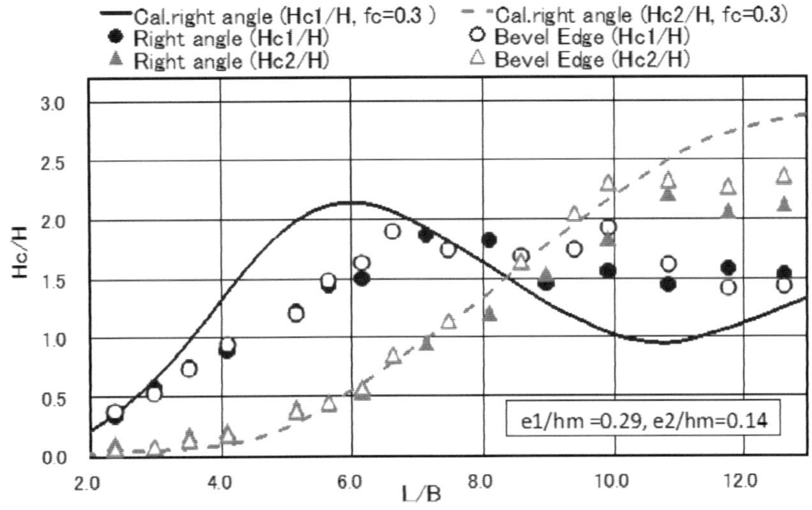

Figure 8. Wave amplification of right angle and bevel edge shaped in first and second water chamber

Figure 9 shows the results of the wave amplifications in the two water chambers without the horizontal plate. The agreement between the experimental and computational results is not good, especially for the second water chamber. It may be caused by the strong wave energy dissipation around the lower edge of the second vertical wall. Comparing the result with horizontal plate case (see figure 8), the peak values of the wave amplification for without horizontal plate in the two water chambers become lower significantly.

Figure 10 shows the influence of an incident wave height on the wave amplification in the two water chambers of the double water chamber type seawall with the horizontal plate. As seen in the figure, the wave amplification in the second water chamber varies significantly with the incident wave height, especially under the resonant condition of the second water chamber.

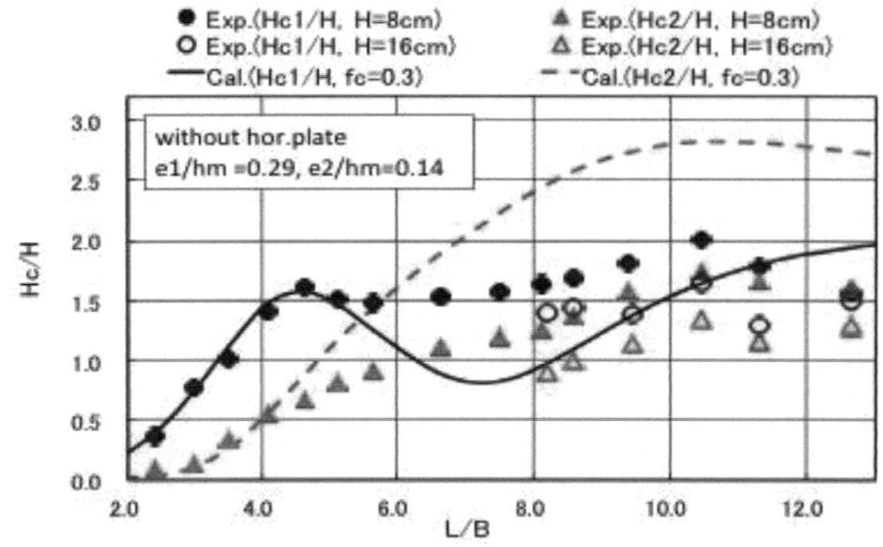

Figure 9. Wave amplification result in first and second chamber without horizontal plate

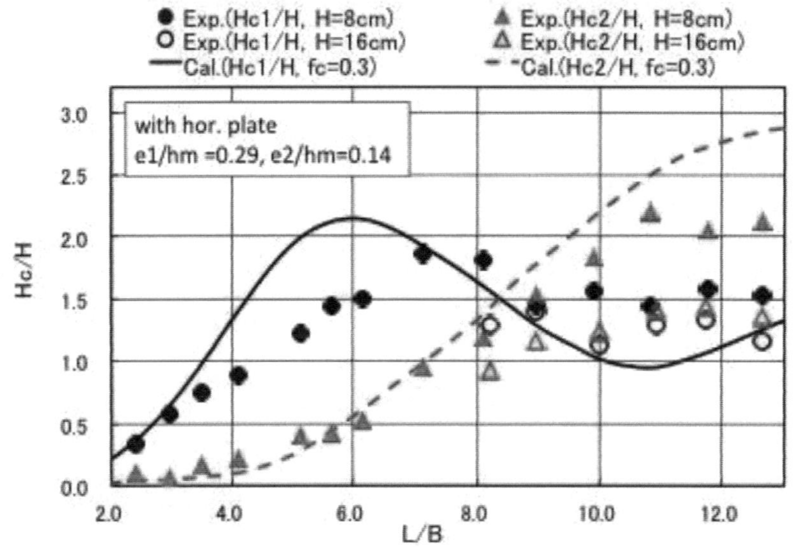

Figure 10. Wave amplification result in first and second chamber with a horizontal plate (water channel height e2=5cm, H=8 and 16cm)

Figure 11 shows the result of wave amplifications in the first and second water chambers in case of e2=10cm (the channel height under the horizontal e2=10cm). As compared with figure 8 which corresponds to the result of e2=5cm, the wave amplification in the second water chamber increases for a larger channel height e2 especially under the resonant wave period condition.

In order to magnify the wave amplification especially in the second water chamber of the double water chamber type, the installation of the horizontal plate under the first water chamber is very effective. Further, if the channel height e2 is comparatively large, we can expect much larger wave amplifications in the second water chamber.

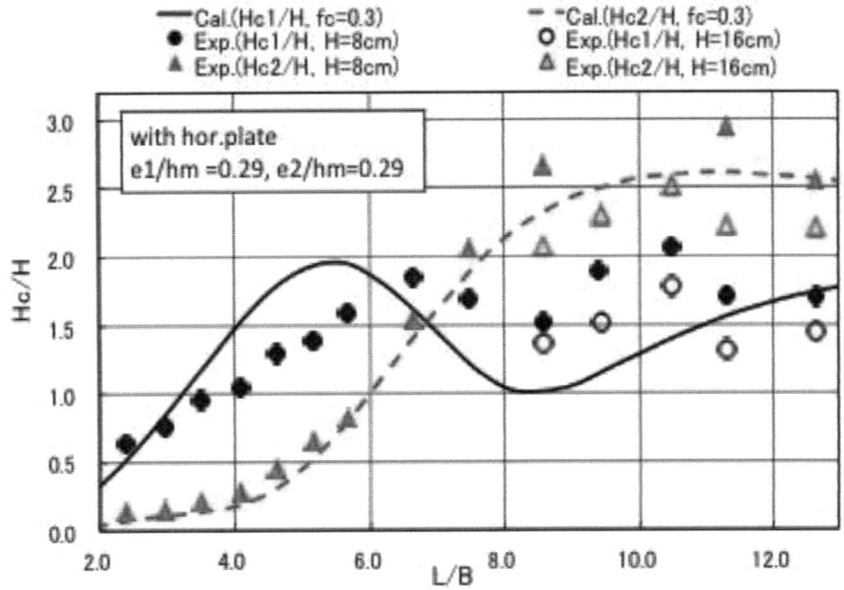

Figure 11. Effect increase of water channel height on wave amplification in the first and second chamber with horizontal plate (e1=10cm and e2=10cm)

5. Conclusion

- By installed a horizontal plate in the lower end of second vertical wall, performance of seawall water chamber is more effective to reduce wave reflections and also wave amplifications in the two chambers as compared to the original model without a horizontal plate.
- Increasing the bottom channel height under the horizontal plate, effective range of wave frequency for the dissipation of reflected waves becomes narrow. However, wave height amplifications in the water chambers, especially in the second water chamber, magnifies significantly.

Based on the experimental result of a double water chamber type with a horizontal plate, the wave amplification value shows good result. Therefore, it possible to continue to extract wave power using a double water chamber type with a horizontal plate as a substructure by installed an air turbine or water turbine in the future work.

Reference

[1] Tanimoto, K. and Yoshimoto, Y. 1982. Theoretical and experimental study of reflection coefficient for wave dissipating caisson with a permeable front wall, Report of Port and Harbor Research Institute, Vol.21, No.3, pp. 43-78

[2] Cox, R. J., Horton, P. R., and Bettington, S. H. 1998. Double walled low reflection wave barrier, *Proc., 26th, Coastal Eng. Conf.*, ASCE, pp. 2221-2234

[3] Nakamura, T., Kohno, T., Makimoto, K. And Kamikawa, H. 2000. Enhancement of wave energy dissipation by a double curtain-walled breakwater with different drafts, *Proc., of Int., Conf. Coastal Structures 99*, Vol. 1, pp. 533-540.

[4] Nakamura, T. and Ide, Y. 1997. Analysis on wave transformations and wave forces about an angular body considering wave energy dissipations, *Proc., of Civil Eng. in the Ocean*, Vol.13, pp. 177-182 (in Japanese)

[5] Sollit, C. K. & Cross, R. H. (1972): Wave transmission through permeable breakwater, *Proc., 13th., Coastal Engineering Conf.*, ASCE, pp. 1827-1846

Effects of stern-foil submerged elevation on the lift and drag of a hydrofoil craft

K Suastika[1] and Apriansyah[1]

[1]Department of Naval Architecture and Shipbuilding Engineering, Faculty of Marine Technology, Institut Teknologi Sepuluh Nopember (ITS), Kampus ITS Sukolilo, Surabaya 60111, Indonesia

Email: k_suastika@na.its.ac.id

Abstract. Effects of the stern-foil submerged elevation on the lift and drag of a hydrofoil craft are studied by using computational fluid dynamics (CFD) and by considering three alternative stern-foil submerged elevations. The submerged elevation of the front foil is kept constant in all the alternatives. From among the alternatives, the deepest stern-foil placement results in the highest stern-foil lift with the highest foil's lift-to-drag ratio. However, considering the lift-to-drag ratio of the whole foil-strut-hull system, the shallowest stern-foil placement results in the highest lift-to-drag ratio. The struts and the foil's submerged elevation significantly affects the drag of the whole foil-strut-hull system.

1. Introduction

The use of a hydrofoil on a boat is particularly aimed at reducing the viscous resistance of the boat while moving in the water. At relatively high speed the hydrofoil can produce a sufficient lifting force so that the hull can be lifted out of the water, thus reducing the ship's wetted surface area (WSA) and the ship's resistance. Besnard et al. [1] conducted research to gain the maximum lift-to-drag ratio of hydrofoils to be applied to fast boats. They used a panel method to calculate the 3-D flow field, taking into account free-surface effects. The numerical tool is integrated into a multi-criteria optimization method applied for the design of a single and biplane foils. Considering vessel's maneuvering, Latorre and Teerasin [2] calculated the take-off speed of hydrofoil vessels with different foil size and angle of attack. Calculation results show that different regions of the foil are exposed to a large impact at the take-off. Putnam et al. [3] performed tests to select the most optimum type of foil that will be used for hydrofoil boats. As parameter for best performance is the maximum lift-to-drag ratio. Different foil types were studied and they recommended Eppler-396 foil for the stern foil and Eppler-420 for the front foil of the hydrofoil craft.

Saputro and Suastika [4] conducted model experiments of a hydrofoil craft in a towing tank and varied the positions of the front- and stern foils in the longitudinal direction. They reported an optimum placement of the foils with the smallest resistance but the lifting of the bow was not followed by the stern. It is suggested that this is due to the alignment of the two foils relative to the water line, which was at the same elevation. This results in that the flow of water that passes the stern foil has been affected by the turbulence generated by the front foil such that the lifting capacity of the stern foil is markedly reduced.

The purpose of present study is to investigate effects of the stern-foil submerged elevation of a hydrofoil craft on the lift and drag. The same vessel is considered as in [4]. The vessel's principal dimensions are summarized in table 1. The submerged elevation of the front foil is kept constant but that for the stern foil is varied in the vertical direction, for which three alternatives are investigated (see figure 1 and table 2). Following Saputro and Suastika [4], both the front and the stern foils are NACA 64(1)212 foils while the struts are NACA 63012 symmetrical foils.

To obtain the most optimum foil size, CFD-simulations were done of foil alone with varying foil's aspect ratio (AR) and angle of attack. The aspect ratio is defined as [5]:

$$AR = \frac{(b)^2}{S} = \frac{b}{c} \tag{1}$$

$$AR = \frac{b}{c}\left[1 + \left(\frac{a}{b}\right)^3\right]\frac{h}{b} \tag{2}$$

where a is the breadth of the craft, b is the foil span, c is the chord length, h is the submerged elevation of the foil below the water line and S is the foil area. Equation (1) is used for calculating the foil aspect ratio (to lift the vessel) while equation (2) for the strut aspect ratio. The range of recommended aspect ratio for a hydrofoil is between 4 and 10 [6]. Then, simulations with foils attached to the hull are done to study effects of the stern-foil submerged elevation on the lift and drag of the hydrofoil craft.

Table 1. Vessel's principal dimensions.

Length overall (L_{OA})	15.85	m
Length of water line (L_{WL})	14.48	m
Beam (B)	4.00	m
Draft (T)	0.70	m
Height (H)	1.80	m
Displacement (∇)	18.75	ton

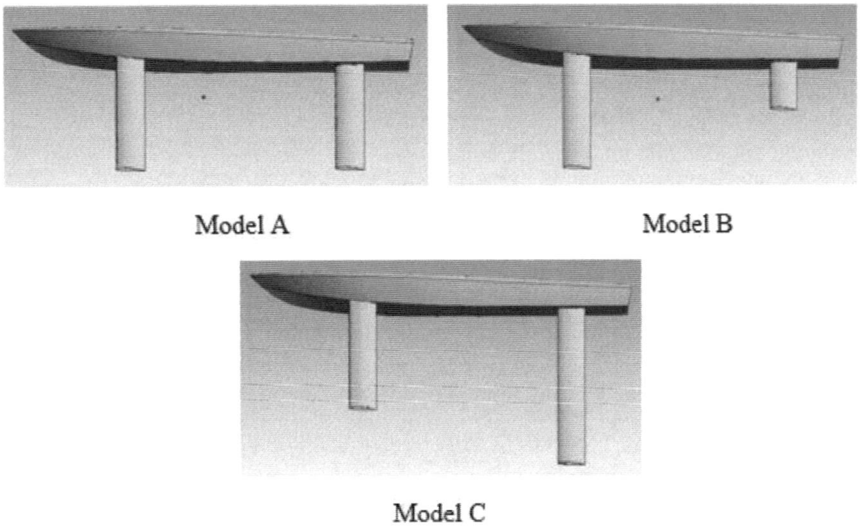

Figure 1. Three alternatives for the stern-foil submerged elevation with constant front-foil elevation.

Table 2. Submerged elevation of the front and stern foils below the water surface.

	Front foil	Stern foil
Model A	6.37 T = 4.46 m	6.37 T = 4.46 m
Model B	6.37 T = 4.46 m	3.19 T = 2.23 m
Model C	6.37 T = 4.46 m	9.56 T = 6.69 m

2. Foil-size optimization

IOP Conf. Series: Earth and Environmental Science **135** (2018) 012003 doi:10.1088/1755-1315/135/1/012003

Based on the location of the center of gravity (LCG) and the positions of the front- and stern foils in the longitudinal direction, the required lift for both the front and stern foils can be calculated. Calculation results show that the required lift for the front foil is 85.66 kN while that for the stern foil is 98.09 kN. To obtain the optimum foil size, CFD-simulations of foil alone were done. Two foil spans are considered, namely 4.0 m (the same as the vessel's beam) and 5.0 m. The Froude number used in the simulations is $Fr = 0.8$, corresponding to a vessel's speed of 9.53 m/s (18.53 knots).

2.1. Meshing, boundary conditions and the turbulence model

Figure 2 shows a mesh of the NACA 64(1)212 foil and figure 3 the position of the hydrofoil model in the computational domain. Grid-independence tests [7] were done to obtain the most optimum grid size. Test results show that the optimum number of elements for the simulations is 1,074,984. The root mean square (rms) error criterion with a residual target value of 10^{-6} is used as the criterion for the convergence of the numerical solutions.

Figure 2. Mesh of NACA 64(1)212 foil used for the hydrofoil craft.

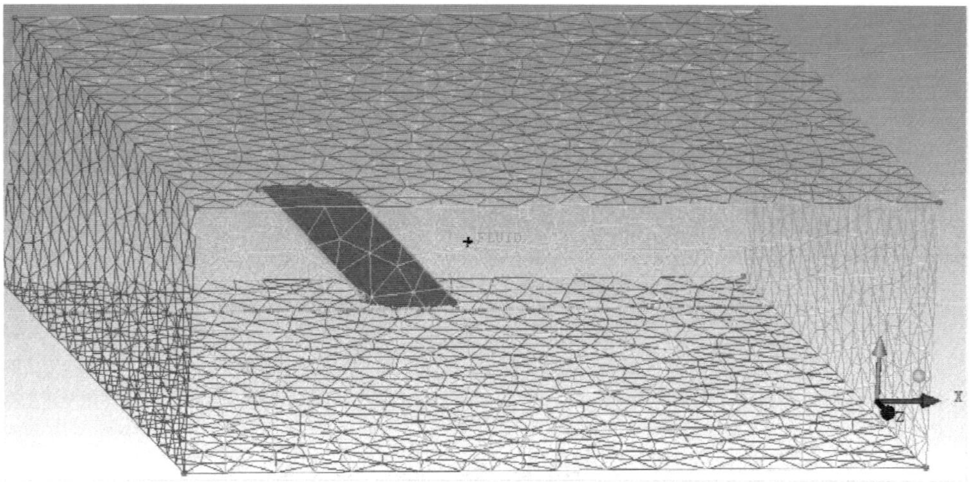

Figure 3. Position of the foil in the computational domain.

The boundary conditions of the computational domain are as follows [8]. The condition at the inlet boundary, located 2-*c* upstream from the leading edge (*c* is the chord's length), is defined as a uniform flow with velocity equalling the vessel's speed (in the simulations, the vessel is at rest while the fluid is flowing). The boundary condition at the outlet, located 5-*c* downstream from the trailing edge, is defined such that the pressure is equal to the undisturb static pressure. The boundary condition on the surface (wall) of the foil is defined as no-slip boundary (smooth wall). The boundary condition at the side walls is defined as a symmetric pressure condition, that is, the pressure on the inside wall is equal to that on the outside wall. Finally, the boundary conditions at the top (2-*c* above the upper surface of the model) and the bottom (2-*c* below the lower surface of the model) surfaces are defined as free-slip boundaries. Free surface effects (generation of waves) are not modeled in this case (foil alone).

The turbulence model used in the simulations of foil alone is the k-ε model [9, 10].

2.2. Results of foil-alone simulations

Figure 4 shows plots of lift-to-drag ratio as function of angle of attack (α) for foils with 4-m span and different chord lengths, while figure 5 shows lift-to-drag ratio plots for foils with 5-m span. In all cases, the aspect ratio is kept within the recommended range, namely, between 4 and 10 [6] (see tables 3 and 4). Figures 4 and 5 show that the lift-to-drag ratio increases monotonically with increasing angle of attack and then decreases. The maximum value of lift-to-drag ratio (stall) takes place at the angle of attack of approximately 3.5 degrees. The most optimum foil with 4-m span is that with chord's length of 0.43 m (AR = 9.39) while the most optimum foil with 5-m span is that with chord's length of 1.14 m (AR = 4.40).

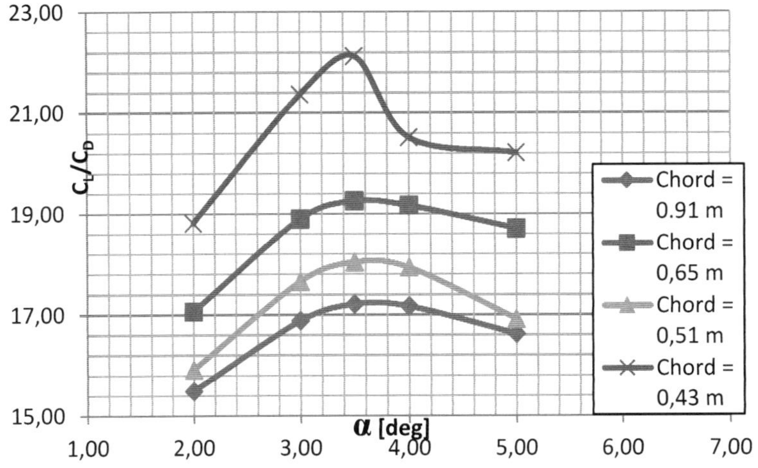

Figure 4. Lift-to-drag ratio as function of angle of attack for foils with 4-m span and different chord's lengths.

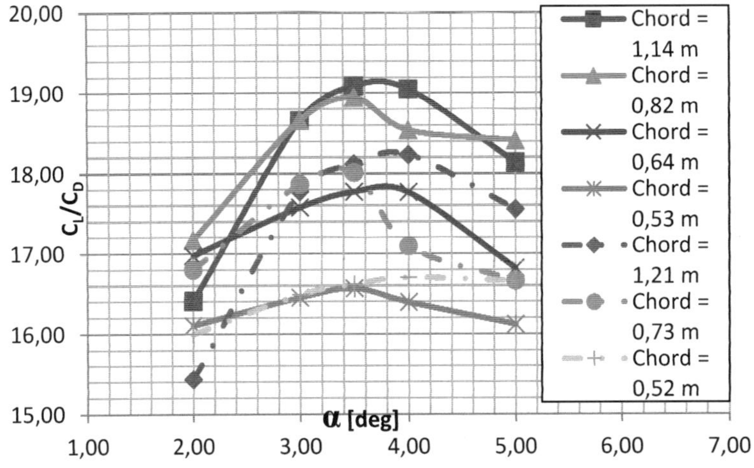

Figure 5. Lift-to-drag ratio as function of angle of attack for foils with 5-m span and different chord's lengths.

Table 3. Foil's lift at $Fr = 0.8$ and $\alpha = 3.5°$ for 4-m foil's span.

Chord [m]	AR	Lift [kN]
0.91	4.40	60.22
0.65	6.13	49.52
0.51	7.83	40.52
0.43	9.39	35.77

Table 4. Foil's lift at $Fr = 0.8$ and $\alpha = 3.5°$ for 5-m foil's span.

Chord [m]	AR	Lift [kN]
1.14	4.40	83.22
1.21	4.14	82.68
0.82	6.13	81.61
0.73	6.87	79.28
0.64	7.83	79.28
0.53	9.39	84.03
0.52	9.57	76.27

Tables 3 and 4 summarize results of lift force calculations with $Fr = 0.8$ and $\alpha = 3.5°$ for 4-m foil's span and for 5-m foil's span, respectively. Tables 3 and 4 show that the lifts generated by the foils with 4-m span are smaller than those generated by the 5-m span foils, as expected. In both cases (4-m and 5-m spans), the generated lifts are smaller than the required lifts (85.66 kN for the front foil and 98.09 kN for the stern foil).

To provide a sufficient lift for foilborne, the foil's span can be increased but AR must be kept within the recommended range. Another option is to keep the foil's span remaining 5.0 m but foilborne is expected to take place at a larger Froude number. The latter option is chosen in the present study. Thus, the foil to be used is NACA 64(1)212 foil with 5-m span and chord of 1.14 m (AR = 4.40).

3. Foils attached to hull

To study effects of the stern-foil submerged elevation on the lift and drag of the hydrofoil craft, simulations were done with foils attached to the hull (by using struts). In the simulations, the foil's angle of attack (both the front and stern foils) was set 3.5° (see section 2). Furthermore, different Froude numbers are considered, namely, $Fr = 0.8$, 0.9, 1.0, 1.1 and 1.2.

3.1. Meshing, boundary conditions and the turbulence model

Figure 6 shows a mesh of the hydrofoil craft for model B while figure 7 shows the position of the hydrofoil-craft model in the computational domain. Grid-independence tests [7] were done to obtain the most optimum grid size. Test results show that the optimum number of elements for the simulations is 1,791,588. The root mean square (rms) error criterion with a residual target value of 10^{-4} is used as the criterion for the convergence of the numerical solutions.

The boundary conditions are as follows [8]. At the inlet, located at 2-*L* upstream from the vessel (*L* is the vessel's length), the condition is defined as a uniform flow with velocity equalling the vessel's speed (in the simulations, the vessel is at rest while the fluid is flowing). The boundary at the outlet (5-*L* downstream from the vessel) is defined such that the pressure is equal the undisturbed static pressure. The boundary condition on the wall of the vessel is defined as no-slip boundary (smooth wall). The boundary conditions at the bottom wall (2-*L* below the vessel) and the top wall (2-*L* above the vessel) are defined as free-slip condition. Free-surface effects (interface between water and air; generation of waves) are modeled in this case (foils attached to hull). The viscous-flow field is solved using Reynold's averaged Navier-Stokes (RANS) solver for incompressible flow.

IOP Conf. Series: Earth and Environmental Science **135** (2018) 012003 doi:10.1088/1755-1315/135/1/012003

Figure 6. Mesh of the hydrofoil craft for model B.

Figure 7. Position of the hydrofoil-craft model in the computational domain.

The turbulence model used is the SST k-ω model [9]. This turbulence model, which is a hybrid model utilizing a k-ω model for the turbulent boundary layer flow and a k-ε model for the flow outside the boundary layer, has been succesfully applied in many applications [10, 11].

3.2. Results of foils attached to hull

Tables 5, 6 and 7 summarize the lift forces for the front and stern foils for Fr = 0.8, 0.9 and 1.0, respectively. Recall that the required lift for the front foil is 85.66 kN while that for the stern foil is 98.09 kN. At Fr = 0.8 and 0.9, the lift forces of the front and/or the stern foils are smaller than the respectively required lifts. At Fr = 1.0, models B and C provide sufficient lift for the vessel to foilborne. The highest lift-to-drag ratio of the stern foil in the range of Fr between 0.8 and 1.1 is given by model C (see figure 8). However, looking at the lift-to-drag ratio of the whole foil-strut-hull system, model B gives the highest lift-to-drag ratio (see figure 9). If the lift-to-drag ratio of the whole foil-strut-hull is considered as parameter for best performance, model B is the best alternative, for which the vessel is expected to foilborne at Fr approximately 1.0 (see table 7).

Table 5. Lift forces of the front and stern foils (Fr = 0.8).

Model	Front Foil [kN]	Stern Foil [kN]	Difference [%]
A	84.87	66.84	26.98
B	80.11	72.59	10.36
C	81.21	76.16	6.64

Table 6. Lift forces of the front and stern foils ($Fr = 0.9$).

Model	Front Foil [kN]	Stern Foil [kN]	Difference [%]
A	106.21	79.42	33.73
B	100.61	85.89	17.15
C	103.58	92.66	11.78

Table 7. Lift forces of the front and stern foils ($Fr = 1.0$).

Model	Front Foil [kN]	Stern Foil [kN]	Difference [%]
A	129.83	95.05	36.58
B	123.06	104.61	17.63
C	127.38	110.26	15.53

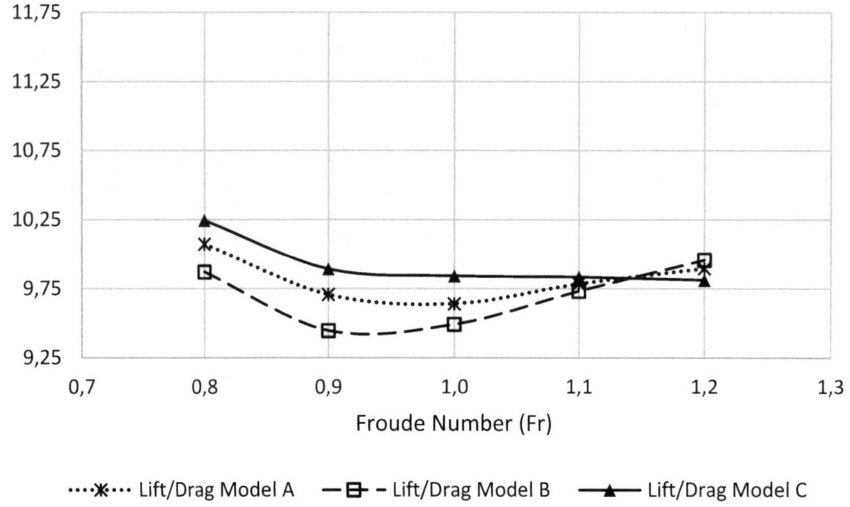

Figure 8. Lift-to-drag ratio of the stern foil for model A, B and C.

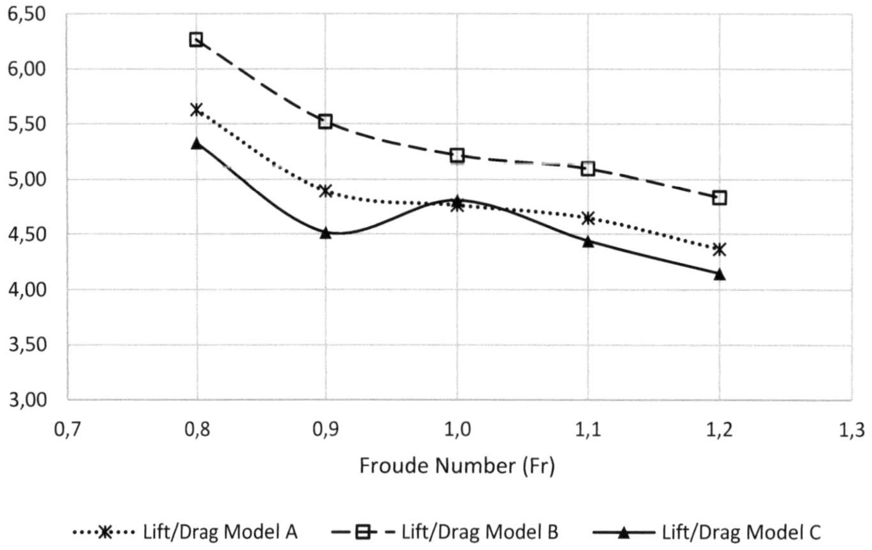

Figure 9. Lift-to-drag ratio of the hydrofoil craft for model A, B and C.

4. Conclusions

Based on the required lift, the recommended foil's aspect ratio and the highest lift-to-drag ratio, the foil chosen for the hydrofoil craft is NACA 64(1)212 with 5-m span and chord of 1.14 m (AR = 4.40). Simulation results of foils attached to the hull show that the stern-foil submerged elevation significantly influences the lift generated by the stern foil. Setting the stern foil deeper than the front foil (model C) or shallower than the front foil (model B) result in a larger lift for the stern foil. The above result support the hypothesis that the front foil significantly affects the flow upstream from the stern foil that can result in a reduction of the lifting capacity of the stern foil. Among the three alternatives considered, model C results in the highest lift and the highest lift-to-drag ratio for the stern foil. However, considering the whole foil-strut-hull system, model B with the shortest strut gives the highest lift-to-drag ratio of the hydrofoil craft. Clearly, the struts and the foil's submerged elevation significantly affects the total drag of the system. If the lift-to-drag ratio of the whole system is used as parameter for best performance, then model B is the most optimum alternative.

Acknowledgment

This research project was financially supported by the Indonesian Ministry of Research and Higher Education (RISTEKDIKTI), under the grant: Penelitian Kerjasama Industri 2017.

References

[1] Besnard E, Schmitz A, Kaups K, Tzong G, Hefazi H, Kural O, Chen H and Cebeci, T 1998 Hydrofoil design and optimization for fast ships, *Proc. 1998 ASME Int. Congress Exhibition* (Anaheim, USA)

[2] Latorre R and Teerasin S 1992 Calculation of hydrofoil craft take-off speed including influence of foil size, foil angle and propeller shaft angle *Ocean Engng* **9**(2) 183

[3] Putnam N, Dickert G and Wagner C 2014 On the Design, Construction, and Testing of a Fully-Submerged Canard Hydrofoil System for a Low-speed Solar Boat, Elmer W. Engstrom Department of Engineering and Computer Science, Cedarville University, Cedarville

[4] Saputro S A and Suastika K 2012 Kajian eksperimental pengaruh posisi perletakan hydrofoil pendukung terhadap hambatan kapal *J Teknik ITS* **1**(1) G-51

[5] Du Cane P 1974 *High Speed Small Craft* (Portland: David & Charles)

[6] van Walree F 1999 *Computational Methods for Hydrofoil Craft in Steady and Unsteady Flow*, Ph.D.-thesis, Delft University of Technology, The Netherlands

[7] Anderson, J D 1995 *Computational Fluid Dynamics: The Basics with Applications* (New York: McGraw-Hill, Inc.)

[8] Versteeg H K and Malalasekera W 2007 *An Intoduction to Computational Fluid Dynamics: The Finite Volume Method* (United Kingdom: Longman Scientific)

[9] Menter, F R 1994 Two-equation eddy-viscosity turbulence models for engineering applications *AIAA J* **32**(8) 1598

[10] Mordhorst C J 2001 Investigation of Open Source CFD Software on Shipyards, M.Sc.-thesis, Chalmers University of Technology, Sweden

[11] Swennberg S U 2000 A test of turbulence models for steady flows around ship, *Proc. Gothenburg Workshop on Numerical Ship Hydrodynamics* (Gothenburg)

The Assessment of the Ultimate Hull Girder Strength of RO-RO Ship after Damages

Muhammad Zubair Muis Alie[1], Ganding Sitepu[1] and Samuel Izaak Latumahin[1]

[1]Department of Naval Architect and Ocean Engineering, Engineering Faculty, Hasanuddin University, Indonesia

Email : zubair.m@eng.unhas.ac.id

Abstract. Many accidents of Ro-Ro ships happen in Indonesia such as collision and grounding. When the collision or grounding takes place on the Ro-Ro ship, the ultimate strength of hull structure after damage becomes decrease. Car and passenger decks are critical location since collision and/or grounding occur. In the present study, the assessment of the ultimate hull girder strength is conducted. The cross section of Ro-Ro ship is taken to be analyzed. The collision and grounding damages are assumed to be palced on the side and bottom area, respectively. The damages are created by removing the element from the side shell and bottom part. Finally, the result obtained is compared with one another.

1. Introduction

Collision or grounding damages may occur on the ship's hull so that the safety of ship and environment would be threatened. Many ship accidents takes place in Indonesia especially for Ro-Ro ship. In spite of human error, the structural degradation after collision and grounding damages gives impact to the ultimate strength of ship's hull. This must be stated in the local regulation.

The analysis of the ultimate hull girder strength has been assessed by some researchers particularly for damage cases. Paik [1] developed a rapid procedure to identify the possibility of hull girder failure after collision and grounding damages based on the closed-form formulae of the ultimate hull girder strength and section modulus after damages. Muis Alie [2] presented a n incremental fromulation of the progressive bending collapse behavior of ship hull girder with damages based on the Smith's method taking the rotation of the neutral axis into account. The proposed method was applied to the residual strength analysis of bulk carrier and tanker having the collision damage at the side structure. Soares [3] evaluated the ability of simplified structural analysis methods based on the smith's formulation to predict the ultimate strength of damage ship's hull. Ohtsubo [4] showed the experimental and numerical work on the ship structural damages due to collision and grounding. This is one of the first attempts to apply the explicit finite element method codes such as LS DYNA and DYTRAN to the collision and grounding problems of ships. Muis Alie [5] investigate the influence of superstructure on the longitudinal ultimate strength of a Ro-Ro ship. To investigate the ultimate strength, the Smith's method was adopted and implemented into the thin-walled beam. The cross section of Ro-Ro shp was taken to be analyzed. Muis Alie [6] analyzed the hull girder ultimate strength of asymmetrically damaged ships using Finite Element Method. The collision damage was modeled by removing the plate and stiffened plate element. The result obtained by FE analysis was compared to the analytical solution and the progressive collapse behavior of ship's hull was presented. Also, the progressive collapse analysis of ship hull girder based on Smith's method was developed by [7]. The previous studies showed that the ultimate strength of ship

IOP Conf. Series: Earth and Environmental Science **135** (2018) 012004 doi:10.1088/1755-1315/135/1/012004

hull girder was very essential particularly for the assessment of the ultimate strength after damage. The assessment of the ultimate strength must be clear in the local regulation especially for Ro-Ro ship.

In the present study, the assessment of the ultimate hull girder strength is performed. The cross section of Ro-Ro ship is taken and it is quite new to be analyzed for the ultimate strength assessment due to damages in Indonesia. The collision and grounding damages are assumed to be palced on the side shell and bottom area, respectively. The damages are created by removing the element from the side shell and bottom part. The result obtained by analytical formulation both hogging and sagging condition for intact and damages are compared with one another.

2. Analytical Solution

The previous study only consider the collision damage of bulk carrier and tanker taking the rotation of the neutral axis into account (Muis Alie, M.Z. et al 2012) [2]. In the present study, the influence of the rotation of the neutral axis is not considered. The ultimate hull girder strength of Ro-Ro ship is analyzed with collsion and grounding damages. It should be noticed that since the collision or grounding takes place at asymmetric position, the assumption of the cross section remained plane may be applied, the axial strain $\varepsilon_i(y_i, z_i)$ at the i-th structural element caused by the horizontal curvature ϕ_H and vertical curvature ϕ_V can be expressed as

$$\varepsilon_i(y_i, z_i) = \varepsilon_0 + y_i\phi_H + z_i\phi_V \tag{1}$$

where ε_0 is the axial strain at the origin of the bottom keel and the axial stress σ_i corresponding to the axial strain ε_i is given by the average stress-average strain relationship is calculated for the individual elements as illustrated in Fig. 1.

To obtain the ultimate strength both hogging and sagging condition, the equilibrium condition must be required, i.e. that the stress where the on neutral axis position is zero and it is a functioned of the strain. In this regard, the relationship is generally a nonlinear function by considering buckling and yielding and it is expressed as

$$\sigma = f_i(\varepsilon) \tag{2}$$

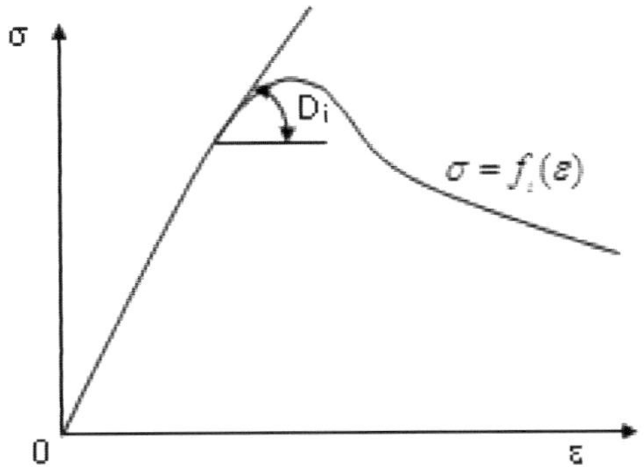

Fig.1 Average stress-average strain relationship of a structural element

Where $f_i(0) = 0$ and integrating the axial stresses to all the intact part of cross section, the axial force P, the vertical bending moment M_V, and the horizontal bending moment M_H can be obtained as

$$P = \sum_{i=1}^{N} \sigma_i A_i \equiv 0 \tag{3}$$

$$M_H = \sum_{i=1}^{N} \sigma_i y_i A_i \tag{4}$$

IOP Conf. Series: Earth and Environmental Science **135** (2018) 012004 doi:10.1088/1755-1315/135/1/012004

$$M_V = \sum_{i=1}^{N} \sigma_i z_i A_i \tag{5}$$

where N is the number of intact elements and A_i is a cross section of the i-th element.

$$\Delta\sigma = D_i \Delta\varepsilon \quad \left(D_i = \frac{df_i}{d\varepsilon} \right) \tag{6}$$

The relationship of axial force, horizontal and vertical bending moment can be given by

$$\begin{Bmatrix} \Delta P = 0 \\ \Delta M_H \\ \Delta M_V \end{Bmatrix} = \begin{bmatrix} \bar{D}_{AA} & \bar{D}_{AH} & \bar{D}_{AV} \\ \bar{D}_{HA} & \bar{D}_{HH} & \bar{D}_{HV} \\ \bar{D}_{VA} & \bar{D}_{VH} & \bar{D}_{VV} \end{bmatrix} \begin{Bmatrix} \Delta\varepsilon_0 \\ \Delta\phi_H \\ \Delta\phi_V \end{Bmatrix} \tag{7}$$

Where

$$\begin{aligned} &\bar{D}_{AA} = \sum_{i=1}^{N} D_i A_i, \quad && \bar{D}_{AH} = \bar{D}_{HA} = \sum_{i=1}^{N} D_i y_i A_i \\ &\bar{D}_{HH} = \sum_{i=1}^{N} D_i y_i^2 A_i \quad && \bar{D}_{AV} = \bar{D}_{VA} = \sum_{i=1}^{N} D_i z_i A_i \\ &\bar{D}_{VV} = \sum_{i=1}^{N} D_i z_i^2 A_i \quad && \bar{D}_{HV} = \bar{D}_{VH} = \sum_{i=1}^{N} D_i y_i z_i A_i \end{aligned} \tag{8}$$

The stiffness equation, Eq. 8, can be simplified by the formulation with respect to the variables defined for the instantaneous neutral axes as described in the following.

The expression of the axial force increment ΔP of Eq. 8 can be rearranged in the form

$$\begin{aligned} \Delta P &= \bar{D}_{AA}\Delta\varepsilon_0 + \bar{D}_{AH}\Delta\phi_H + \bar{D}_{AV}\Delta\phi_V \\ &= \sum_{i=1}^{N} D_i \left(\Delta\varepsilon_0 + y_i\Delta\phi_H + z_i\Delta\phi_V \right) A_i \\ &= \sum_{i=1}^{N} D_i \left\{ \Delta\varepsilon_G + (y_i - y_G)\Delta\phi_H + (z_i - z_G)\Delta\phi_V \right\} A_i \end{aligned} \tag{9}$$

where

$$\Delta\varepsilon_G = \Delta\varepsilon_0 + y_G\Delta\phi_H + z_G\Delta\phi_V \tag{10}$$

y_G and z_G are the coordinates and $\Delta\varepsilon_G$ is the axial strain increment at the neutral axis plane caused by $\Delta\varepsilon_0$, $\Delta\phi_H$ and $\Delta\phi_V$. In this regard y_G and z_G are given by

$$y_G = \left(\sum_{i=1}^{N} y_i D_i A_i \right) / \left(\sum_{i=1}^{N} D_i A_i \right) \tag{11}$$

$$z_G = \left(\sum_{i=1}^{N} z_i D_i A_i \right) / \left(\sum_{i=1}^{N} D_i A_i \right) \tag{12}$$

And the Eq. 9 can be simplified as

$$\Delta P = \left(\sum_{i=1}^{N} D_i A_i \right) \Delta\varepsilon_G \tag{13}$$

Eq. 13, when ΔP =0, there is no axial strain at the neutral axis is produced for any combination of horizontal and vertical curvature increments. Therefore, the relationship becomes

$$\begin{Bmatrix} \Delta P = 0 \\ \Delta M_H \\ \Delta M_V \end{Bmatrix} = \begin{bmatrix} D_{AA} & 0 & 0 \\ 0 & D_{HH} & D_{HV} \\ 0 & D_{VH} & D_{VV} \end{bmatrix} \begin{Bmatrix} \Delta \varepsilon_G \\ \Delta \phi_H \\ \Delta \phi_V \end{Bmatrix} \quad (14)$$

Where

$$\begin{aligned} D_{AA} &= \sum_{i=1}^{N} D_i A_i, \quad D_{HV} = D_{VH} = \sum_{i=1}^{N} D_i (y_i - y_G)(z_i - z_G) A_i \\ D_{HH} &= \sum_{i=1}^{N} D_i (y_i - y_G)^2 A_i, \quad D_{VV} = \sum_{i=1}^{N} D_i (z_i - z_G)^2 A_i \end{aligned} \quad (15)$$

Using the coefficients of Eq. (15), the relationship of the bending moments and curvatures can be given by

$$\begin{Bmatrix} \Delta M_H \\ \Delta M_V \end{Bmatrix} = \begin{bmatrix} D_{HH} & D_{HV} \\ D_{VH} & D_{VV} \end{bmatrix} \begin{Bmatrix} \Delta \phi_H \\ \Delta \phi_V \end{Bmatrix} \quad (16)$$

3. Methodology

The ultimate strength analysis of ship hull girder with collision and grounding damages are performed using analytical formulation. The cross section of Ro-Ro ship is taken to be analyzed. The breadth of ship is 15000 mm and the depth is 10693 mm. The longitudinal direction is represented as one-frame space which has 550 mm in length. The boundary condition is assumed to be simply supported at both sides of the cross section as shown in Fig. 2. The element subdivision consists of plate and stiffener element for Ro-Ro ship is illustrated in Fig. 3.

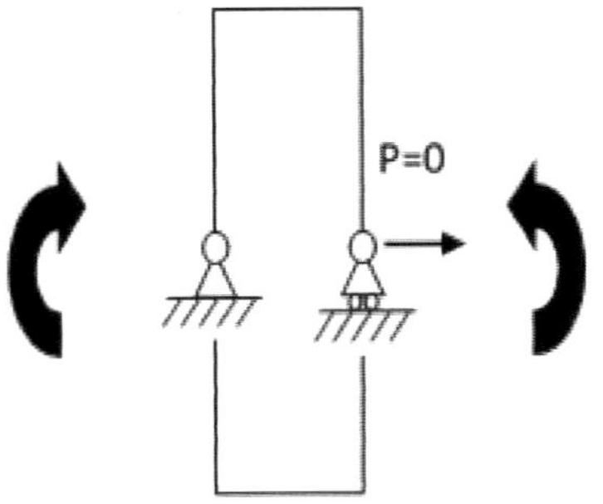

Fig. 2 Boundary condition

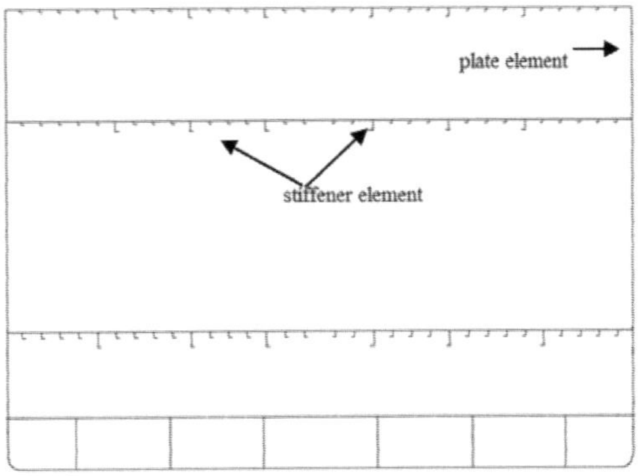

Fig. 3 Element subdivision on Ro-Ro ship

The collision and grounding damages are placed to be located at the side shell and bottom part area, respectively. The collision represented by vertical damage extent and grounding represented by horizontal damage extent. The vertical damage extent is taken as the percentage of ship's depth, while for horizontal represented by ship's breadth. Both collision and grounding damages consider two cases; those are 50% and 70%. Additional damage extent is also applied when collision takes place at deck part, that is B/16, and it is constant for 50% and 70% damages. The collision and grounding damages are illustrated in Fig. 4 and Fig. 5.

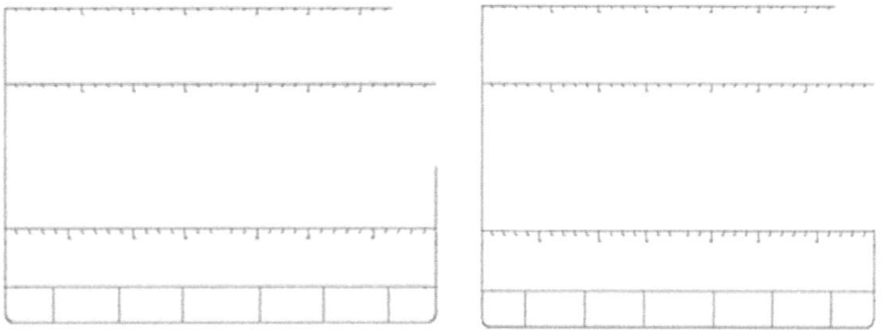

(a) 50% (b) 70%

Fig. 4 Collision damages

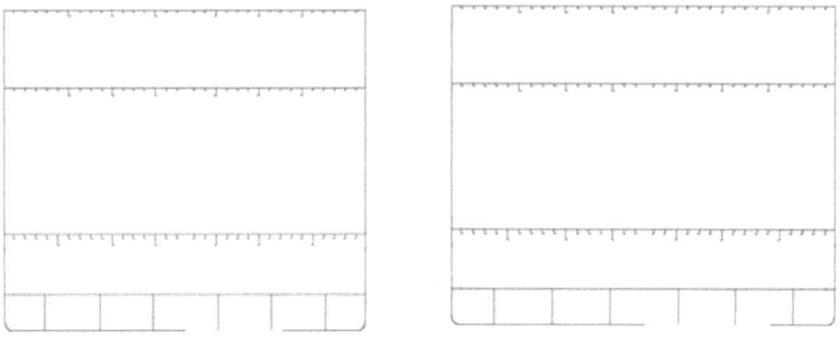

(a) 50% (b) 70%

Fig. 5 Grounding damages

4. Results and Discussion

The ultimate hull girder strength analysis of Ro-Ro ship is performed using the Smith method implemented into the program code HULLST developed by Yao and Nikolov (1992). Fig. 6 and Fig. 7 show the comparison of the ultimate strength in hogging and sagging condition for 70% and 50% damages between collision and grounding.

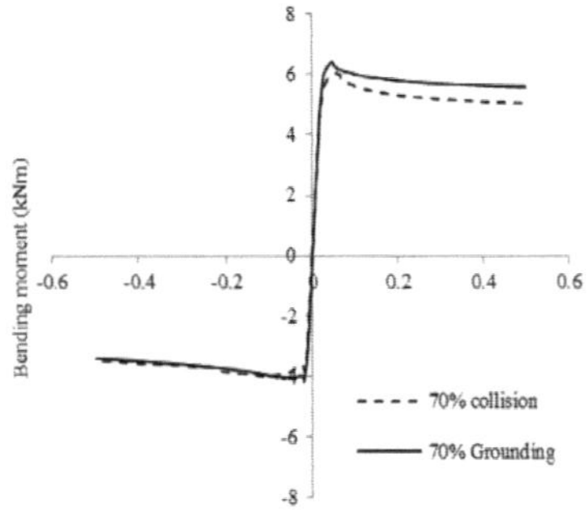

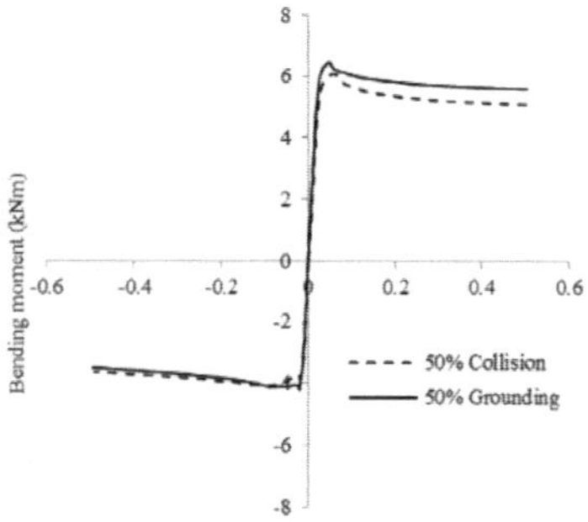

Fig. 6 Comparison of ultimate strength for 70% damages
Fig. 7 Comparison of ultimate strength for 50% damages

Table 1 Ultimate strength in hogging

Damages (m)	Moment (kNm)	
	Collision	Grounding
50%	6,10	6,47
70%	6,05	6,42

Table 2 Ultimate strength in sagging

Damages	Moment (kNm)	
	Collision	Grounding
50%	4,10	4,22
70%	4,04	4,18

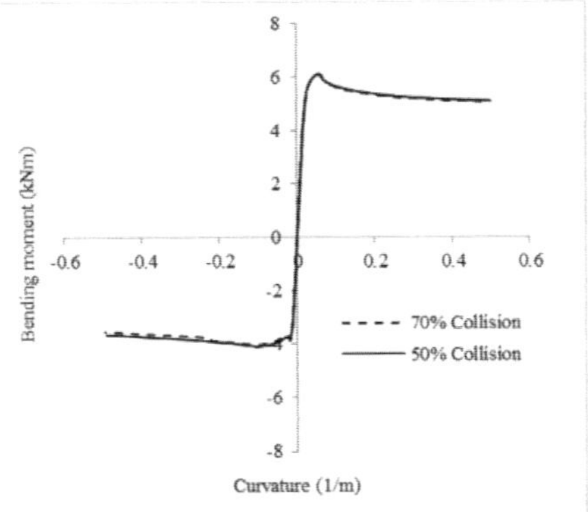

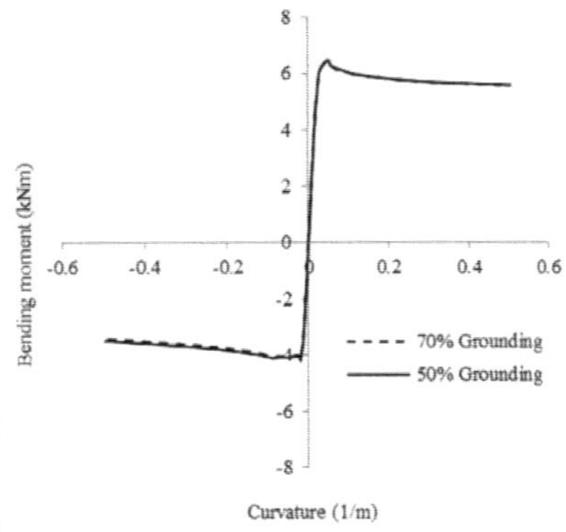

Fig. 8 Comparison of ultimate strength for collision damages
Fig. 9 Comparison of ultimate strength for grounding damages

The result of the ultimate hull girder strength obtained in Fig. 6 and Fig. 7 are summarized in Table 1 for hogging and Table 2 for sagging conditions. According to Table 1 that the ultimate bending moment capacity of the hull girder in hogging condition for collision and grounding are decreased from 50% to 70% damages. This result is also as expected when compared to Table 2, where the ultimate bending moment capacity of the hull girder in sagging condition reduced due to damage extension from 50% to 70%.

Fig. 8 and Fig. 9 express the comparison of the ultimate strength for collision and grounding damages in hogging and sagging condition. The different of the ultimate bending moment capacity for collision and grounding damages from 50% to 70% is 0,05 or 5% in hogging condition. While, the different of the ultimate bending moment capacity for collision damage from 50% to 70% is 0,06 or 6% and grounding damage from 50% to 70% is 0,04 or 4% in sagging condition. The influence of rotation of neutral axis when the collision or grounding damage is considered is not significant. This may be due to the Ro-Ro ship consists of several decks like main, car, passenger and bridge decks.

This may be also caused by the neutral axis still below the car deck and the maximum collision damage (70%) is still above the car deck. But the different of the ultimate bending moment capacity takes place when it is compared for the damage percentage of ship dimension, according to Fig. 6 and Fig. 7 especially in hogging condition. Probably, this behaviour is influenced by the structural component on the bottom and deck part of Ro-Ro ship, where the cross sectional area is quite different between deck and bottom part measured from the neutral axis up to deck and bottom area. It is observed that the deck part consists of many stiffened plate like car, passenger and bridge deck. On the other hand, plate is only found on the bottom part.

Fig. 10 Stress distribution in hogging condition for 70% collision damage

Fig. 11 Stress distribution in sagging condition for 70% collision damage

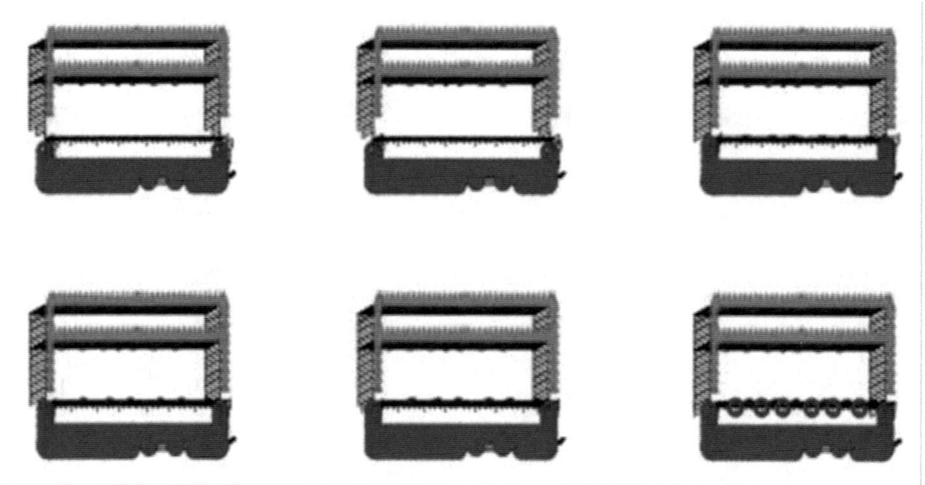

Fig. 12 Stress distribution in hogging condition for 70% grounding damage

Fig. 13 Stress distribution in sagging condition for 70% grounding damage

Fig. 10~13 shows the stress distribution in hogging and sagging condition for 70% collision and grounding damages. The red colour (triangle) indicates collapse elements in tension, and circle one (blue colour) illustrated collapse elements in compression. It is observed that the influence of the rotation of the neutral axis is not significant toward the ultimate strength; it is clear from the stress distribution for the maximum damage (70%).

The stress distribution in hogging condition for 70% of collsion damage is different copared to grounding one. Yielding (red color) distributed to all members above the neutral axis for grounding with 70% damage. While, for the collsion with 70% damage, the yielding spreads only on the intact part of side shell, car, pessanger and bridge decks. Buckling fully takes place on the bottom part in hogging condition for 70% collision damages. For 70% grounding damage, buckling is not fully distrubuted on the bottom part.

For sagging condition, buckling fully occurs at both side shell including pessanger and bridge decks with 70% grounding damage. But yielding is not fully spreaded on bottom area. In case of 70% collision damage, buckling is not totally distributed on the deck member components, but yielding spreads on bottom area.

5. Conclusion

The ultimate hull girder strength analysis of Ro-Ro ship after collision and grounding damages have been performed based on the Smith's method. The following conclusions are; the effect of the rotation of neutral axis on the ultimate bending moment capacity of hull girder is not significant. This may be due to the Ro-Ro ship consists of several decks such as car, passenger and bridge decks. But the different of the ultimate bending moment capacity takes place when it is compared for the damage percentage of ship dimension.

6. References

[1] Paik, J.K. et. al. 1998. *J. Marine Technology*, 35, 38-54.
[2] Muis Alie, M.Z, et. Al. 2012 *Proc 22nd Int Offshore and Polar Eng. Conf.* Greece, ISOPE, 4, 831–8389
[3] Soares, C.G. et al. 2008 *J Intl. Shipbuilding Progress,* 55, 87-107.
[4] Ohtsubo, H. et. al. *J Nuclear Engineering and Design*, 150, 385-396.
[5] Muis Alie, M.Z, et. al. 2016 *Proc 26th Int Offshore and Polar Eng Conf*, Greece, ISOPE, 4, 1022–1029.
[6] Muis Alie, M.Z, et al 2016 *Proc 35th Int Conf. on Ocean, Offshore and Arctic Engineering*, OMAE, Korea.
[7] Yao, T. and Nikolov, P.I. 1992 *J Soc. Naval Arch. of Japan*, Vol 172, 437-446.

The Influences of Stirring and Cow Manure Added on Biogas Production From Vegetable Waste Using Anaerobic Digester

N O Abdullah[1], E S Pandebesie[1]

[1]Departement of Environmental Engineering, Sepuluh Nopember Institute of Technology, Indonesia

Email : nurjannahoktorina@gmail.com

Abstract. Based on Indonesian Government Regulation number 18, 2008, solid waste management should be conducted from the source to minimize the amount of waste. The process includes the waste from domestic, commercial, and institution. This also includes in 3R program (reduce, reuse, and recycle). Vegetable waste from market is a potential material to produce biogas due to its chemical composition (hemi-cellulose, cellulose, and lignin) which transform the biomass to be the raw material of biogas. Acid substance of vegetable becomes an obstacle in process of producing biogas. There has to be *buffer* material which can improve the performance of biogas process. Cow manure is a material which can be easily obtained as *buffer*. This research used 24 biogas reactor in volume 6 L by batch method. Biogas volume is measured by checking the preferment in manometer. Methane measurement is conducted by using *Gas Chromatography* (GC) Hewlett Packard (HP-series 6890) in day 15 and 30. The research was started by sample characterization, sample test by total solid analysis, volatile solid, lignin, ratio C/N, ammonium, and ash. Analysis of pH, temperature, and biogas volume is conducted every day.

1. Introduction

Approximately 1,5 billion ton/year of vegetable and fruit were produced worldwide and 45% of them were wasted. Based on Indonesian Government Regulation number 18, 2008, solid waste management should be conducted from the source to minimize the amount of waste. The process includes the waste from domestic, commercial, and institution. This also includes in 3R program (reduce, reuse, and recycle). Biogas is an environmentally advantageous energy source which is mostly comprised of methane (60%) and carbon dioxide (35-40%). Moreover, biogas contains a low quantity of other gasses such as ammonia (NH_3), hydrogen sulfide (H_2S), hydrogen (H_2), oxygen (O_2), nitrogen (N_2), and carbon monoxide (CO) [1,2]. Biogas is the gas evolved from a process known as anaerobic digestion (AD). AD is known as the degradation of organic compounds to simple substances by microorganisms which live as syntrophy under the lack of oxygen with releasing biogas [3,4]. Vegetable waste from market is a potential material to produce biogas due to its chemical composition 40-50% cellulose, 20%-35% hemi-cellulose, and 15%-30% lignins which transform the biomass to be the raw material of biogas. Acid substance of vegetable becomes an obstacle in process of producing biogas. There has to be *buffer* material which can improve the performance of biogas process. Cow manure is a material which can be easily obtained as *buffer*.

Another factor to produce biogas is mixing. The importance of mixing in achieving efficient substrate conversion has been reported by several researchers [5]. The main factors affecting digester mixing are the mixing strategy, intensity, duration, and the location of the mixer. Adequate mixing was shown to improve the distribution of substrates, enzymes, and microorganism throughout the digester [6], whereas inadequate mixing was shown to result in stratification and formation of floating layer of solids. Continuous mixing was shown to improve biogas production compared to unmixed. Nevertheless, intermediate mixing appears to be the most optimal for substrate conversion. Mixing intensity was also shown to affect digester performance and biogas production [7]. The effect of mixing on aerobic digestion of manure was evaluated in lab-scale and pilot scale experiments at 55°C. They investigated the effect of continuous (control), minimal (mixing for 10 min prior to extraction/feeding), and intermittent mixing (with holding mixing for 2 h prior to extraction/feeding) on methane production in three lab-scale continuously stirred tank reactors. They found that the intermittent and minimal mixing strategies improved methane productions by 1,3% and 12,5%, respectively as compared with continuous mixing.

2. Methods

The digester experiments were carried out in a batch system with a total volume of 6.000 mL. The digester was supported by the mixer and gas sampler. Gas sampling was located at positions corresponding to the top of digester contents. The digester had one outlet at the bottom. The contents of the digester were mixed as controlled by manually 8 times/day which was carried out for 5 minutes. Digester was run at temperature of mesophilic. As for the samples conducted in this experiment were cow manure and vegetable wastes as substrates. The cow manure was taken from Pegirian slaughterhouse located in Surabaya, Indonesia. Vegetable wastes have been collected from local market. It was manually screened to remove any inorganic material like bones, stones, polythene bags, and citrus material like lemon residues. Experiments were operated in batch system. The biogas produced was measured daily by calculating increase water level in the manometer. Measurement of methane rate in the biogas using Gas Chromatography (GC) Hewlett Packard (HP-series 6890). The schematic diagram of experimental laboratory set up as shown in Figure 1.

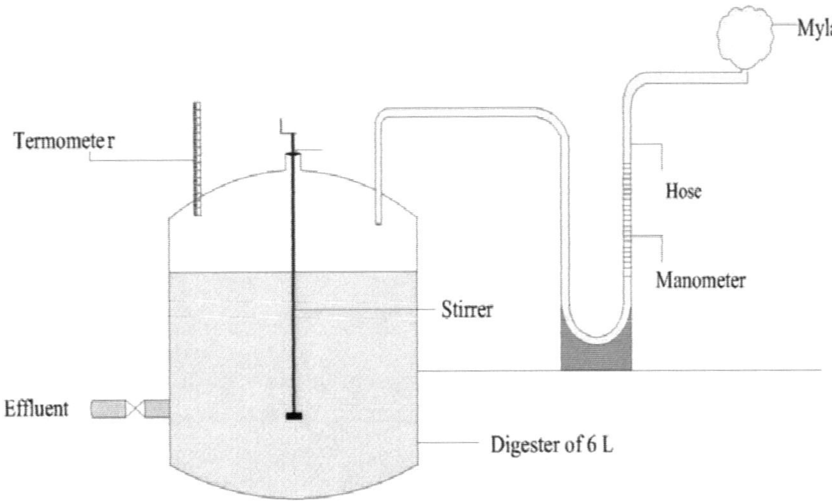

Figure 1. Design of Digester

Figure 2. Plastic Anaerobic Digester

This research used a variable of stirring time, cow manure added, 10% and 20% of total solid. The stirring time was 8 times for 5 minutes, 4 times for 5 minutes, and without stirring. Vegetable waste were added by cow manure. After that, vegetable waste pH was neutralized using NaOH until the pH becomes neutral. The concentration of total solids of the mixture variation in this study was 10% and 20%. Table 1 showed the variables used in this study. The variables used in this study was the influence of the stirring time, the effect of cow manure, and the total effect of solid.

Table 1. Research Variable

No.	Stirring Time (%) (A)	Cow Manure	
		Total Solid 10% (B)	Total Solid 20% (B)
1	8 Times	A1B1	A1B2
2	4 Times	A2B1	A2B2
3	0 Times	KK1	KK2

A1B1: Total solids 10% and stirring 8 times/day
A2B1: Total solids 10% and stirring 4 times/day
KK1 : control and without stirring times
A1B2: Total solids 20% and stirring 8 times/day
A2B2: Total solids 20% and stirring 4 times/day
KK2 : control stirring times

3. Results

3.1 Characterization of Samples

The samples has been characterized and the results have been tabulated as follows. The batch digestion in lab-scale was carried out in 6.000 mL with working volume of 5.000 mL at temperature of 35 ± 1 °C. Characteristics of substrates are important parameters in designing and operating anaerobic reactors. The initial characteristics of substrates strongly affect the biogas yield and anaerobic process stability. Substrates used in this study were individually analyzed for their initial physicochemical characteristics and the results are presented in Table 2. The initial pH of vegetable wastes was lower than the optimum pH required for anaerobic digestion, while the pH of cow manure used for this experiment was comparatively high i.e., 7,9. When substrate was mixed with individual substrate mixtures, pH of the substrate mixtures increased, hence pH was not adjusted artificially. The pH of digestion mixtures A1B1, A2B1, A1B2, and A2B2 after mixing with cow manure was 7,1; 6,9; 7,2; 7,1 respectively. C/N ratios of individual substrates used in this study were either greater or lesser than that of reported optimum range of C/N ratio for anaerobic digestion. C/N ratio of vegetable wastes relatively higher as compared to cow manure, which is due to high nitrogen content of vegetable wastes mainly organic form like protein [5]. Such types of organic wastes like vegetable wastes with relatively high content of lignocelluloses always have low biogas potential [6].

Table 2. Characteristics of vegetable wastes and cow manure

Parameters	Vegetable Wastes	Cow Manure
Water content (%)	90,49	57,4
Volatile solids (%)	93,90	81,66
Carbon (%)	52,17	45,37
Hydrogen (%)	6,26	5,44
Oxygen (%)	38,24	33,81
Nitrogen (%)	1,58	2,06
Ash (%)	0,61	11,8
P-Total (%)	1,14	1,52
Lignin	6,09	9,74
C/N	32,81	21,92
pH	5,6	7,9

For co-digestion of vegetable wastes and cow manure, C/N ratio of 32,81 was not desirable, as it yielded 38% lesser biogas than a mixture with C/N ratio 20. Lower C/N ratio of feedstock or higher organic nitrogen content leads to accumulation of ammonia nitrogen and high pH of 8,5 or more which can inhibit methanogenesis. While, on the other hand at higher C/N ratio of feedstock or lesser organic nitrogen content, methanogens consume nitrogen more rapidly resulting in nitrogen deficiency and hence low biogas yield [7]. The C/N ratio of substrate in range of 20-30 is considered optimum for anaerobic digestion [7]. High C/N ratio of vegetable wastes was used to adjust the C/N ratio of cow manure to 20-30 by mixing both the co-substrates in different ratios without addition of any external nutrients.

3.2 The Effect of Stirring Process on The Production of Biogas

The daily biogas production and methane content were determined. The obtained data revealed that the different stirring periods had the positive effect on the biogas production. It gave a higher biogas yield and biogas productivity. Moreover, the increase the stirring periods increase the biogas yield with about 54,29%; 41,05%; 56,06%; 48,12% at stirring periods of 8 times/day and 4 times/day for 5 minutes respectively. The obtained results, indicated that the stirring periods of 8 times/day was gave the highest value of biogas productivity at stirring period of 5 minutes. These highest values may be due to the complete homogenous of the digestion materials which led to increase the exposed surface area resulted in increasing the microbial growth and consequently increase the totals methanoginic bacteria. The frequency of the best stirring to the formation of biogas contained in the reactor with stirring 8 times for 5 minutes and the addition of cow manure on the total solid 20%. Reactor with stirring frequency 8 times creates a homogeneous substrate preventing stratification and formation of a surface crust, and ensures solids remain in suspension. Further, mixing also enables heat transfer, particle size reduction as digestion progresses and release of produced biogas from digester contents.

IOP Conf. Series: Earth and Environmental Science **135** (2018) 012005 doi:10.1088/1755-1315/135/1/012005

Table 3. Biogas Volume Production

Total Solid	Reactor	Stirring Time	Volume of Biogas (mL)
10%	A1B1	8 times	5.407
	A2B1	4 times	4.192
	KK1	0 times	2.471
20%	A1B2	8 times	8.743
	A2B2	4 times	7.403
	KK2	0 times	3.841

Low solid is the percentage of the mixture with a water content of as much as (4-10: 96-90% w/w). Medium solid is the percentage of the mixture with a water content of as much (15-20: 85-80% w/w) [7,8]. Figure 3 showed the highest cumulative biogas volume and methane levels of biogas generated by a mixture of vegetable waste and cow manure 70:30 (%w/w) with total solid 10% and 20%. The cumulative biogas volume produced at reactor A1B1 was 5.607 mL. The cumulative biogas volume and methane levels produced in the same mixture with 20% total solid at reactor A1B2 was 8.743 mL.

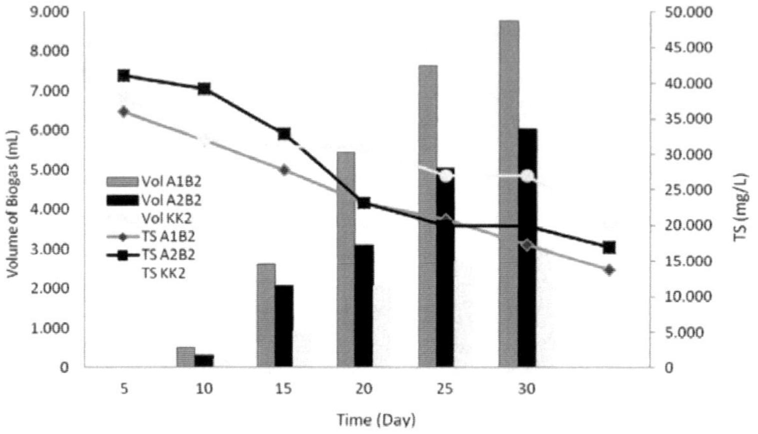

(a)

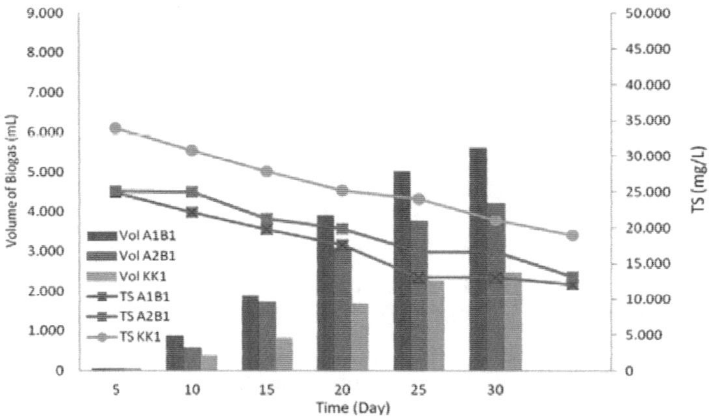

(b)

Figure 3. Total Solids from Six Substrate Mixtures

Total solids and volatile solids reduction from six substrate mixtures is shown in Figure 3. Highest TS and VS reduction was observed for substrate mixture A1B2, while these were the lowest for KK1.

Higher volatile solids reduction yielded higher biogas as trend in VS reduction was similar to biogas yield. Substrate mixture A1B2, that produced the highest biogas also showed the highest VS reduction than all other substrate mixtures showing a relationship between biogas production and VS removal. These results are in accordance with another study where the same trends of TS and VS reduction and relationship between VS reduction and biogas yield have been reported [9]. Significant difference was found among solids reduction from six substrate mixtures. Volatile solids reduction from A1B2 was 1,39 times higher and TS reduction was 1,2 times higher than A1B1 which is total solid 10%. It might be due to higher fraction of vegetable wastes used in total solid 20% as compared to total solid 10%. Similar results have also been reported by several authors where higher VS reduction was achieved with high fractions of vegetable wastes used in substrate mixtures [9].

3.3 The Effect of Cow Manure Added on The Production of Biogas

The anaerobic digestion process is a set of biological reactions under the bacteria action, organized into several microbial populations, each of them playing successive roles for the transformation of complex organic mass into methane and carbon dioxide. All reactions, multiple, and complex, that take place in the digester can be divided into four main stages characterizing the action of different bacteria groups: hydrolysis, acidogenesis, acetogenesis, and methanogenesis [9]. The addition of cow manure as a bio-activator in the anaerobic digester process was expected to increase the activity of the microorganisms to decompose organic matter and produce methane gas. In Figure 4 could be seen the results cumulative biogas formed by the addition of cow manure.

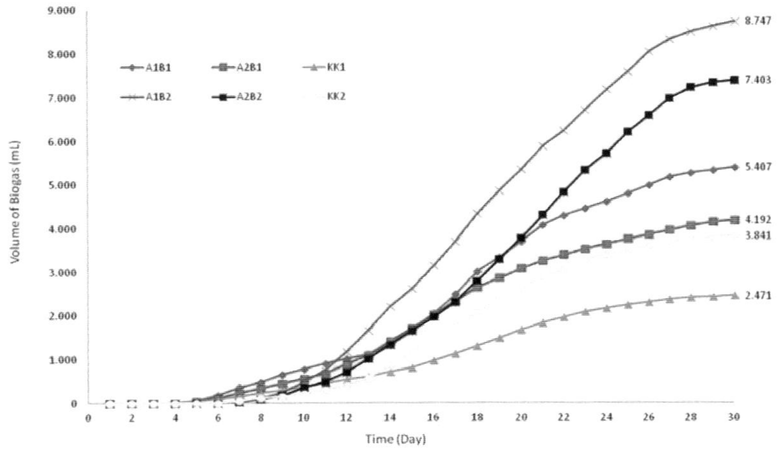

Figure 4. Cumulative of Biogas Volume

To determine the effect of cow manure on the formation of biogas is necessary comparison between the reactor. Best reactor with the addition of cow manure A1B2 contained in the reactor amounted to 8.768 mL. The influence of the addition of cow manure to the formation of biogas contained in the reactor with stirring frequency of 8 times and the addition of cow manure on the total solid 20%.

4. Conclusion

The highest cumulative biogas volume and methane levels generated by stirring 8 times and cow manure added. The highest cumulative volume produced in the reactor amounted to 8.743 mL A1B2 and methane percentage of 61,09%. The volume of biogas produced by the reactor without cow manure is lower than the reactor added cow manure.

References

[1] Abdeshahian, P., Lim, J. S., Ho, W. S., and Hashim, H. 2016. *Potential of Biogas Production from Farm Animal Waste in Malaysia.* Renewable and Sustainable Energy Reviews, 60(1), pp. 714-723.

[2] Alibardi, L. dan Cossu, R. 2015. *Compotition Variability of the Organic Fraction of Municipal Solid Waste and Effects on Hydrogen and Methane Production Potentials.* Waste Management, 23(1), pp. 147-155.

[3] Tchobanoglous G., Theisen H., and Vigil S. *Integrated Solid Waste Management: Engineering Princiles and Management Issues.* Singapore. McGraw-Hill, Inc. 1993.

[4] Cristina, R., Duda, M., Leite, D., and Oliveira, A. 2016. *Anaerobic Co-Digestion of Vegetable Waste and Swine Wastewater in High-Rate Horizontal Reactors with Fixed Bed.* Waste Management, 52(1), pp. 112-121.

[5] Ghosh S., Henry M. P., & Christoper R. W., *Hemicellulose Conversion by Anaerobic Digestion.* Institute of Gas Technology and United Gas Line Company. USA, 1984, 257-258.

[6] Harun M.Y., Radiah D.A.B., Abidin Z.Z., and Yunus R., *Effect of Physical Pretreatment on Dilute Acid Hydrolysis of Water Hyacinth (Eichhornia crassipes).* Bioresource Technology, 2011, 5193-5199.

[7] Lindmark, J., Eriksson, P., dan Thorin, E. 2014. *The Effects of Different Mixing Intensities During Anaerobic Digestion of the Organic Fraction of Municipal Solid Waste.* Waste Management, 78(2), pp. 78-85.

[8] Luthfianto, D., Mahajoeno, E., dan Sunarto. 2012. *Pengaruh Macam Limbah Organik dan Pengenceran Terhadap Produksi Biogas dari Bahan Biomassa Limbah Peternakan Ayam.* Bioteknologi, 9(1), pp. 18-25.

[9] Mazareli, R., C., D., S., Duda, R., M., Leite, V., D., dan Oliveira, D., R., A., D. 2016. *Anaerobic Co-Digestion of Vegetable Waste and Swine Wastewater in High-Rate Horizontal Reactors with Fixed Bed.* Waste Management, 52(1), pp. 112-121.

The Evaluation of Cone Capsule as an Alternative Hull form for Portable Tsunami Lifeboat to Support Evacuation System in the Coastal Regions and Small Islands

Ahmad Fauzan Zakki[1], Suharto[2], Aulia Windyandari[2]

[1]Naval Architecture Department, Engineering Faculty, Diponegoro University, Semarang, Indonesia
[2]Industrial Technology Department, Vocational School, Diponegoro University, Semarang, Indonesia

Email : ahmadfzakki@undip.ac.id

Abstract. Several attempts have been made to reduce the risk of tsunami disasters such as the development of early warning systems, evacuation procedures training, coastal protection and coastal spatial planning. Although many efforts to mitigate the impact of the tsunami in Indonesia was made, no one has developed a portable disaster rescue vehicle/shelter as well as a lifeboat on ships and offshore building, which is always available when a disaster occurs. The aim of the paper is to evaluate the performance of cone capsule shaped hull form that would be used for the portable tsunami lifeboat. The investigation of the boat resistance, intact stability, and seakeeping characteristics was made. The numerical analysis results indicate that the cone capsule is reliable as an alternative hull form for the portable tsunami lifeboat.

1. Introduction

Tsunamis are waves that propagate in all directions and occur because of an impulsive disturbance on the seabed. The impulsive disorders occur because of changes in the shape of the geological structure of the seabed vertically and in a short time. The changes were caused by three main sources, namely tectonic earthquakes, volcanic eruptions, or a landslide that occurred on the seabed. Based on the three sources, the main cause of the tsunami in Indonesia is a tectonic earthquake.

Indonesia is a country prone to tsunamis, especially the coastal areas which deal directly with the meeting of the Eurasian Plate, the Indo-Australian and Pacific, among others, the western part of Sumatra Island, southern part of Java, Nusa Tenggara, the northern part of Papua, Sulawesi and Maluku, and the eastern part of the Borneo Island, [1].

Tsunami in Aceh on December 26, 2004, killed about a quarter million lives around the Indian Ocean region. History tsunami disaster within the last twenty years shows that at least ten tsunami disaster in Indonesia (see Figure 1), [1]. Nine tsunamis of them are destructive tsunami and caused casualties and material, the tsunami in Flores on December 12, 1992 which claimed more than 2000 lives, the tsunami in Banyuwangi, East Java (1994), Biak (1996), Maluku (1998), Banggai, Sulawesi Utara (2000), Ransiki, West Papua (2002), a large tsunami in Aceh (December 2004), the tsunami in Nias (2005), West Java (2006), Bengkulu (2007), and Mentawai (2010), [1]. Impact damage caused by the tsunami has destroyed residential areas and the number of casualties.

Based on data from the Meteorology and Geophysics, it shows that arrival time of the tsunami that occurred in Indonesia, generally between 10-60 minutes, [2]. It is determined that tsunamis occurring in

IOP Conf. Series: Earth and Environmental Science **135** (2018) 012006 doi:10.1088/1755-1315/135/1/012006

Indonesia are a local tsunami. Very short tsunami arrival time will influence the evacuation process and time. The evacuation process should be done rapidly and be supported by the ease of access to the safety equipment.

Accordingly, the application of lifeboat technology which is used on ships and the offshore building is developed as a solution in providing protection and tsunami rescue equipment (rescue disaster-shelter). The design of portable tsunami lifeboat is proposed by adopting the cone capsule as an alternative hull form of the portable tsunami lifeboat. Therefore, the main focus of this paper is to evaluate the performance of cone capsule shaped hull form that would be used for the portable tsunami lifeboat. The investigation of the boat resistance, intact stability, and seakeeping characteristics is made. The results of this study are expected able to provide a positive contribution to the development of technology disaster mitigation in the coastal regions and small islands.

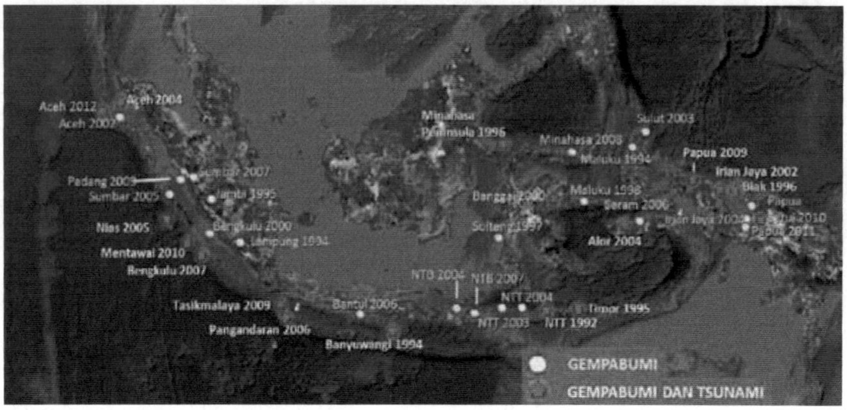

Figure 1. The distribution of tectonic earthquake and tsunami damage between years 1991-2010

2. Literature Reviews

In the study of tsunami lifeboat technology development, the majority of reviewed articles were related to freefall lifeboat technologies, collision and water entry phenomena. It is due to the tsunami lifeboat usability and operational conditions similar to freefall lifeboat that is experiencing an excessive impact load/slamming during water entry phase.

E. Schreck, [3], conducted a study on the development of lifeboat launching simulation in storm conditions. In this simulation, it is considered some variables of detail geometry, waves, and the wind. Coupled simulation of fluid flow and flow induced motion of rigid bodies are used to solve the problems of simulation. The results showed that good agreement between simulation and experimental results.

D. M. Bae, H. S. Kim, A. F. Zakki, [4], conducted a study on the structural response simulations freefall lifeboat for 25 passengers in the event of impact with the water. The use of explicit methods on Arbitrary Lagrangian Eulerian ALE Fluid-Structure Interaction FSI analysis is used to solve this problem. In another paper Zakki, A. F., [5] conduct a study on the assessment of impact load on freefall lifeboat for 25 person launching system using HYBRID III dummy model. Response to motion of Hybrid III Dummy Model was assessed to determine the mechanism of injury occurrence and the effectiveness of the occupant's restraint system in freefall lifeboat.

D. M. Bae, A. F. Zakki, H. S. Kim, J. G. Kim, [6], perform numerical studies on the estimated response of acceleration in freefall lifeboat with FSI Analysis using LS-DYNA Code. The results of this study show that the estimation of the response acceleration has a good agreement between the experimental results with simulation analysis.

D. M. Bae, A. F. Zakki, [7], in another study describes a comparison between the method of multi-material ALE (MMALE) and single material ALE (SMALE) for estimation of the response acceleration in freefall lifeboat. The results showed that MMALE requires computing time is longer than SMALE. However, the use SMALE also be accepted at the design stage of freefall lifeboat development.

H. Luo, H. Wang, C. G. Soares, [8], conduct numerical and experimental studies of elastic response and hydrodynamic collision when a wedge with stiffened panel free fall. In this study, Wagner Uncoupled Method by combining theory and the finite element method is used to analyze slamming 3D structure. Matched asymptotic theory developed to predict the movement and slamming pressure on the free-drop rigid body. Slamming pressures are then added to the finite element models to predict the transient response of the structure. Comparison between the numerical and experimental results shows good agreement.

Olav Aagaard, [9], conducted a study on the hydroelastic analysis on flexible wedges. The simulation model was developed to analyze water entry of the flexible wedges. Effect of viscosity and compressibility is considered in the investigation. The results showed a good agreement between the simulation and experiment.

Espen Larsen, [10], conducted a study on the impact load on Circular Cylinder. 2D simulations using the STAR-CCM + has been carried out at a constant speed and freefall cylinder. The results showed that there was good agreement between the simulation and experiment.

M. Shito, [11], conducted numerical and experimental studies on the tsunami for the design of coastal and offshore structures. Numerical simulation of tsunami propagation, wave impact load and hydraulic experimental techniques to reduce the wavelength of a tsunami was studied in depth in this study. The study shows that it takes more detailed hydraulic experiments to examine every aspect of the tsunami wave loads.

A. F. Zakki, A. Windyandari, D. M. Bae, [12], conducted the development of new type free-fall lifeboat using FSI Analysis. The results of the study provide the parameters for numerical analysis of the acceleration response induced by slamming load during water entry phase in the launching process.

A. R. Prabowo, et al., [13], examined the effects of the selected collision parameter values on the characteristics of collision energy in several ship collision scenarios. The angle position between the two objects during the impact process significantly contributed to the damage pattern on the side hull. The final results also indicated that the cross-section of the target point's location influenced the observed parameters.

D. M. Bae, et al., [14], investigated the several parameters used in the numerical simulation of a collision between two ships. The element model study performed indicates that the Belytschko-Tsay element formulation should be recommended for use in virtual experiments. It is recommended that the real value of the friction coefficient for materials involved is applied in simulations.

D. Chrismianto, et. al., [15], the use of optimization analysis and computational fluid dynamics was applied to the development of the submarine hull form. The parametric model of submarine hull form was proposed using the cubic bezier curve and curve-plane intersection method. The results show that the submarine hull shape might be generated with some variation of the input parameters.

3. The Principal Dimension of Cone Capsule Shaped Hullform

Generally, in freefall lifeboat design planning, there are several requirements provided by the buyer (owner and mission requirements) such as the height of the offshore platform, service speed (Vs), and occupants capacity, in addition to the requirements related to safety, comfort, and beauty. The mission requirements became important information that will be used as a basis by the lifeboat planners to determine the lifeboat's principal dimensions are compatible regarding technical and economical.

In determining the principal dimension of the tsunami lifeboat, it is determined that the design configurations of the lifeboat dimension have a fixed draught of 500mm, the lifeboat heights of 2500 mm and 3500 mm, and the lifeboat breadth/length of 4000 mm. Furthermore, the estimated displacement of the design configurations was used as a reference for the cone capsule shaped tsunami lifeboat dimension.

Since the principal dimensions of the tsunami lifeboat were made, furthermore the shape of the cone capsule shaped hull form is determined by the configuration of geometry parameters, see **Table 1**. The radius of bilge, the ratio of the height of cone and radius of the cone and the depth of floor was considered for the shape configuration of cone capsule shaped hull form see **Figure 2**. The detail shapes of cone capsule shaped hull form for tsunami lifeboat are presented in **Figure 3**.

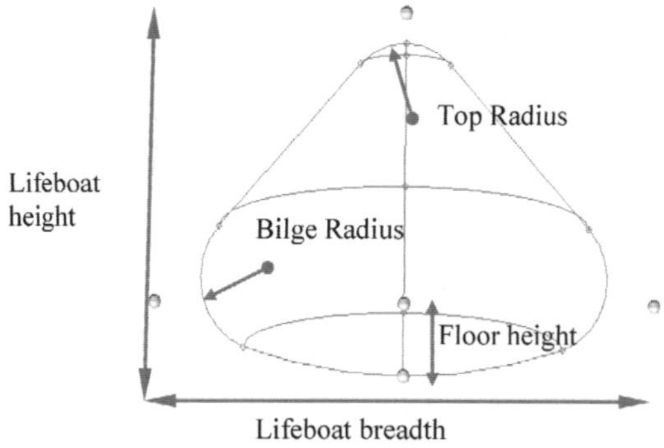

Figure 2. The geometry parameters of cone capsule

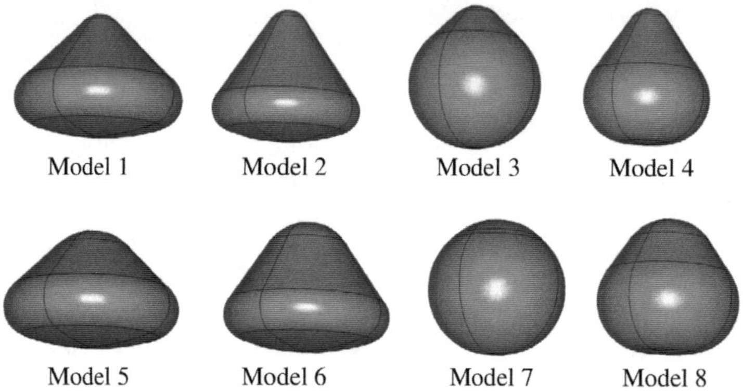

Figure 3. The detailed shape of the configurations of cone capsule shaped hull form

4. Results and discussion

4.1. Intact Stability Performances

The stability curves of the proposed cone capsule shaped hull form are presented in **Figure 4**. Comparison of stability curves of the design configurations of the cone capsule shaped hull form in each of loading conditions has been plotted jointly. Regarding the stability curves calculated for each loading conditions, all conditions have positive GZ. It is indicated that the positive intact stability was conducted by the tsunami lifeboat hull form.

All the case studies fulfilled all applicable IMO stability criteria; see **Table 1**. In the case of area 0° to 40°, the model 6 has the largest area than others. It is indicated that the model 6 have a better righting moment during the heel interval from 0° to 40°. The model 6 also shows better performance in the heel angle 30° to 40°. In the criterion of GZ at 30°, the model 6 and model 2 have the same GZ length of 0.85m which is the largest GZ length amongst others. Accordingly, the numerical results show that the model 6 has an intact stability performance better than the other models.

Table 1. The design configurations of cone capsule shaped hull form for tsunami lifeboat.

Hullform Design	Radius of top [mm]	Radius of bilge [mm]	Draught [mm]	Lifeboat Height [mm]
Model 1	500	500	500	2500
Model 2	500	500	500	3500
Model 3	500	1000	500	2500
Model 4	500	1000	500	3500
Model 5	1000	500	500	2500
Model 6	1000	500	500	3500
Model 7	1000	1000	500	2500
Model 8	1000	1000	500	3500

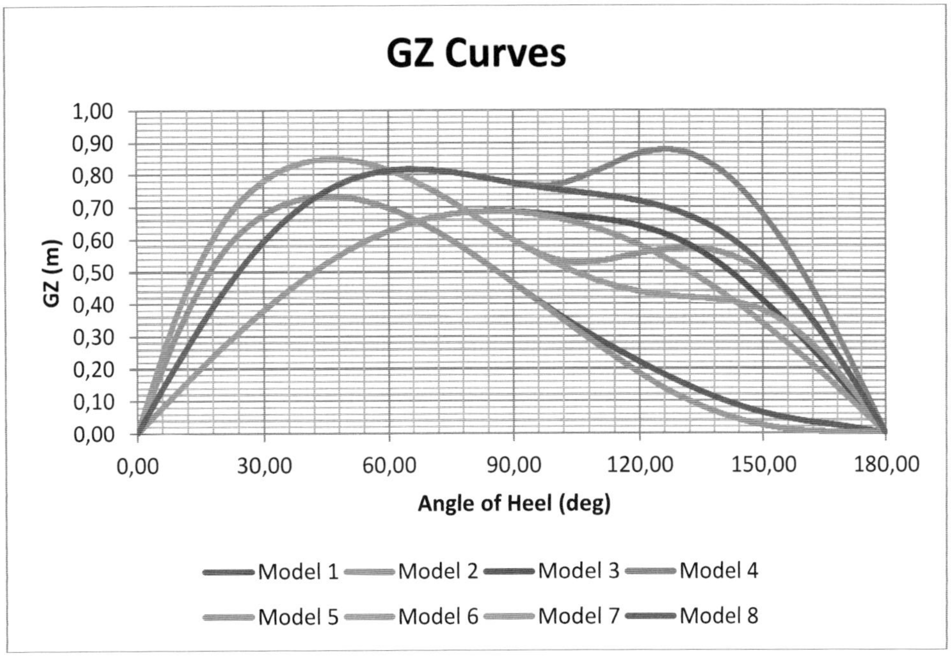

Figure 4. The stability curves of the configuration of cone capsule shaped hull form

4.2. The Seakeeping Performances

In ship design, it is important to pre-determine the behavior of the ship or floating structure when it is subjected to waves. It can be calculated, found through physical model testing and ultimately measured on board the vessel. The calculations can be performed analytically for simple shapes like rectangular barges, but need to be calculated by a computer for any realistically shaped ship. The results of some of these calculations or model tests are transfer functions called Response Amplitude Operators (RAO)

Seakeeping analysis is performed using the wave and wind characteristic of the Java Sea which is defined as the high, rough and phenomenal conditions. The environment conditions were made as an equivalent environment for the tsunami incidence. The rough conditions have the sea state of 5 and the significant wave height of 4 m while the high conditions have a sea state of 7 with the significant wave height of 9m. The most severe condition was defined as phenomenal conditions have the sea state of 9 and the significant wave height of 15 m. Considering the Java Sea characteristics, the ITTC spectra was chosen for the seakeeping analysis. Calculations were made at 0 knots for the lifeboat speed, Head Sea and the ITTC wave spectrum with significant wave height of the rough, hard and phenomenal conditions.

IOP Conf. Series: Earth and Environmental Science **135** (2018) 012006 doi:10.1088/1755-1315/135/1/012006

Table 2. Tsunami lifeboat intact stability and IMO criteria

Cone Capsule Shaped Tsunami Lifeboat									
Criteria	Required	Model							
		1	2	3	4	5	6	7	8
Area 0° to 30°	3,15 m.deg	12.33	14.42	5.89	9.57	12.33	14.43	5.89	9.57
Area 0° to 40°. or Downflooding point	5,16 m.deg	19.40	22.56	10.22	16.12	19.40	22.57	10.21	16.12
Area 30° to 40°. or Downflooding point	1,719 m.deg	7.07	8.14	4.33	6.55	7.07	8.15	4.33	6.55
GZ at 30°. or greater	0,2 m	0.734	0.85	0.687	0.82	0.734	0.85	0.69	0.82
Angle of GZ max	25 deg	45.5	46.4	85.5	126.4	45.5	46.4	85	65.9
GM	0,15 m	1.93	2.33	0.78	1.31	1.93	2.34	0.78	1.31

Seakeeping analysis is calculated by using commercial software with frequency domain strip theory. The result of the analysis is Response Amplitude Operator (RAO). RAO is the transfer function expressing the relation between wave spectrums in response to movement of the vessel (ship response spectrum). The response characteristics of the movement of the proposed hull form can be seen from RAO. On the RAO diagram, the magnitude of the response sensitivity of the lifeboat against the excitation force caused by ocean waves might be seen. The RAO on the heave motion of the developed tsunami lifeboat hull form in the Head Sea might be seen in **Figure 5**. Furthermore, the motion response of the tsunami lifeboat was shown in **Table 2**.

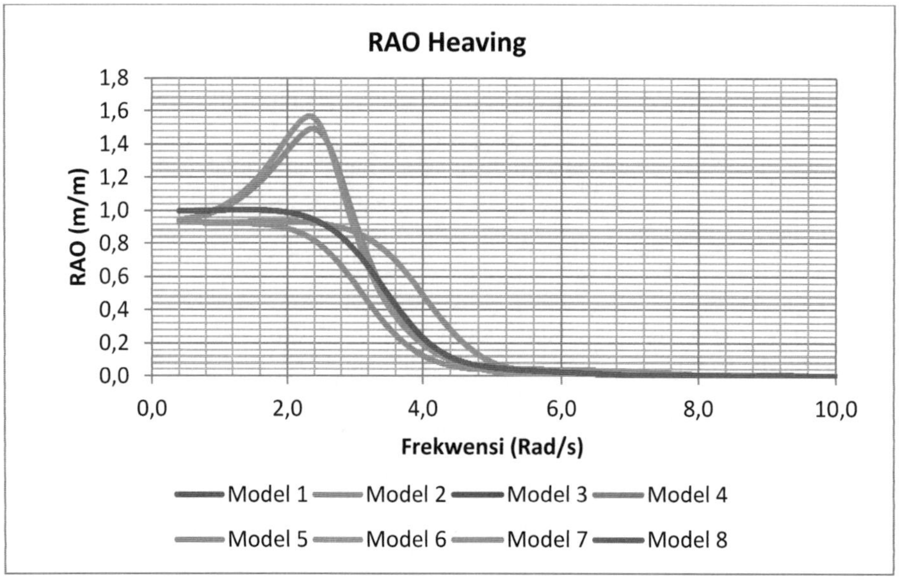

Figure 5. RAO of cone capsule shaped models heave in head sea

Based on the results obtained from the calculation (**Figure 5** and **Table 2**), it appears that the heave motion response of the entire lifeboat hull forms was similar. The amplitude of the heave of the lifeboat hull forms is 0.93m-1m, 2.15m-2.25m and 3.49m-3.75m for the rough, high and phenomenal conditions, respectively. The smallest heave motion amplitudes are shown by model 2. Otherwise, the model 4 and model 8 have shown the largest amplitude of the heave motion as high as 3.75m. Accordingly, it is indicated that the high cone with the relatively flat bottom able to reduce the heave motion response. The large bottom area has a damping effect to the heave motion of the lifeboat. Furthermore, the heave velocity and acceleration response appeared in the range of 0.79-1.70m/s and 0.93-1.44m/s^2, respectively. The minimum heave velocity and acceleration response were also made by the model 2. On the other hand, the model 8 has the maximum heave velocity, and the model 6 has the maximum heave acceleration.

In the case of pitch motion, it appears that the minimum pitch motion response was made by the model 2. The amplitude of the pitch motion of the lifeboat hull forms is 6.12°-8.90°, 6.93°-9.57° and 7.32°-9.90° for the rough, high and phenomenal conditions, respectively. The smallest amplitude of pitch motion was made by the model 2 of 6.12° in the rough conditions. Otherwise, the maximum amplitude is shown by the model 6 of 9.90° in the phenomenal sea conditions. The pitch velocity and accelerations of the lifeboat hull forms were obtained as 0.19rad/s -0.30rad/s and 0.41rad/s2-0.80rad/s2, respectively. The minimum pitch velocity and acceleration are shown by the model 2. Accordingly, it may be concluded that the model 2 have the best seakeeping performance than the other hull forms.

Table 3. Tsunami lifeboat motion behavior in the rough, high and phenomenal environment

H significant 4 m (Sea State 5 - Rough)								
	Model 1	Model 2	Model 3	Model 4	Model 5	Model 6	Model 7	Model 8
Heave motion (m)	0.99	0.93	0.93	1.00	0.99	1.00	0.93	1.00
Pitch motion (deg)	8.73	6.12	7.02	6.58	8.73	8.90	7.02	6.58
Heave velocity (m/s)	0.91	0.79	0.81	0.86	0.91	0.93	0.81	0.86
Pitch velocity (rad/s)	0.29	0.19	0.27	0.23	0.29	0.29	0.27	0.23
Heave acceleration (m/s2)	1.24	0.93	1.03	1.04	1.24	1.27	1.03	1.04
Pitch acceleration (rad/s2)	0.68	0.47	0.80	0.60	0.68	0.66	0.80	0.60

H significant 9 m (Sea State 7 - High)								
	Model 1	Model 2	Model 3	Model 4	Model 5	Model 6	Model 7	Model 8
Heave motion (m)	2.15	2.09	2.10	2.25	2.15	2.16	2.10	2.25
Pitch motion (deg)	9.40	6.93	7.71	7.34	9.40	9.57	7.71	7.34
Heave velocity (m/s)	1.31	1.21	1.23	1.31	1.31	1.33	1.23	1.31
Pitch velocity (rad/s)	0.30	0.20	0.26	0.23	0.30	0.30	0.26	0.23
Heave acceleration (m/s2)	1.36	1.06	1.15	1.18	1.36	1.38	1.15	1.18
Pitch acceleration (rad/s2)	0.68	0.47	0.74	0.59	0.68	0.66	0.74	0.59

H significant 15 m (Sea State 9 - Phenomenal)								
	Model 1	Model 2	Model 3	Model 4	Model 5	Model 6	Model 7	Model 8
Heave motion (m)	3.56	3.49	3.50	3.75	3.56	3.57	3.50	3.75
Pitch motion (deg)	9.72	7.32	7.81	7.64	9.72	9.90	7.81	7.64
Heave velocity (m/s)	1.66	1.58	1.59	1.70	1.66	1.68	1.59	1.70
Pitch velocity (rad/s)	0.29	0.19	0.22	0.21	0.29	0.29	0.22	0.21
Heave acceleration (m/s2)	1.41	1.14	1.19	1.25	1.41	1.44	1.19	1.25
Pitch acceleration (rad/s2)	0.64	0.41	0.53	0.48	0.64	0.63	0.53	0.48

IOP Conf. Series: Earth and Environmental Science **135** (2018) 012006 doi:10.1088/1755-1315/135/1/012006

5. Conclusion

The evaluation of the cone capsule shaped hull forms was made. The design configurations were developed through the modifications of the cone capsule geometry parameters such as the radius of bilge, the height of the boat, and the radius of the top. It was determined for the tsunami lifeboats draught is 0.50m. The modifications may produce the hull form models that was used as an alternative hull form for the tsunami lifeboat

Based on the results of numerical analysis, it is indicated that the entire hull forms model was accepted by the IMO Criteria. However, the model 6 has shown a better intact stability performance than the other hull form models. The small radius of bilge and the large radius of the top may improve the intact stability performance for the cone capsule shaped hull form. In the case of seakeeping analysis, the model 2 shows the better seakeeping performance than the others. The minimum amplitude of the motions was obtained as 3.49m for heave motion and 7.32° for pitch motion on the phenomenal sea condition.

Although the numerical analyses have shown that the entire hull form models were reliable for the tsunami lifeboat on the intact stability and seakeeping performance, however, the experimental studies should be made as a requirement of the classification regulations for the development of a new lifeboat hull forms.

References

[1] BMKG 2012 *Pedoman pelayanan peringatan dini tsunami* (Indonesia: Pusat gempa bumi dan tsunami kedeputian bidang geofisika Badan Meteorologi Klimatologi dan Geofisika)
[2] BMKG 2010 *Katalog Tsunami*
[3] Schreck E, 2008 *Proc. 6th Int. Conf. on CFD in Oil & Gas (Trondheim)* p 1
[4] Bae D M, Kim H S and Zakki A F 2009 *Proc. of the Ann. Aut. Meet. SNAK (Mungyeong)* p 929
[5] Bae D M, Kim H S and Zakki A F 2009 *Proc. of the Ann. Aut. Meet. SNAK (Mungyeong)* p 938
[6] Bae D M, Zakki A F, Kim H S, and Kim J G 2010 *J. Soc. Nav. Archit. Korea* **47** 681
[7] Bae D M and Zakki A F 2011 *J. Soc. Nav. Archit. Korea* **48** 552
[8] Luo H, Wang H, and Soares C G 2012 *Ocean Eng.* **40** 1
[9] Aagaard O 2013 *Hydroelastic Analysis of Flexible Wedges* (NTNU)
[10] Larsen E 2013 *Impact Loads on Circular Cylinders* (NTNU)
[11] Shito M, Inuzuka I, Amaya I, Saito H, and Kurata J 2014 *IHI Eng. Rev.* **46** 21
[12] Zakki A F, Windyandari A and Bae D M 2016 *J Mar Sci and Tech-Taiw* **24** 575
[13] Prabowo A R, Bae D M, Sohn J M and Zakki A F 2016 *IJTech* **4** 592
[14] Bae D M, Prabowo A R, Cao B, Zakki A F and Haryadi G D 2016 *J Mar Sci Appl* **15** 63
[15] Crismianto D, Zakki A F, Adietya B A, Kim D. J. 2015 *J Mar Sci Appl* **14** 399

IOP Conf. Series: Earth and Environmental Science 135 (2018) 012007 doi:10.1088/1755-1315/135/1/012007

The Study on Stability and Seakeeping Characteristics of the Glass Bottom Boat Trimaran in Karimunjawa Island

Berlian Arswendo Adietya[1], Aulia Windyandari[2], Ahmad Fauzan Zakki[1]

[1]Naval Architecture Department, Engineering Faculty, Diponegoro University, Semarang, Indonesia
[2]Industrial Technology Department, Vocational School, Diponegoro University, Semarang, Indonesia

Email : berlian@undip.ac.id

Abstract. Recently the diversity of fish populations in the waters Karimunjawa Island is only appreciated by those who have the ability to play diving and snorkeling. It is due to the unavailability of a vehicle that is specially made to delight in the fascination of the underwater panorama. One of the alternative solutions is using the glass bottom boat technology which is using transparent bottom that might look out the underwater scenery instead of swimming and snorkeling. The paper has focused on the study of intact stability and seakeeping characteristics of glass bottom boat trimaran in Karimunjawa Island. The intact stability characteristics will be investigated at the various load cases and weight distribution configurations which are influenced by the passenger positions and fuel tank condition. Regarding the seakeeping performance analysis, the ITTC-Bretschneider will be adopted as the wave spectrum at the wave parameters defined from the operational environment. The influence of the parameters on the stability and seakeeping of the glass bottom boat trimaran are presented and discussed.

1. Introduction

Jepara regency is one of the regencies in Central Java province. The Jepara regency district borders are the Java Sea in the west and the north, Pati and Kudus Regency in the east, as well as Demak district in the south. Jepara regency also includes the Karimunjawa Island, located in the Java Sea. Jepara district located in the eastern north coast of Central Java, where the western and northern parts bounded by the sea. The eastern part of the District is a mountainous region.

Jepara regency also includes the Karimunjawa Islands, which is a cluster of islands in the Java Sea. Two of its largest island is the Karimun and Kemujan Island. Most of the area is protected under Karimunjawa Marine Conservation. The public transportation to the islands served by a ferry that departed from the Port of Jepara and the small airport is available for the small plane that may be used in private or emergency conditions.

Karimunjawa, a group of islands in the Java Sea, a few miles northwest of the city of Jepara, Central Java, is a paradise for coral reefs, mangroves, and coastal forests. Karimun Islands, which has a land area of 1,500 hectares and waters covering an area of 110,000 hectares, is an excellent habitat for nearly 400 species of marine fauna. Ecological environment in Karimun until now relatively still preserved. Since the 27 islands in the archipelago of Karimunjawa, only five are inhabited, while 22 other islands are original, beautiful and a virgin. The inhabited islands are Karimunjawa, Kemujan, Nyamuk, Parang, and Genting. On March 15, 2001, The Karimunjawa Island was declared as a National Park by the

Government of Karimunjawa Jepara regency to protect the natural wealth of coral reefs, mangroves, coastal forests and marine fauna of the archipelago. Some of the endangered species is live comfortably in Karimunjawa Islands, including white-bellied sea eagle, hawksbill and green turtles, as well as approximately 242 species of ornamental fish.

Panorama and the natural resources are owned by the Karimun Islands often tempt tourists to come to the island despite having a long journey. They usually visit the island of Menjangan. Menjangan Besar is an island that is used for breeding center sharks, turtles, and eagles. The visitors who have more guts can also swim or do snorkeling accompanied by dozens of sharks, and watch the beautiful underwater panorama, see **Figure 1**.

Figure 1. Tourist attraction in Karimunjawa

Nowadays the beauty of the underwater panorama is usually only enjoyed by those who can swim, snorkel and dive. It is due to the unavailability of a vehicle that is specially made to enjoy the beautiful underwater panorama, for tourists who cannot swim and dive. One of the alternative solutions is using the glass bottom boat technology which is using transparent bottom that might look out the underwater scenery instead of swimming and snorkeling. Therefore the glass bottom trimaran design is proposed and studied to have the optimum hull form and seakeeping performance to support the tourism activities in Karimunjawa.

Based on the condition, the paper has focused on the study of intact stability and seakeeping characteristics of glass bottom boat trimaran in Karimunjawa Island. The intact stability characteristics will be investigated at the various load cases and weight distribution configurations which are influenced by the passenger positions and fuel tank condition. Regarding the seakeeping performance analysis, the ITTC-Bretschneider will be adopted as the wave spectrum at the wave parameters defined from the operational environment.

2. Literature Reviews

Hullform trimaran is a design concept that has been developed from a boat equipped with two outriggers on the right and left aside. Trimaran design advantage is endurance capability that allows the ship to be driven at high speeds even faster than a speedboat. The trimaran also offers the ability to deal with the conflicting requirement to provide sufficient hydrostatic restoring moment to achieve adequate transverse stability, which is the primary function of the outriggers or side hulls, especially, vital for a ship when damaged and partially flooded. However, the main advantage trimaran design that is considered as a glass bottom hull form is a larger area and the flexibility for the design of the particular main deck and bottom part than a monohull. Additionally, trimaran design is possible to provide wider view access at the bottom part which is not blocked by demi-hull like on the catamaran hull form. The trimaran proposed design may be seen in **Figure 2**.

IOP Conf. Series: Earth and Environmental Science **135** (2018) 012007 doi:10.1088/1755-1315/135/1/012007

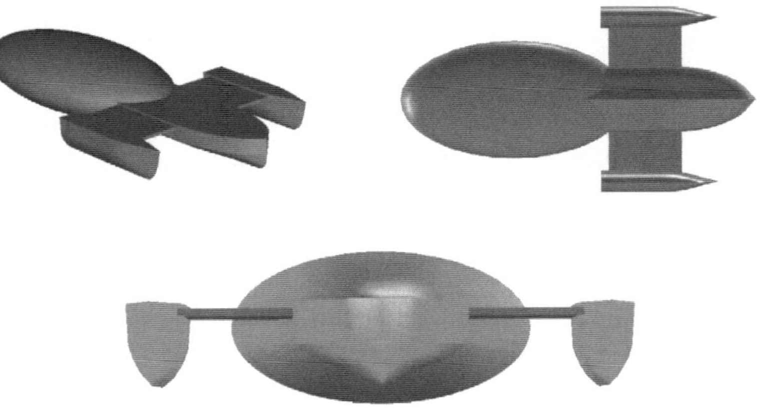

Figure 2. The trimaran hull form for the glass bottom boat

In this study of glass bottom technology development, the commonly reviewed articles were related to the trimaran technology/multihull and hull form development procedure. It is due to the glass bottom boat that would be developed using the trimaran hull form. Konstantin I. Matveev, [1], conducted a study on the aerodynamic characteristics of the hybrid model trimaran. Investigation of the aerodynamic behavior of the hybrid trimaran which consists of three wave-piercing hull planning and wing-shaped superstructure was made. The use of interceptors on the pressure side of the wing has significantly augmented the aerodynamic lift force.

Ming-Chung Fang, [2], performs the parametric studies of the trimaran wave loads on ships. Analysis based wave spectra carried out to determine the setting side of the hull during sailing trimaran wave conditions. The magnitude of the significant wave loads, including shear force, bending moments and moments of torque in each different location in the main hull and the cross deck with consideration stagger and clearance were derived using spectral analysis. This study provides information for setting hull side by considering the wave loads on ships trimaran.

Xu Min, [3], conducted a study on the numerical optimization to ship trimaran hull side. Based on the Green's function and potential theory of three-dimensional, this study provides significant value to the wave loads, including shear forces and bending moments vertically at each different location on the main hull, as well as bending moments transverse to the connection structure and the main hull. Parametric optimization on wave loads for the layout and hull length different side also performed. The results of this study are very useful for the design and optimization of a new type of trimaran.

Hafez, K. A., [4], to investigate the effect of variation comparison stagger at high-speed trimaran hydrodynamic interference. Methods slender body which is the software used to calculate Hullspeed symmetry series resistance trimaran on the water surface calm and deep waters. Each series consists of 4681 trimaran configuration derived by considering 151 staggers. Three series trimaran being studied are AMECRC, Wigley, and NPL.

Su Yu-Min, [5], conducts studies and numerical experiments on hydrodynamic performance of the trimaran planing type channel. Numerical studies were made using RANS-VOF solver to analyze the performance of hydrodynamic channel type trimaran planing. A series of experiments on the hydrodynamic towing tank was made, to get the movement behavior and performance barriers trimaran sailing. The study indicates that the resistance was decreased significantly due to the influence of a combination of hydrodynamic and aerodynamic lift, while the peak of high-speed resistance is reached.

Lv Jian, [6], undertakes the analytical studies on the response of the bottom panel due to the slamming loads. The response of bottom panel was studied analytically by using elastic Euler-Bernoulli beam as a cross-sectional representation of the bottom panel. The slamming pressure was modeled as a high-intensity peak which is followed by a constant lower pressure, at a constant velocity along the beam. The study shows that the maximum deflection and bending moment occurs when slamming loads on transverse beams approaching the lowest natural frequency of the beam. At high-speed slamming, the bottom panel response was decreased. The high-intensity peak pressure is not the cause of the structural response to be enlarged. However, the total peak load slamming has a significant effect on the response of the structure. The slamming load on the low motion was not influenced significantly. However, the load case was influenced substantially by the rapid motion.

A. F. Zakki, A. Windyandari, D. M. Bae, [7], the development of new type free-fall lifeboat using FSI Analysis. The results of the study provide the parameters for numerical analysis of the acceleration response induced by slamming load during water entry phase in the launching process.

D. Chrismianto, et. al., [8], the use of optimization analysis and computational fluid dynamics was applied to the development of the submarine hull form. The parametric model of submarine hull form was proposed using the cubic bezier curve and curve-plane intersection method. The results show that the submarine hull shape might be generated with some variation of the input parameters.

3. Intact Stability and Loading Conditions

The Archimedes principle states that a body immersed in fluid experiences an upthrust equal to the weight of the fluid displaced. Since the glass bottom boat trimaran is a floating body, the boat floating freely in still water experiences a downward force due to gravity is known as the weight. The boat is in equilibrium, while the weight has the same magnitude and the same line action with the buoyancy force. Otherwise, the boat would move until the equilibrium position achieved.

The boat might be inclined in any direction. Any inclination might be considered as made up of an inclination in the transverse plane and inclination in the longitudinal plane. In equilibrium and stability calculations the transverse inclination is known as the heel, and the longitudinal inclination called trim. The equilibrium position and intact stability characteristics of the glass bottom trimaran boat were estimated, considering the loading conditions that describe the operational conditions of the boat.

The loading conditions of the trimaran boat were determined based on the number of occupants and the configuration of the occupant positions during the evacuation process. Therefore, the loading conditions in the equilibrium and intact stability calculations are:

a. Condition I: Departure Full Occupant (LCG =-0.287 m from midship)
b. Condition II: Departure 50% Occupant (LCG =-0.059m from midship)
c. Condition III: Arrival 100% Occupant (LCG =-0.426 m from AP)
d. Condition IV: Arrival 50% Occupant (LCG =-0.119 m from AP)
e. Condition V: Lightweight Condition (LCG=0.223 m from midship)

The intact stability calculations have been made using Krylov's method, considering the same load case with the loading condition that was used for the equilibrium calculations. No free surfaces in boat units were considered. The center of gravity for the lightweight/empty condition was considered at midship. For the cases of the calculations using the loading conditions, the center gravity was adapted following the weights distribution of the occupant positions.

4. Wave spectrum on Seakeeping Analysis

The ship response for the certain speed in the regular wave is described in Response Amplitude Operator (RAO) curve. Strip Theory Method has used this study with several speeds 0 to 6 knot and 180o for the heading of the wave. For calculation of ship response to the irregular wave can describe with Response Spectrum as shown in Equation 1.

$$S\zeta r\,(\omega) = RAO2 \times S\zeta\,(\omega) \tag{1}$$

$S\zeta\,(\omega)$ is noted as Wave Spectrum. The Bretschneider or ITTC formula with two parameters is used as shown in Equation 2. The first parameter is Significant Wave Height (Hs) and the second parameter is Average Period (Tav). The several Hs and Tav are shown in **Table 1**.

$$S_{ITTC\varsigma}(\omega) = \frac{A}{\omega^5} EXP\left(\frac{-B}{\omega^4}\right) \quad (2)$$

where :

ω = wave frequency (rad/sec) $A = 172{,}75 \frac{{H_s}^2}{{T_{ave}}^4}$ $B = \frac{691}{{T_{ave}}^4}$

Table 1. The height of Significant Wave (Hs) and Average Period (Tav).

Hs (m)	Tav (s)
0,25	1,931
0,50	2,731
1,00	3,862

The area under the Response Spectrum is called m0 as shown in Equation 3. The average and height of significant ship motion are described in Eq. 4 and Eq.5. The calculation of the average and height of significant velocity and acceleration are available in Eq. 6 and Eq. 7 by combining with Equation 4 and 5 with substituting the value of $m_{0,2,4}$.

$$m_0 = \int_0^{\infty} S_{\varsigma r}(\omega)\ d\omega \quad (3)$$

$$(\zeta)av = 1{,}253 \sqrt{m_0} \quad (4)$$

$$(\zeta)s = 2 \sqrt{m_0} \quad (5)$$

$$m_2 = \int_0^{\infty} \omega^2\ S_{\varsigma r}(\omega)\ d\omega \quad (6)$$

$$m_4 = \int_0^{\infty} \omega^4\ S_{\varsigma r}(\omega)\ d\omega \quad (7)$$

The results of seakeeping calculation are evaluated according to criteria. The criteria of seakeeping that used are General Criteria for Fast Small Craft [9] as showed in **Table 2**.

Table 2. General Criteria for Fast Small Craft, [9]

No	Standard	Value
1	Single amplitude average pitch, (θ)av	3°
2	Significant heave acceleration, $(a_z)s$ for no people working on deck	≤ 0,4g

5. Results and discussion

5.1. The Glass Bottom Boat Trimaran Stability

The results GZ curve for each load case is showed in **Figure 3**. The criteria evaluation is showed in **Table 3**. The GZ curve is a moment arm that use to return the ship to an initial condition after roll motion. The positive value of GZ indicates the ship can return to the initial condition. On the other hand, the negative value of GZ indicates the ship cannot return to the initial condition.

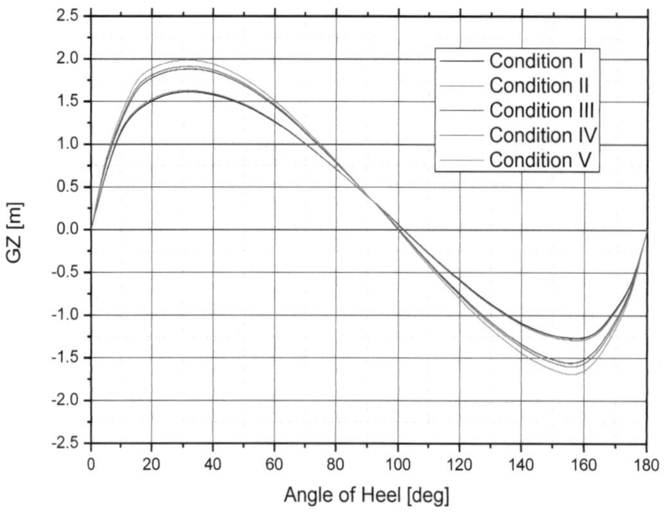

Figure 3. GZ curves of the glass bottom boat trimaran

Table 3. Criteria Evaluation for GZ Curve

Criteria	Value	Condition I	Condition II	Condition III	Condition IV	Condition V
Angle of Maximum GZ	25 deg	31.8 deg	31.4 deg	32.3 deg	32.3 deg	31.4 deg
Initial GMt	0.15 m	7.73 m	7.89 m	9.10 m	9.37 m	9.89 m

As showed in Figure 3 and Table 3, the boat has the positive value of GZ over 90o of heel angle in every load case. It indicates that the boat has a good stability. For criteria evaluation, the boat passes the criteria

5.2. The Seakeeping Characteristics.

Ship response in the regular ship is described in RAO Curve. The minimum response for the ship occurs when the peak of RAO curve is low. **Figure 4** and **Figure 5** show the heave and pitch RAO curve with the different speed.

RAO for each motion is constant with speed. The increasing speed of the vessel, the peak RAO has increased as well. The results of criteria evaluation were available in **Table 4** and **Tablc 5**.

Table 4. Criteria Evaluation for Single Amplitude Average Pitch (Degree)

Hs [m]	Standard	0 knot	1.5 knot	3 knot	6 knot	12 knot
0.25	< 3	2.7×10-4	2.3×10-4	2.0×10-4	1.5×10-4	1.0×10-4
0.50	< 3	9.4×10-4	8.6×10-4	7.9×10-4	6.8×10-4	5.1×10-4
1.00	< 3	2.1×10-3	2.0×10-3	1.9×10-3	1.8×10-3	1.7×10-3

Table 5. Criteria Evaluation for Significant Heave Acceleration (g) for No People Working on Deck

Hs [m]	Standard	0 knot	1.5 knot	3 knot	6 knot	12 knot
0.25	< 0.4	1.02×10-5	2.04×10-5	2.04×10-5	1.02×10-5	1.02×10-5
0.50	< 0.4	2.04×10-5	2.04×10-5	2.04×10-5	2.04×10-5	3.06×10-5
1.00	< 0.4	3.06×10-5	4.08×10-5	4.08×10-5	5.10×10-5	8.16×10-5

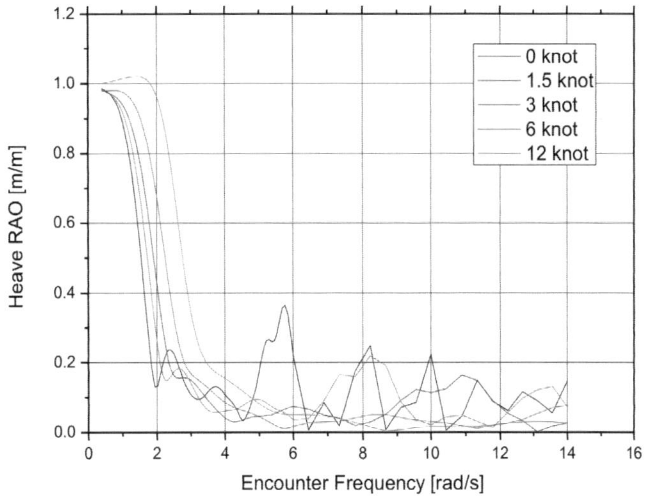

Figure 4. Heave RAO

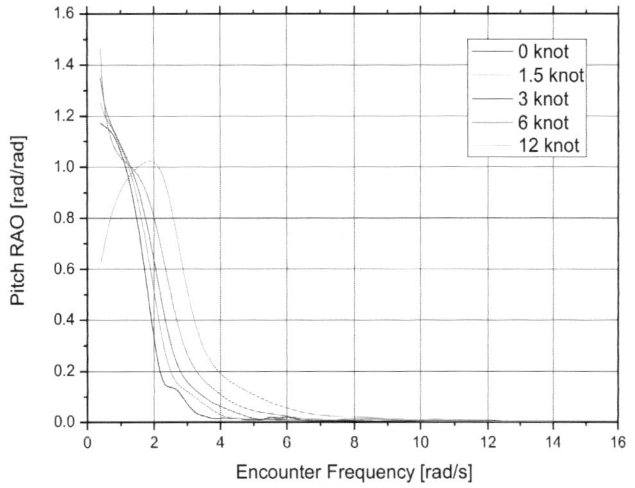

Figure 5. Pitch RAO

As showed in **Table 4** and **Table 5**, for average amplitude pitch the trimaran boat is passed the criteria for condition 0-12 knot and height of significant wave 0.25 – 1 m. Based on the seakeeping analysis, it is concluded that the trimaran boat is reliable for the Glass Bottom Hullform to support tourism activities in Karimunjawa Islands.

6. Conclusion

The calculation of stability analysis shows that trimaran boat has the positive value of GZ over 90o of heel angle in every load case. The trimaran boat can return to the initial condition in roll motion. It indicates that the trimaran boat has a good stability.

The seakeeping analysis shows that for average amplitude pitch the trimaran boat was accepted by the criteria. It indicates that trimaran boat might be used as an alternative hull form for Glass Bottom Boat to support tourism activities in Karimunjawa islands.

IOP Conf. Series: Earth and Environmental Science **135** (2018) 012007 doi:10.1088/1755-1315/135/1/012007

References

[1] Matveev K and Dubrovsky V A 2007 *Ocean Eng.* **34** 616
[2] Fang M C and Chen T Y 2008 *Ocean Eng.* **35** 749
[3] Xu M and Zhang S 2011 *J. Hydrodyn. Ser. B* **23** 265
[4] Hafez K A and El-Kot A A 2012 *Alexandria Eng. J.* **51** 153
[5] Su Y, Wang S, Shen H, and Du X 2014 *J. Hydrodyn. Ser. B* **26** 549
[6] Lv J and Grenestedt J L 2015 *Ocean Eng.* **94** 116
[7] Zakki A F, Windyandari A, and Bae D M 2016 *J Mar Sci Tech-Taiw* **24** 575
[8] Crismianto D, Zakki A F, Adietya B A, Kim D. J. 2015 *J Marine Sci Appl* **14** 399
[9] NORDFORSK 1987 General Operability Limiting Criteria for Ships

Analysis Of Resistance And Effective Wake Friction Due To Addition Of Stern Tunnels On Passenger Ship Using Cfd

D Chrismianto[1], Tuswan[1] and P Manik[1]

[1]Department of Naval Architecture, Faculty of Engineering, Diponegoro University, Semarang, Indonesia

E-mail: *deddychrismianto@yahoo.co.id*

Abstract. In this study, the stern tunnel to improve the efficiency of ship propulsion system is analysed. Stern tunnels installed on the two sides of the ship stern. Analysis of ship resistance and wake friction of the ship using CFD are carried out. The stern tunnel height (Hw) and length (L) are implemented to find the better stern tunnel form of the ship. The result of analysis showed that model has a high stern tunnels (Hw) of 1,444 m or additional high stern tunnels ratio of 16% and stern long tunnels (L) about 7 m is a model that has the smallest resistance about 1.1137 N or able to make reduction of resistance amount 11.2582%. While, the model with the addition of height of 0.2 m and a length of 9 m of stern tunnel is a model that has the better advanced speed about 4,927% in increase, and better wake friction about 30.4% in reduce.

1. Introduction

This research was carried out on the passenger ship of MV Tropical Princess Cruises that has V hullform using two propulsor. In this passenger ship will be added stern tunnels on both sides of the ship stern.

Stern tunnels are one of hullform modification that aims to centralize the flow of water towards the propeller which improves speed advanced tunnels stern of the ship so it can add thrust and propeller efficiency as well as reducing the wake value of ship. The study is conducted to modify the shape of the stern tunnel on passenger ships. The parameters of stern tunnel is considerd to be modified, that is the height of stern tunnel (Hw) and the length of stern tunnel (L).

Some of the studies CAD-integrated CFD and optimization method can be used simultaneously to solve any problems, including the optimization of hull shape. Deddy, C [2, 3] has developed the parametric bulbous bow design and submarine design are constructed using the cubic Bezier curve method and the curve-plane intersection method, then it can be used to generate a parametric bulbous bow shape automatically using solid modeling for the optimization calculation in the effort to minimize the ship's resistance.

For ship stern tunnel form that have been done in which the secondary parameter is varied to modify the ship stern tunnels by height stern tunnels (Hw) using 3(three) variations of 1.1 m - 1.5 m on a cargo ship of BM-Duisburg to analyze the ship's performance. The result showed that the improvement of Hw tends to reduce wake friction and increase the thrust deduction factor [1].

Refer to this fact, beside the height of stern tunnel (Hw), it is also essential to do the assessment on the parameter of the length of stern tunnels (L) on passenger vessels with various variations in order to get the larger advanced speed value, the largest and the smaller wake friction value.

2. Stern Tunnel Parameters

Stern Tunnels is a modified form of the stern of the ship to make a convex curve so allowing the installation of larger diameter of propeller. The advantage of this one is a higher efficiency of the wake friction due to the combination of the decline of the shaft angle and increase the propeller dimension. Figure 1 shows the dimensions of stern tunnel is used in this study.

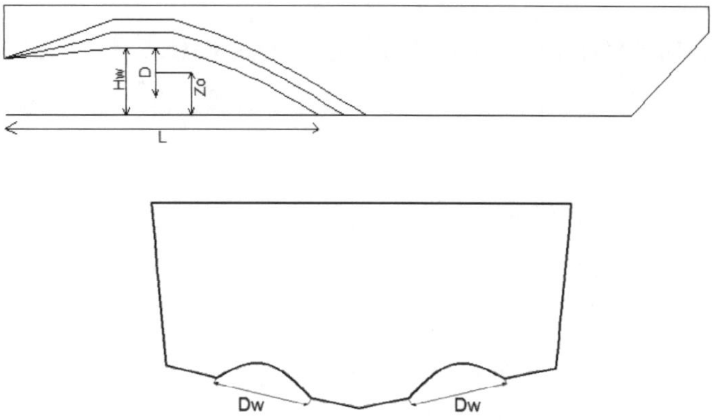

Figure 1. Stern Tunnel Parameters

where:

L : The length of stern tunnel
Hw : The height of stern tunnel
D : Propeller diameter
Dw : Stern tunnel diameter
Zo : Distance of hub propeller to baseline.

3. Resistance And Wake Calculation Using Cfd

3D Modeling of ship is created to analyze ship resistance and effective wake friction using CFD (Computational Fluid Dynamic) method then it will be validated with empirical method.

This research is focused to obtain a hullform has a small resistance and wake of ship comparing with original hullform of ship after changing of stern form using adding of stern tunnel. Stern tunnel lenght (L) and stern tunnel height (Hw) as parameters are applied to create stern form of models.

Ship model uses a scale of 1:40. This model is used to a CFD simulation process acccording with total number of mesh elements (Table 1).

Table 1. Ship principle dimensions (original and CFD model scale)

No.	Items	Original Scale	CFD Model Scale	Unit
1	*Length Over All (LOA)*	30.66	0.7665	m
2	*Length of Waterline (Lwl)*	28.655	0.7164	m
3	*Breadth (B)*	6.08	0.152	m
4	*Depth (H)*	3.3	0.0825	m
5	*Draft (T)*	1.52	0.038	m
6	*Wetted Surface Area (WSA)*	170.953	0.10676	m^2
7	*Speed (Vs)*	7.716	1.2188	m/s
8	*Froude Number (Fn)*	0.46	0.46	
9	*Coefficient Block (Cb)*	0.415	0.415	

3.1. Ship Model Variations

Table 2 shows variation of stern form based on 2(two) parameters, such as: stern tunnel lenght (L) and stern tunnel height (Hw). Correspondence one to one is implemented to obtain Model B to Model M.

Table 2. Variation of models based on stern tunnel height (Hw) and Lenght (L)

Models	Hw	Ratio of Hw added (%)	L	Ratio of L added (%)
Model A	1,244 m	-	-	
Model B	1,344 m	8	7 m	23 %
Model C	1,344 m	8	8 m	26 %
Model D	1,344 m	8	9 m	30%
Model E	1,394 m	12	7 m	23 %
Model F	1,394 m	12	8 m	26 %
Model G	1,394 m	12	9 m	30%
Model H	1,444 m	16	7 m	23 %
Model I	1,444 m	16	8 m	26 %
Model J	1,444 m	16	9 m	30%
Model K	1,494 m	20	7 m	23 %
Model L	1,494 m	20	8 m	26 %
Model M	1,494 m	20	9 m	30%

Table 3 shows value of wetted surface area after adding the stern tunnel on the ship stern. The Percentage of WSA difference on the model M is 0,925% larger than original model.

Table 3. Wetted Surface Area on each model

Model	WSA (m^2)	WSA Difference to Model A (%)
Model A	0.106760	0.000
Model B	0.107125	0.342
Model C	0.107199	0.412
Model D	0.107252	0.461
Model E	0.107331	0.535
Model F	0.107378	0.579
Model G	0.107392	0.592
Model H	0.107398	0.598
Model I	0.107430	0.628
Model J	0.107563	0.752
Model K	0.107582	0.771
Model L	0.107679	0.861
Model M	0.107748	0.925

3.2. CFD Process

Phase mesh is the most complicated stage. When an error is occured in the process working, so mesh production will be stoped and failed. Then the mesh step must be repeated, or if the error due to the form of the model then needs to be fixed before the new model do mesh process again. So, it is advisable to be careful and meticulous. The greater of the number of mesh elements, so the mesh results will be more delicate and can get more accurate of results. If the number of elements is too much, the numerical simulation process will be more severe and more consumable time. Figure 2 shows the result of mesh elements for ship model.

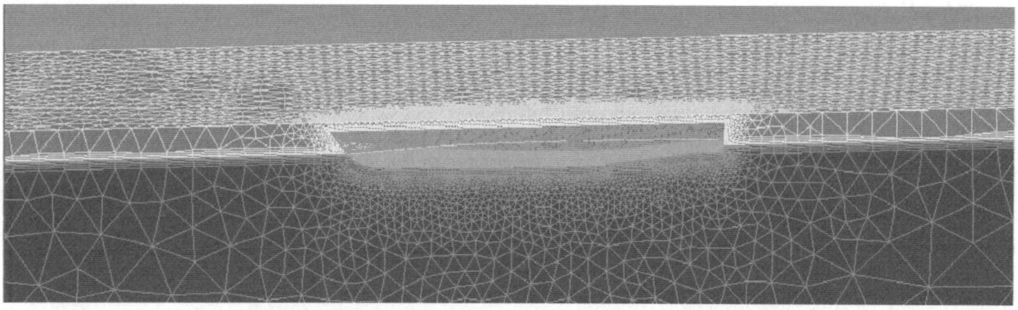

Figure 2. Mesh process and total number of elements

Meshing process time depends on total number of elements. In this research, total number of mesh elements about 600,000 elements for each models. Consumable time of mesh production abou 5 minutes. Table 4 and 5 show the result of mesh statistics for each models.

Table 4. Mesh statistics for each models

Mesh Statistics	Model A	Model B	Model C	Model D	Model E	Model F	Model G
Total number of nodes	137131	142196	136908	153808	156063	153617	149487
Total number of tetrahedra	387984	404299	425586	446661	450143	481215	474254
Total number of pyramids	3295	2301	3658	3423	3775	4304	4460
Total number of prisms	119819	125116	105975	131891	134623	118185	112737
Total number of elements	511098	531716	535219	581975	588541	603704	591451

Table 5. The Remaining mesh statistics models

Mesh Statistics	Model H	Model I	Model J	Model K	Model L	Model M
Total number of nodes	153617	151140	169472	172532	165995	176283
Total number of tetrahedra	481215	479633	496213	501138	528640	525142
Total number of pyramids	4304	4601	3835	4342	5779	4542
Total number of prisms	118185	114450	144537	147967	124482	147110
Total number of elements	598684	603704	644585	653447	658901	676794

IOP Conf. Series: Earth and Environmental Science **135** (2018) 012008 doi:10.1088/1755-1315/135/1/012008

Model setup stage is a important step in numerical simulation based on Computational Fluid Dynamic (CFD) because choice of boundary condition has any effects into the good result. The wrong input of boundary condition value will cause an error of simulation process.

In this simulation process, boundary condition of bottom uses wall type and free slip wall, it means fluid motion did not occure any slip. Boundary condition of the hull uses the wall type, the same as the bottom. Mass and momentum use the free slip wall which means that the fluid moves no slip. While the wall roughness applies a smooth wall. For inlet boundary condition, the mass and momentum uses Cartesian velocity components. This inlet fluid will move and the inlet fluid velocity is also determined. Determination of velocity at the inlet is filled using three coordinates, namely speed U, V, W (X, Y, Z). Turbulence in the inlet is 1% (low). The boundary condition of the oulet isoutlet type which its function as a discharge of the fluid. The flow is a subsonic at the outlet. Side boundary is set up as a wall because of its function as a closed wall. Then the mass and momentum uses free slip wall like as the bottom. The boundary condition of symmetry is located on the ship hull. The boundary type uses symmetry. Then the boundary condition of top uses an opening because its function is as an open field and subsonic fluid flow. Mass and momentum use an entrainment with relative pressure of 0 (zero).

Calculation of total resistance of ship is performed by simulating stern form using CFD analysis. CFD simulation results are validated with empirical calculations to determine the level of calculation error about ± 5%. Table 6 shows comparison of empirical and numerical calculations.

Table 6. Comparison of empirical and numerical calculations

Metode	Rf (N)	Rv (N)	Rw (N)	Rt (N)
Empiris	0.398	0,5065	0.739	1,2463
Numerik	0.406	0.517	0.738	1.255
Different (%)	2,01	2,07	0,13	0,7

Table 6 shows that the error value of empirical and numerical calculations is less than 5%. This result proves that the numerical calculations based CFD is very closed with empirical calculation.

3.2.1. Ship resistance calculation

Total ship resistance value is obtained from magnitude of force on the ship hull. In the Table 7 are presented the value of total ship resistance in the each models.

Table 7. Total resistance calculation for each models

Models	Coeff. of total resistance (Ct)	Total Resistance (N)	Difference with Model A (%)
Model A	0.01544	1.2550	-
Model B	0.0150829	1.2306	-1.9124
Model C	0.01434	1.1709	-6.2948
Model D	0.01401	1.1426	-7.7230
Model E	0.01417	1.1581	-8.8606
Model F	0.01388	1.1335	-9.7052

Models	Coeff. of total resistance (Ct)	Total Resistance (N)	Difference with Model A (%)
Model G	0.01377	1.1265	-10.6454
Model H	0.01362	1.1137	-11.2582
Model I	0.0136725	1.1182	-10.8980
Model J	0.01404	1.1510	-6.5006
Model K	0.01472	1.2075	-5.3778
Model L	0.0145	1.1875	-3.7624
Model M	0.01489	1.2220	-2.6265

Table 7 shows that the addition of stern tunnels on all of variation models tend to reduce the total ship resistance (Rt) with a range between 1.91 -11.26%. Model H has a high stern tunnels (Hw) of 1,444 m or additional high stern tunnels ratio of 16% and stern long tunnels (L) about 7 m is a model that has the smallest resistance = 1.1137 N with a difference with the model A of 0,1413 N or a reduction of resistance amount 11.2582%.

Figure 3 shows a comparison between the resistance value of each ship models that are: friction resistance (Rf), viscous resistance(Rv), wave resistance (Rw) and total resistance (Rt).

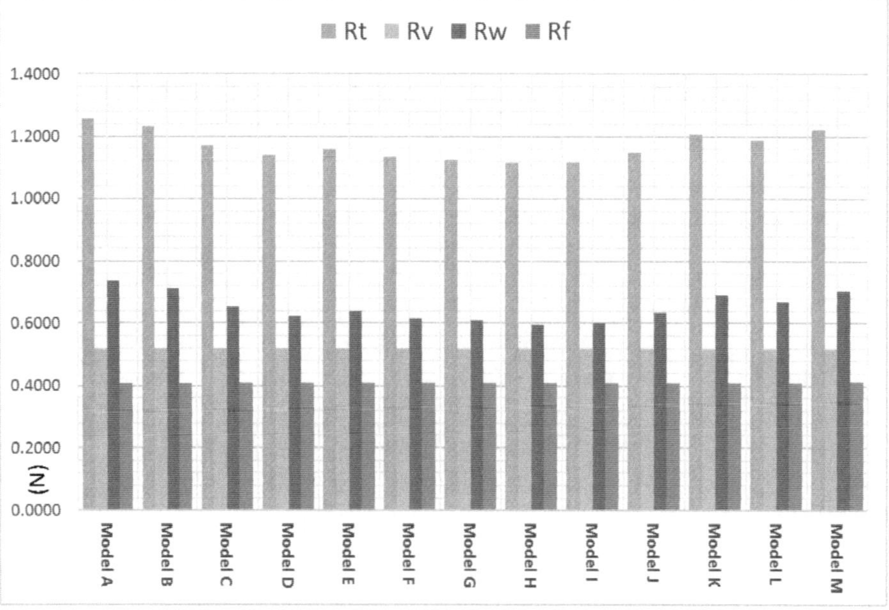

Figure 3. Components of resistance for each models

Based on the results are shown in Figure 3 can be conclude that the addition of stern tunnels on a Froude number of 0.46 is obtained that wave resistance (Rw) is greater than the friction resistance and viscous resistance. The wave resistance has a significant effect in reducing total resistance with addition of the stern tunnels comparing with other resistance components such as friction resistance and viscous resistance that tends to stay constant due to the value of original model resistance. In Figure 4 and Figure 5 presents a wave countour on side of the Model A and Model H.

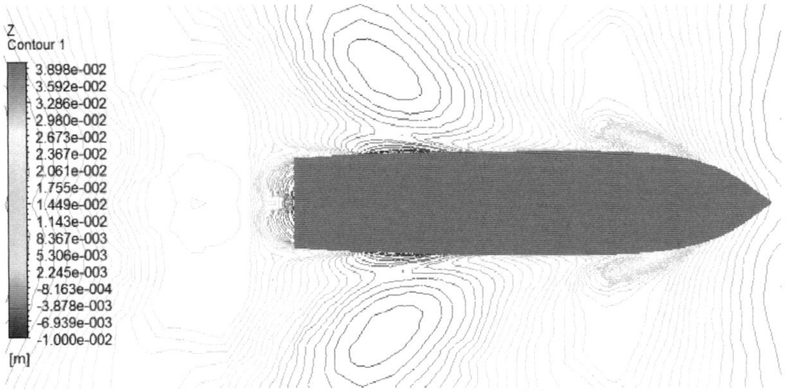

Figure 4. Wave pattern model A

Figure 5. Wave pattern model H

3.2.2. Wake Calculation

In calculating the value of a wake, in this study the average value Va scale model obtained from Ansys CFD-Post is converted into an average Va original model by using the calculation of the Froude number. Here are the results of calculations wake of thirteen models such variations.

Table 8. Comparison advance velocity (Va) and wake for each models

Model	Va [M/S]	Wake	Difference (%)
Model A	6.6407	0.13936	
Model B	6.6913	0.13281	-4.69891
Model C	6.7194	0.12916	-7.31773
Model D	6.7928	0.11965	-14.14364
Model E	6.8413	0.11337	-18.64840
Model F	6.8860	0.10756	-22.81302
Model G	6.8512	0.11207	-19.57719

Model	Va [M/S]	Wake	Difference (%)
Model H	6.8769	0.10875	-21.96265
Model I	6.8951	0.10639	-23.65843
Model J	6.9364	0.10104	-27.49690
Model K	6.8773	0.10870	-22.00055
Model L	6.9155	0.10374	-25.55509
Model M	6.9679	0.09695	-30.43011

From the all variation models are analyzed based on CFD can be stated that the addition of stern tunnels can concentrate the flow of water leading to the propeller that cause advanced speed (Va) is increased or with another word the value of Va is increases and the value of ship wake-friction becomes decrease. From the results is obtained that the model M with high stern tunnels 1,494 m (stern tunnels add rate of 20%) and long stern tunnels 9 m is a model that has the smallest value that is equal to 0.09695 wake with a percentage of the difference in value wake the original model was reduced by 30.4%.

4. Conclusions

In this study, Model H has a high stern tunnels (Hw) of 1,444 m or additional high stern tunnels ratio of 16% and stern long tunnels (L) about 7 m is a model that has the smallest resistance = 1.1137 N with a difference with the model A of 0,1413 N or a reduction of resistance amount 11.2582%.

While Model M has a stern tunnel height about 1,494 m (increase by 20%) and stern tunnel length about 9 m are model that have an advanced speed value is larger than other model is equal to 6.9679 m/s with the difference in Va value comparing the original model increased by 0,3272 m/s or 4,927%.

Model M has a stern tunnel height about 1,494 m (increase by 20%) and stern tunnel length about 9 m are model that have an effective wake friction value is smaller than other model is equal to 0.0969 by the difference in wake value comparing the original model reduced by 0.0424 or 30.4%. The addition of stern tunnels can increase the advanced speed of ship and reduce the effective wake-friction ship.

REFERENCES

[1] Kulczyk J and Tabaczek 2014 Coefficients of propeller- hull interaction in propulsion system of inland waterway vessel with stern tunnels *TransNav the International Journal on Marine Navigation and Safety of Sea Transportation* Vol 8 No 3 p 377-384

[2] Chrismianto D and Dong J K 2014 Parametric bulbous bow design using the cubic Bezier curve and curve-plane intersection method for the minimization of ship resistance in CFD *J Mar Sci Technol* Vol 19 p 479-492

[3] Chrismianto D, Zakki A F, Arswendo B and Dong J K 2015 Development of Cubic Bezier Curve and Curve-Plane Intersection Method for Parametric Submarine Hullform Design in order to Optimize the Hull Resistance by Using CFD *Journal of Marine Science and Application* Vol 14 No 4 p 399-405

[4] Lewis, Edward V 1988 Society of Naval Architects and Marine Engineers (SNAME) *Principles of Naval Architecture Vol. II, Resistance and Propulsion* (New Jersey)

[5] Molland F A, Turnock R S and Hudson AD 2011 *Ship Resistance and Propulsion: practical estimation of ship propulsive power* (New York: Cambridge University Press)

[6] Setiawan A Y 2014 *Modifikasi Bentuk Buritan Pada Shallow Draft Bulk Carrier Untuk Meningkatkan Efisiensi Sistem Propulsi* (ITS Undergraduate paper) p 1-6

[7] Harvald S S 1983 *Resistance and Propulsion of Ships* (New York: John Wiley and Sons)

Study of Water Quality Changes due to Offshore Dike Development Plan at Semarang Bay

M Wibowo[1] and B A Hakim[1]

[1] Center for Port Infrastucture and Coastal Dynamic Technology (BTIPDP) – BPPT, Indonesia

Email : mardi.wibowo@bppt.go.id

Abstract. Now, coast of Semarang Gulf is experiencing rapid growth because Semarang as a center economic growth in Central Java. On the other hand, coast of Gulf Semarang also experience a variety of very complex problems, such as tidal flood, land subsidence, as well as coastal damage due to erosion and sedimentation process. To overcome these problems BPPT and other institutions proposed construction of offshore dike. Construction of the offshore dike is a technology intervention to the marine environment that will certainly affect the hydrodynamic balance in coastal water including water quality in the Gulf of Semarang. Therefore, to determine changes in water quality that will happen is necessary to study the water quality modeling. The study was conducted by using a computational modeling software MIKE-21 Eco Lab Module from DHI. Based on this study result knowed that development offshore dike will change water quality in the west and east dam that formed. In west dam the average value of the DO decline 81.56% - 93.32 % and the average value of BOD rise from 22.01 to 31.19% and in the east dam, there is an increase average value DO of 83.19% - 75.80%, while the average value of BOD decrease by 95,04% - 96.01%. To prevent the downward trend in water quality due to the construction of the offshore dike, its necessary precautions at the upstream area before entering the Gulf of Semarang.

1. Introduction

1.1. Background

Coast of Semarang Gulf is area that consist land and water areas along the coast that stretches from the mouth of K. Bodri-Kab. Kendal in the west, the coast of Semarang City up to the estuary of K. Wulan-Kab. Demak in the eastern part. This area has approximately 104 km long coast. Coast of Semarang Gulf is currently experiencing rapid growth. As a consequence of these developments and natural conditions this area, now coast of Semarang Gulf experiencing complex problems.

The first problem that occurred in the Gulf of Semarang city is tidal flood and river flood. Now, if occurrence of high tides, most areas of the Semarang City especially coastal areas (± 3,100 ha) will be floods [1].

The second problem is the occurrence of land subsidence. Land subsidence occurs in the range of 2008-2009 reached -12.4 cm, then increased in 2009-2010 became -20.4 cm and fell back to 10.5 cm in 2010-2011 [2]. According to the Indonesia Geological Agency, land subsidence speed range of 0-1 cm/yr (in the district. Tugu, West Semarang, Semarang, Central and East of Semarang) and 8-9 cm/yr (North Semarang and Genuk) [1].

The third problem is the occurrence of abrasion along the coastline due to the dynamics of the coast. Between the years 1972 - 1992 in some places along the coast of Semarang back about 500 m, between the years 1992-2001 partially some beach experience abrasion large enough [3].

To overcome these problems has been several concept that proposed to regional governments both to Central Java Government and Semarang City Government. Concepts that are currently the most prominent is the concept of offshore dam and belt/ coastal dike along the shoreline and offshore dike concept proposed by BPDP-BPPT. These concepts beside to overcome tidal flood, land subsidence and coastal erosion, also offer solutions of transportation infrastructure development.

Construction of coastal infrastructure such as offshore dikes that proposed by BPPT is a technology intervention to the marine environment that will certainly have an impact on coastal dynamic balance includes the impact on water quality in the Semarang Gulf. Until now the concepts proposed for addressing concerns in the Semarang Gulf has been no systematic study of the impact on existing water in particular the water quality. Therefore this study will be conducted studies on changes in water quality that would occur if the construction of the offshore dike actually built.

1.2. Location and period

The area of this study is along the Semarang Gulf that stretches from the mouth of K. Bodri in Kab. Kendal, the coast of Semarang City until the estuary of K. Wulan in Kab. Demak (figure 1).

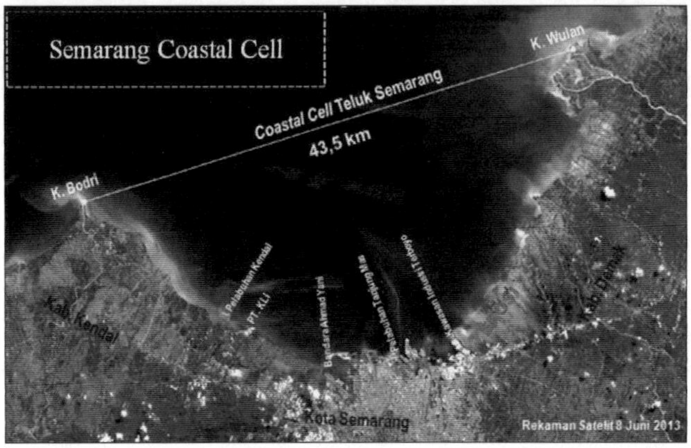

Figure 1. Area of study

The period of this study is the rainy season (January) with consideration the data required for modeling is complete compared to other months.

2. Method

The main step in this study is field surveys and computational modeling.

2.1. Steps of Study

The stages of the activities to be undertaken are as follows:

- Literature and secondary data collection; field surveys, including bathymetric surveys, sampling and basic sediment drift, currents measurements and tidal measurements
- Analysis of secondary, primary data and preparation of scenario computational modeling
- Computational modeling and analysis the modeling results

2.2. Water Quality Modeling

Numerical advection-dispersion modeling was conducted to determine the distribution of water quality parameters that wants to be reviewed, based on hydrodynamic conditions of local waters. The modeling is done by using modeling software MIKE 21 Modules Eco Lab with WQ Simple Template MIKE 21/3. Based on this template is the main parameter DO (Dissolved Oxygen), BOD (Biological Oxygen Demand) and salinity [4].

Builder equation for water quality modeling based on the advection dispersion process. Dynamism of a substance to be modeled in the advection process can be expressed by the equation of its transport, which for non-conservative form can be written as follows [4]:

$$\frac{\partial c}{\partial t} + u\frac{\partial c}{\partial x} + v\frac{\partial c}{\partial y} + w\frac{\partial c}{\partial z} = D_x\frac{\partial^2 c}{\partial x^2} + D_y\frac{\partial^2 c}{dy^2} + Dz\,\frac{\partial^2 c}{dz^2} + |S_c + P_c \quad (1)$$

where :

c	:	Concentration of ECO Lab variabel
u, v, w	:	Current velocity component
Dx, Dy, Dz	:	Coefficient of dispersion
Sc	:	Sources and sinks
Pc	:	Eco Lab processes

2.3. Equipments and Data

Equipments used in this study are : echo sounding SB CEEDUCER for measure bathymetry; sediment grabber for sampling basement sediment; Nansen Bottle or sampling water, software MIKE-21 FM : hydrodynamics and ECO-Lab modul, software Global Mapper 13. The data that used in this study :

- Hydrodynamic data at *boundary*, such as salinity, temperature and tidal level, etc.
- Source : river debit (river at Semarang area [5]; Kendal area [6]; Demak area [7])
- Water quality data at boundary, concentration of BOD and DO that will simulated ((river at Semarang area [8]; Kendal area [9]; Demak area [10])
- ECO Lab forcing and ECO Lab constanta
- ECO Lab Load : temperature, concentration at source (BOD, DO and TSS)

3. Result and Discussion

Discussion of the results of water quality modeling is done per-scenario to try to compare the existing condition and ultimate. Existing condition is a condition that exists today (in the absence of giant dike and islands reclaimed) and the condition after the construction of the offshore dike proposed by BPPT.

3.1. Modeling Scenarios

Generally, scenarios is divided into three major parts of the existing condition and the condition after the construction of the offshore dike according Concept of BPPT 1 and BPPT 2 (see figure 2).

Due to limited data and time available, simulating is only done for the west season (represented by the month of January 2014), because the existing data for the month of January 2014 is more complete compared with the data in September 2014 so expect the resulting condition can approach the actual conditions. Simulation chosen dates 1 to 11 January 2014, with the consideration that this happens timescales low tide and high tide.

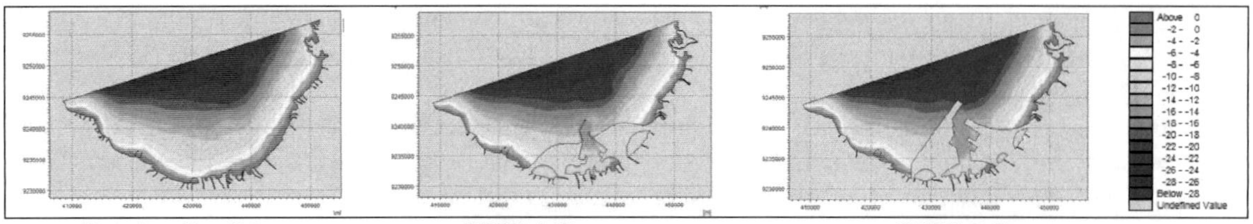

Existing Condition BPPT 1 Concept BPPT 2 Concept

Figure 2. Bathymetry of Semarang Gulf as domain model : existing condition, after construction offshore dike BPPT 1 and BPPT 2 concepts

3.2. Discussion

Based on the results of water quality modeling, the folders are formed and on the waters of the Port of Tanjung Emas shipping lanes are located between the dike changes offshore water quality significantly. The figure to show the changes in water quality over time until the end of the simulation shown in the figure 3 below.

To determine the changes in more detail that caused by infrastructure development like BPPT 1 concept and BPPT 2 concept, it needed extraction at 4 locations, namely : outside the west dam; inside the west dam; in the navigation channel Tanjung Mas; outside the east dam (see figure 4). And to know changes in the value of the DO and BOD during the simulation process (time series) then made extraction at several points for observation (see figure 5).

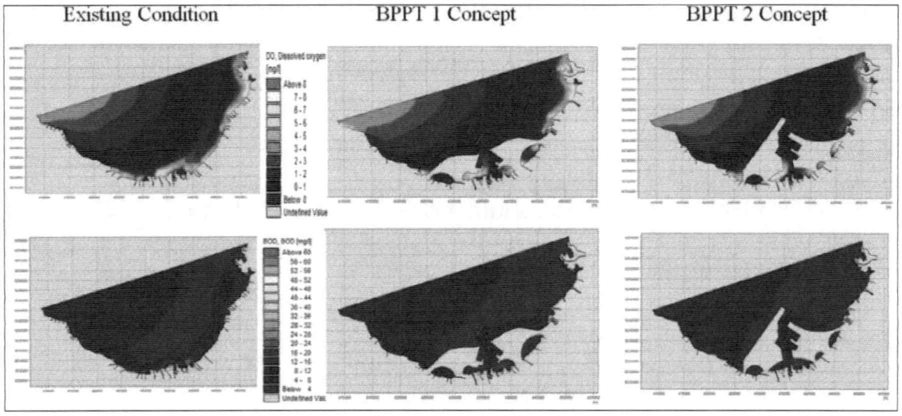

Figure 3. Comparison modeling result between existing condition, BPPT 1 Concept and BPPT 2 Concept d 240 hours simulation (DO : above; BOD : below)

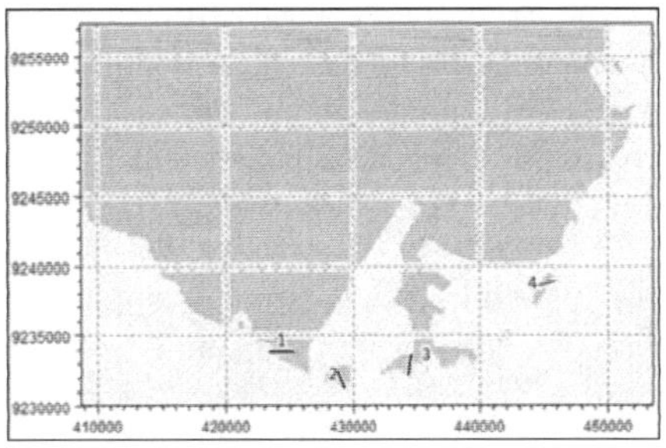

Figure 4. Extraction line to observe water quality changes

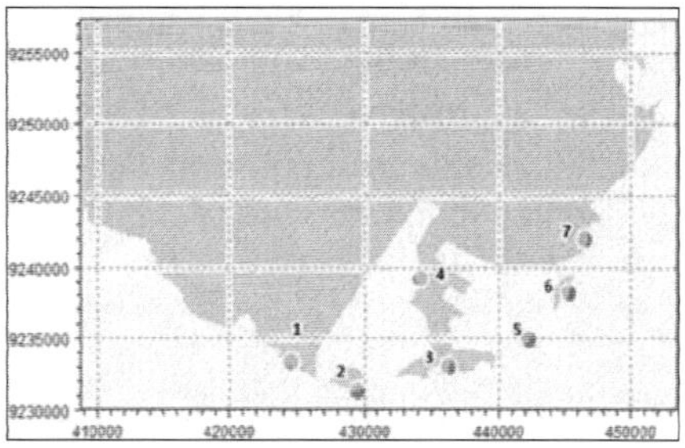

Figure 5. Extraction point to know the water quality changes as a function of time

Based on modeling results in this scenario, generally at the end of the simulation (day 10th) in the west, especially the outside dam with the concept of BPPT 1 and 2 will raise the value of the DO particularly at positions 1 km of the coastline (see figure 6.a). This is probably occur caused of the movement of water due to the lack of infrastructure that can improve the aeration process at locations in the west infrastructure.

The sea waters at the west, that according to the concept of BPPT 1 and 2 will be formed dam, a decline in the average value of DO is significant and bit an increase in the average value of BOD (see figure 6.b). The decreasing DO value occurs because the presence of dam cause the waters became calm because it is not affected by tidal and wave from the offshore waters so the aeration process is much reduced resulting in dissolved oxygen is also reduced. While the BOD value will raise because with the dam organic substances polutant will be concentrated in the dam and can't spread to the open waters). The magnitude of decrease in the average value of the DO in the dam west this amounted to 81.56% (due to the BPPT 1 Concept) and 93.32% (due to BPPT 2 Concept) of the average DO value of the existing condition of 4.42 mg/l. As for the average BOD value increased by 31.19% (due to the BPPT 1 Concept) and amounted to 22.01% (due to the concept of BPPT 2) of the average value of BOD existing condition of 10.45 mg/l.

The waters in the shipping channel Tanjung Mas Port decline in the average value of the DO and the average value of BOD (see figure 6.c). A decrease in the average value of DO is due to the limited contact with the waters of the open sea so will reduce aeration process.. The magnitude of decrease in the average value of the DO in the waters around Tanjung Mas about to 35.79% (due to the BPPT 1 Concept) and 46.51% (due to the concept of BPPT 2) from the existing condition of 5:16 DO mg/l. As for the average value of BOD decrease about to 48.25% (due to the BPPT 1 Concept) and about to 19.79% (due to the concept of BPPT 2) from the existing condition of 23.27 mg/l.

The waters at the east, which according to the concept of BPPT 1 and 2 will be formed dam, an increase in the average value of the DO as well as a decrease in the average value of BOD significantly (see figure 6.d). Increasing DO value in the dam is caused by the high value of the DO of river water into the dam east though the process airation as tidal and wave did not happen because isolated by dikes. While the decline in value caused by the low BOD value of river water into the dam is compared to the value of the initial BOD sea waters before. The magnitude of the increase in the average value of the DO in deep waters east dam is at 83.19% (due to the BPPT 1 Concept) and 75.80% (concept of BPPT 2) of the average DO value of the existing condition of 3.17 mg/l. As for the average BOD value dropped by 95.04% (due to the BPPT 1 Concept) and about to 96.01% (due to the concept of BPPT 2) of the average value of BOD existing condition of 6.25 mg/l.

An example of extraction results at some observation point such as outside the dam (point 1) and inside the dam (point 2) as a function of time will be discuss at below.

At the observation point outside the west dam is seen that the pattern of changes in the value of the DO was not significant (see figure 7.a). At the end of simulation, DO value changes less than 1%. It is caused at this location is still directly connect with the open sea as the existing condition. Slight changes seen in the BOD value (about 1.37%) due to the concept of BPPT 1 because of the shifting of the flow K. Siangker.

At the observation point in the west dam is seen that the pattern of changes are very significant (see figure 7.b). This is probably due to the existing condition is very dynamic beside the influence of the incoming river flow is also due to the tides, waves and wind. So with the concept of BPPT changes in the value of the DO and BOD become more stable does not fluctuate like when it is open to the ocean. At the end of the simulation, DO value decreasing about 97%. While the BOD values fell between 49.4% (due to BPPT 1 concept) and 65.67% (BPPT 2 concept).

At the observation point no. 3 in the shipping channel of Tanjung Mas Port, shows that the pattern of changes in the value of the DO and BOD as a result of infrastructure development is significant (see figure 7.c). This is probably due to that the existing condition upheaval of sea waters at this location is large enough, the opposite occurs when it is already built infrastructure, at this location although it still can be in direct contact with the waters of the open sea, but the condition is relatively quiet because shielded from waves. At the end of the simulation decreasing DO value approximately 53.56% (due to

IOP Conf. Series: Earth and Environmental Science **135** (2018) 012009 doi:10.1088/1755-1315/135/1/012009

the BPPT 1 Concept) and 79.39% (due to BPPT 2 Concept). While the BOD values fell between 63.69% (due to the BPPT 1 Concept) and 68.03% (due to the concept of BPPT 2).

At the observation point no. 4 in the shipping channel of Tanjung Mas Port, shows that the pattern of changes in the value of the DO and BOD as a result of infrastructure development was not significant (see figure 7.d). This is probably due to that the existing condition and has built the infrastructure can still be in direct contact with the open sea waters despite all effect is not large. Thus the fluctuations of DO and BOD values change as a result of this infrastructure is relatively uniform.

At the end of the simulation impaired DO approximately 1.92%. While the BOD values fell between 19.78% (BPPT 1 Concept) and 33.66% (BPPT 2 Concept).

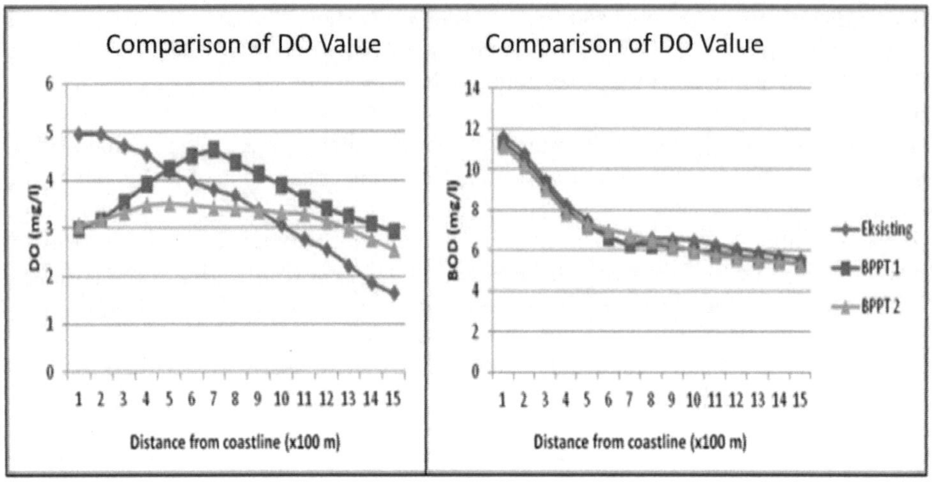

Figure 6. Comparison DO and BOD in January at 10th simulation day (In outside west infrastructure, location 1)

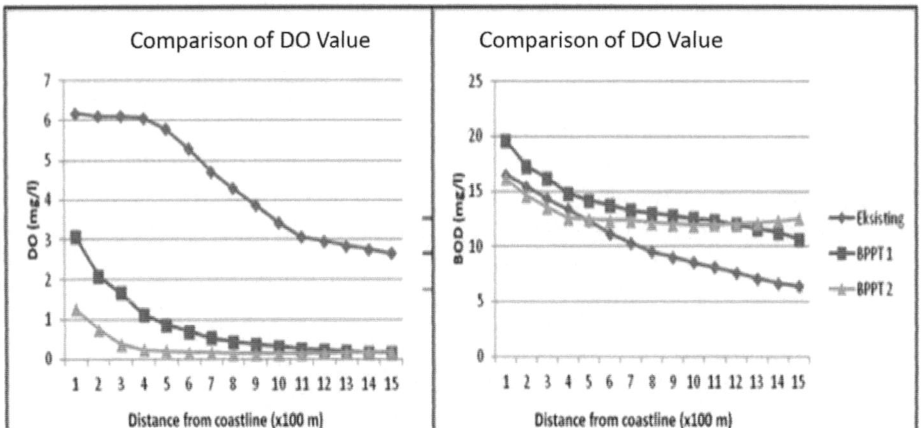

Figure 7. Comparison DO and BOD in January at 10th simulation day (Inside west dam, location 2)

IOP Conf. Series: Earth and Environmental Science 135 (2018) 012009 doi:10.1088/1755-1315/135/1/012009

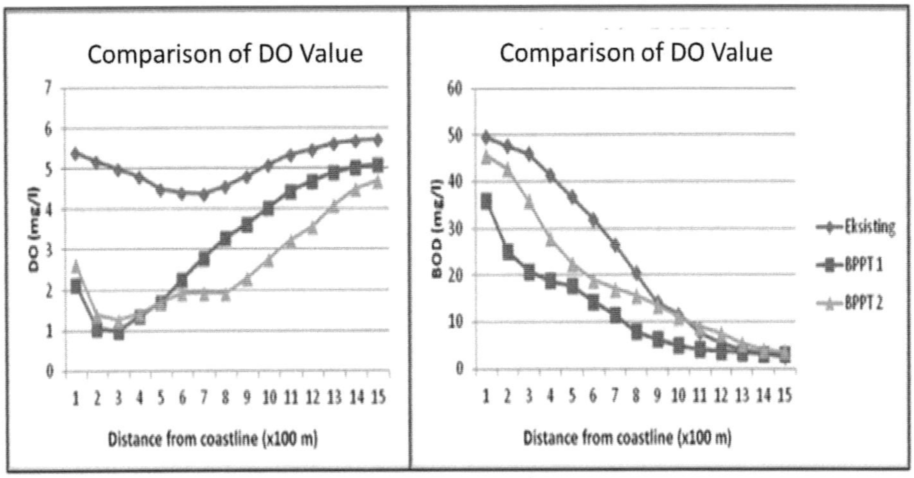

Figure 8. Comparison DO and BOD in January at 10th simulation day (In the shipping channel Tanjung Mas Port, location 3).

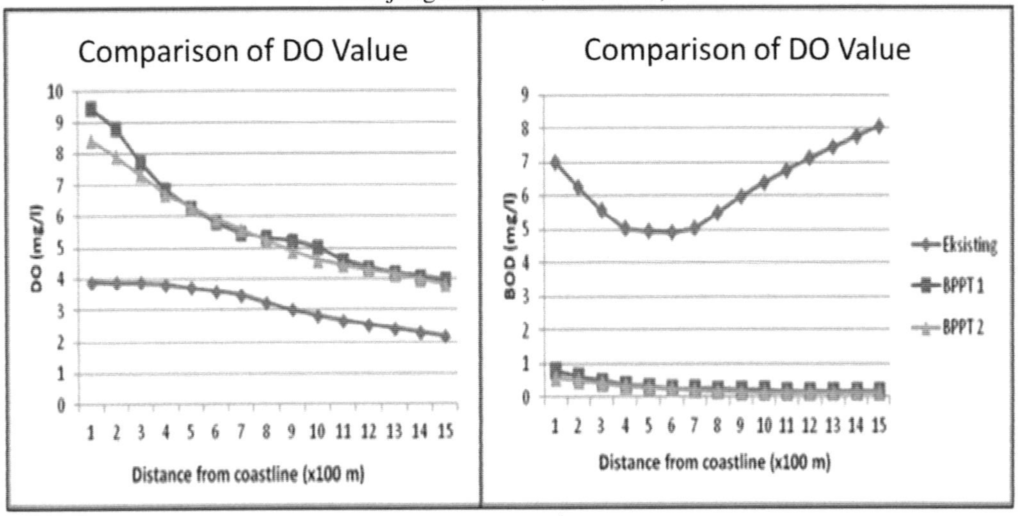

Figure 9. Comparison DO and BOD in January at 10th simulation day (Inside east dam, location 4)

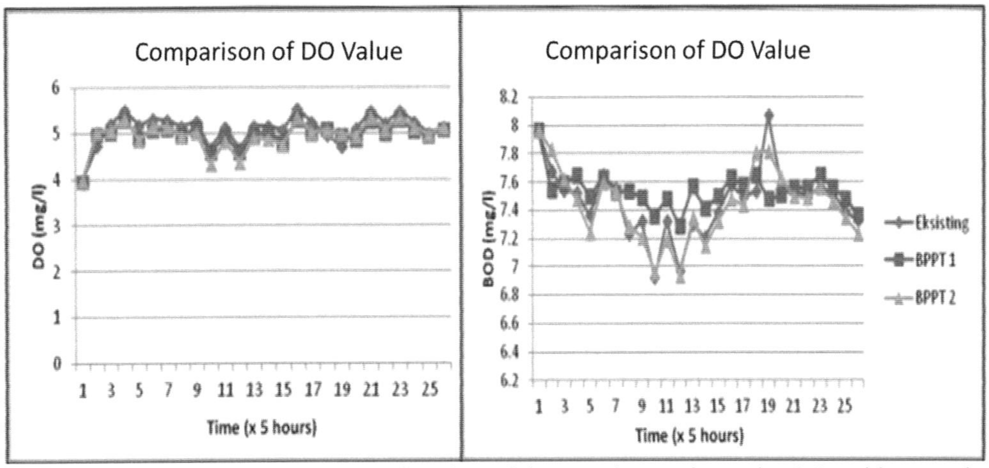

Figure 10. Water quality changes as a function of time at observation point 1 (outside west dam)

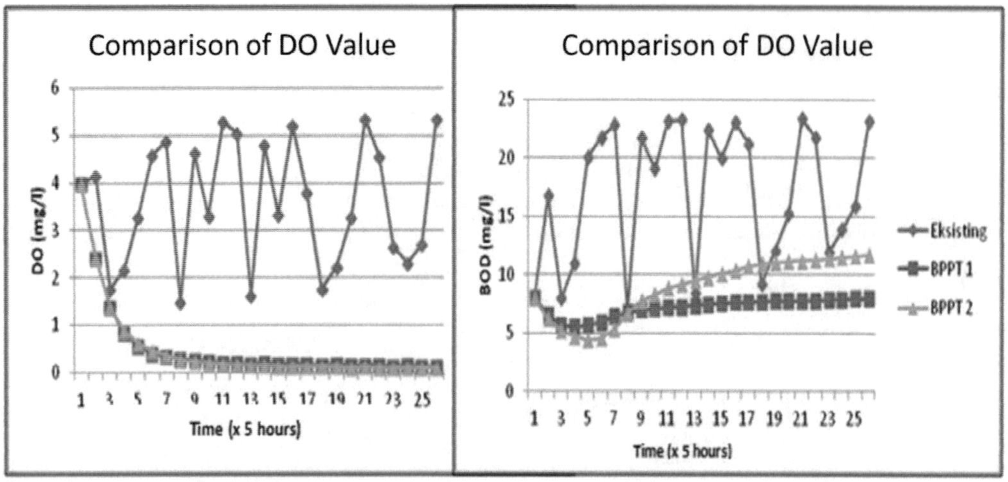

Figure 11. Water quality changes as a function of time at observation point 2 (inside west dam)

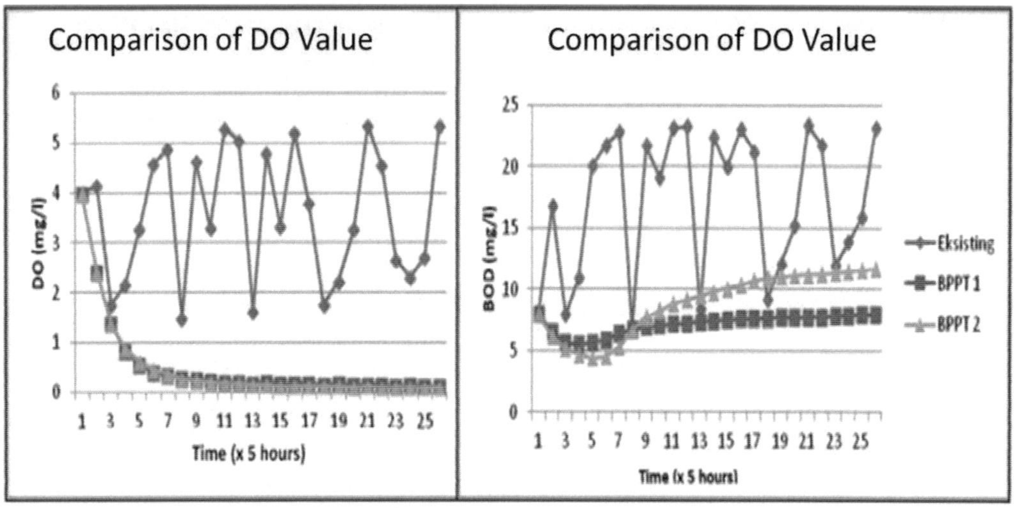

Figure 12. Water quality changes as a function of time at observation point 3 (Tanjung Mas shipping channel)

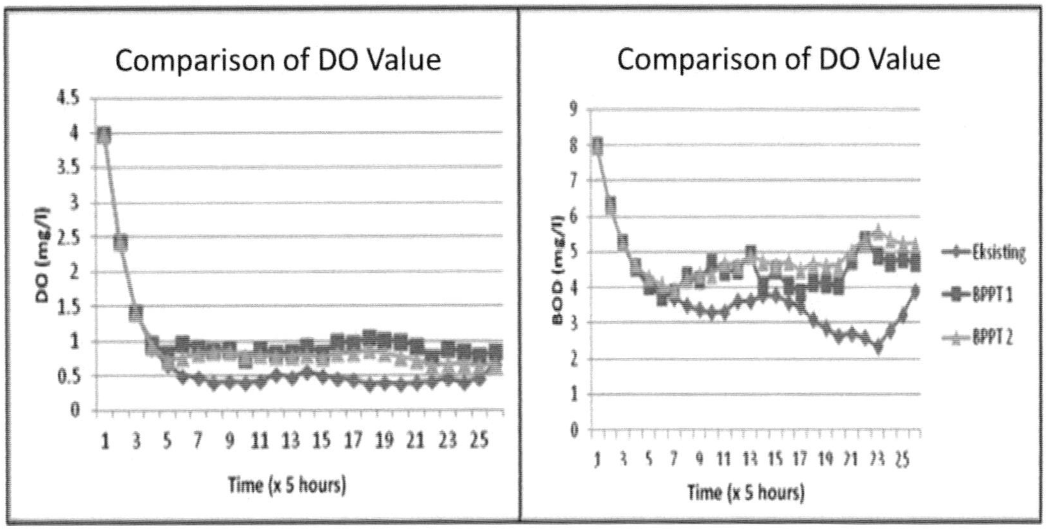

Figure 13. Water quality changes as a function of time at observation point 4 (Tanjung Mas shipping channel)

4. Conclusion

Based on the result of this study, can be concluded as follows :

- In general, the concept of BPPT I and BPPT II will form the dam that isolated from open waters, the input of water from river discharge will affect the quality of water in the dam formed.
- As a result of the construction of the offshore dike according concept of BPPT 1 or BPPT 2, in the dam formed in the west part decline in the average value of the DO 81.56% - 93.32% and the average value of BOD increasing by 22.01 to 31 , 19%.
- In the navigation channel of Tanjung Mas Port, decrease the average DO value of 35.79% - 46.51%, while the average BOD value dropped by 19.79 - 48.25%.
- In the dam to be formed to the east of an increase in the average DO value of 83.19% - 75.80%, while the average BOD value dropped by 95.04% - 96.01%.

Acknowledgements

Thanks to Dr. Rahman Hidayat as the Co-Chair of KJS Program, TIRBR–BPPT 2014, to Dr-Ing. Imam Fachrudin as the head of Center for Port Infrastructure and Coastal Dynamic Technology (BTIPD), to the BoE of BPDP, all engineers involved in the program Technology Clearing House Java Sumatera Connectivity (KJS) – DIPA BPDP BPPT 2014, Semarang Survey Team BPPT 2014, DinasPSDA of Central Java, BAPPEDA of Central Java, BAPPEDA of Semarang city, BLH of Semarang City, BLH of Kendal District, BLH of Demak District and BMKG Meteorological Station A. Yani Semarang 2014.

References

[1] KOICA-BAPPENAS 2012 *Disaster Survei and Thematic Map Building in Semarang Area,* (Jakarta: Koica-Bappenas) p 21

[2] Saraswati S A 2012 *Study of Land Subsidence in Semarang* (Denpasar: Thesis at Environmental Science Faculty, Post Graduate Program, Udayana University) not published p ix

[3] Marfai M A 2011 *GIS Modelling of River and Tidal Flood Hazards in a Waterfront City – Case Study : Semarang City* (Netherllands: International Instutie for Geo-Infromation Science and Earth Observation Enschede) p 59

[4] DHI 2004 *Marine MIKE 2/3 Ecological Modelling- Mike 21/3 Eco Lab FM Short Discreption*, (Horsholm Denmark: DHI) p 4
[5] BLH Semarang City 2013 *Buku Data Status Lingkungan Hidup Daerah Kota Semarang Tahun* (Semarang: Pemerintah Kota Semarang) p 12
[6] BLH Kabupaten Kendal 2013, *Laporan Status Lingkungan Hidup Daerah, Kabupaten Kendal (*Kendal: Pemerintah Kabupaten Kendal) p I-10
[7] Direktorat Jenderal Sumberdaya Air-Departemen Pekerjaan Umum 2010 *Profil Balai Besar Wilayah Sungai Pemali Juana* (Semarang: BBWS Pemali Juana) p 25
[8] BLH Kota Semarang 2014 *Hasil Pemantauan Kualitas Perairan 2013* (Semarang: Pemerintah Kota Semarang)
[9] BLH Kabupaten Kendal 2014 *Hasil Pemantauan Kualitas Perairan 2013* (Kendal: Pemerintah Kabupaten Kendal)
[10] BLH Kabupaten Demak,, 2013, *Buku Laporan Pemantauan Kualitas Air Sungai Kabupaten Demak Tahun 2013* (Demak: Pemerintah Kab. Demak) p I-8

IOP Conf. Series: Earth and Environmental Science **135** (2018) 012010 doi:10.1088/1755-1315/135/1/012010

Evaluating Inundation in Urban Drainage Systems in Tamalanrea District Makassar Based Ecodrainase

Rizky Alfidhdha[1] and Nieke Karnaningroem[2]

[1]Sanitation Environmental Master Program Engineering Department, Faculty of Civil and Planning, Sepuluh Nopember Institute of Technology, Indonesia
[2]Sanitation Environmental Engineering Department, Faculty of Civil and Planning, Sepuluh Nopember Institute of Technology, Indonesia

Email : rizkyalfidhdha@gmail.com

Abstract— Makassar City is one of the major cities in Indonesia with a population of approximately 1.7 million inhabitants, which continues to grow and followed the development of urban infrastructure facilities. The development also resulted in adverse effects on the environment, especially for water catchment area turns into a watertight region resulting changes in surface runoff were greater, especially in the rainy season because the drainage coefficient values are increasing as well. The purpose of this study was to analyze the capacity of the capacity of drainage channels in terms of technical aspects, analyze and formulate efforts to address flooding in a drainage channel system environmentally friendly in terms of environmental aspects, and Calculating the cost and benefit the development of the handling of flooding with a drainage channel system environmentally friendly in terms of financial aspects. The results obtained from the analysis of the technical aspects there are 14 of 41 channel capacity is insufficient accommodation capacity, resulting in the analysis of environmental aspects require 867 infiltration wells, and 3.19 rate of Benefit Cost Ratio (BCR).

1. Introduction

Currently the drainage system has become one of the most important urban infrastructure. The quality management of a city can be seen from the quality of the existing drainage system. Good drainage system can liberate the city from puddles. Stagnant water causes the environment dirty and slovenly, become mosquito breeding, and sources of other diseases, which can degrade the quality of the environment and public health. One way of realizing a good drainage system will require a valid network data associated drainage system from upstream to downstream. The existence of a map of the flow of drainage network can facilitate the identification of the interference problem drainage and land-use planning Sustainable drainage (suistanable) and Environmental (Ecology). The drainage system in the district has a primary drainage Tamalanrea with a length of 14.65 km 33.63 km of secondary channels. The water level reaches 50 cm in Tamalanrea and Tamalanrea Jaya village, 150 cm in the village Tamalanrea Beautiful, Kapasa, Parangloe, and Bira with an average time of low tide more than 2 hours. With an area of inundation varied, but the overall total inundation area in District Tamalanrea of 649.7 ha (BPBDs Makassar, 2014). The number of roads that run into puddles in the district as much as 670 road, or about 40% of the roads in 1708 with an area of 375,017.4 m². Now, then the concept of the drainage needs to be changed from the conventional concept (old concept), a principled drain surface water as soon as possible to the body of water, into a new concept of urban drainage environmentally sound principled drain excess water to accommodate and absorb into the body of water. Therefore, the analysis of the environmental aspects is necessary so that the drainage

IOP Conf. Series: Earth and Environmental Science **135** (2018) 012010 doi:10.1088/1755-1315/135/1/012010

system has the goal of accommodating, soak, drain, and maintain, so that soil water conservation is maintained and can take place continuously and dimensions can be more efficient drainage infrastructure. In addition to technical and environmental analysis, financial analysis needs to be reviewed in this study.

2. Material and Methods

In this study, the method used is descriptive quantitative method to evaluate the condition of existing drainage channels are supported by primary data, the survey results in the determination of alternative ways control inundation accordance with the conditions of research areas. Activities carried out by conducting site surveys, analyze and interpret the data obtained for solutions and solving problems found in the field. Data obtained derived from the information society and institutions.

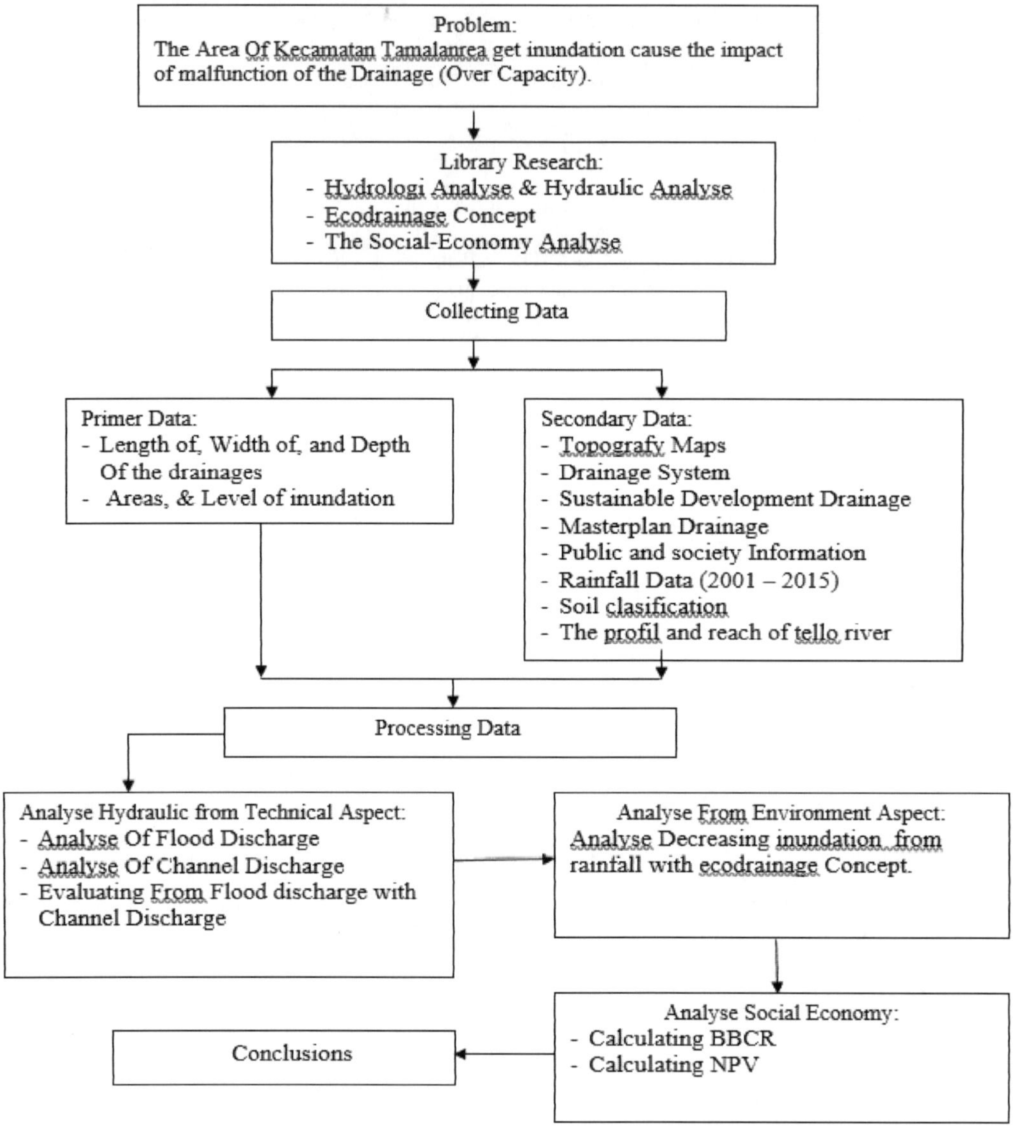

Figure 1. The Flowchart of research step.

3. Result and Discussions

3.1 Technical Aspect

The technical aspects include: rainfall averages region, analysis of rainfall plan, analysis of the frequency of the maximum daily rainfall (HHM) plan, test the suitability of the distribution, analysis of arch rainfall plans, analysis discharge storm water runoff, water analysis household waste, the analysis of the flow rate channel.

3.1.1. Capacity of Drainage Channels

Hydraulics analysis includes the calculation capacity of the existing channels. Hydraulics analysis calculation is hereinafter used as the basis for the evaluation of the capacity of existing channels. Channel capacity is analyzed to determine the amount of water flow capable drained by the channel cross section. Water discharge capable drained by the channel are affected by such broad dimensions, hydraulic radius and velocity of flow or runoff.

Table 1. The capacity of drainage channel

Channel orientation	Location	L	Shape Of Channel	Dimension			S	m	n	A	P	R	V	Capacity
				b_1	b_2	h								
Secondary	1	870	Trapesium	1	0.7	0.8	0.005	1.25	0.023	0.7	3.56	0.191	4.37	2.972
Secondary	2	120	Trapesium	0.8	0.6	0.6	0.071	1.33	0.023	0.4	2.80	0.150	3.26	1.371
Secondary	3	350	Trapesium	0.6	0.5	0.6	0.034	1.00	0.023	0.3	2.30	0.144	2.19	0.722
Secondary	4	870	Trapesium	0.6	0.5	0.6	0.047	1.00	0.023	0.3	2.30	0.144	2.59	0.855
Secondary	2 ke 4	620	Trapesium	1	0.8	1	0.046	1.00	0.023	0.9	3.83	0.235	3.56	3.205
Secondary	A	800	Trapesium	1.5	1	1	0.024	1.50	0.023	1.3	5.11	0.245	2.65	3.312
Secondary	5	170	Trapesium	1.2	0.6	0.8	0.032	1.50	0.023	0.7	4.08	0.176	2.44	1.759
Secondary	6	180	Trapesium	0.7	0.6	0.6	0.013	1.17	0.023	0.4	2.54	0.153	1.42	0.556
Secondary	7	500	Trapesium	0.8	0.5	0.5	0.006	1.60	0.023	0.3	2.69	0.121	0.84	0.274
Secondary	8	470	Trapesium	0.5	0.4	0.4	0.042	1.25	0.023	0.2	1.78	0.101	1.92	0.346
Secondary	9	340	Trapesium	0.6	0.4	0.4	0.031	1.50	0.023	0.2	2.04	0.098	1.62	0.323
Secondary	10	90	Trapesium	0.5	0.4	0.3	0.037	1.67	0.023	0.1	1.67	0.081	1.57	0.212
Secondary	11	100	Trapesium	0.6	0.5	0.5	0.041	1.20	0.023	0.3	2.16	0.127	2.22	0.610
Secondary	12	650	Trapesium	0.6	0.5	0.5	0.071	1.20	0.023	0.3	2.16	0.127	2.92	0.804
Secondary	13	610	Trapesium	0.3	0.2	0.3	0.061	1.00	0.023	0.1	1.15	0.065	1.74	0.130
Secondary	14	310	Trapesium	1	0.5	0.8	0.029	1.25	0.023	0.6	3.56	0.168	2.25	1.352
Secondary	15	200	Trapesium	0.7	0.5	0.6	0.009	1.17	0.023	0.4	2.54	0.142	1.13	0.407
Secondary	16	1330	Trapesium	1.2	0.8	0.8	0.061	1.50	0.023	0.8	4.08	0.196	3.61	2.886
Secondary	17	720	Trapesium	2	1.5	1.2	0.003	1.67	0.023	2.1	6.94	0.303	1.07	2.253
Secondary	18	210	Trapesium	0.4	0.3	0.5	0.010	0.80	0.023	0.2	1.68	0.104	0.96	0.168
Secondary	19	720	Trapesium	1.2	0.8	0.8	0.009	1.50	0.023	0.8	4.08	0.196	1.40	1.124
Secondary	20	320	Trapesium	0.8	0.5	0.5	0.061	1.60	0.023	0.3	2.69	0.121	2.62	0.851
Secondary	21	310	Trapesium	0.7	0.5	0.5	0.037	1.40	0.023	0.3	2.42	0.124	2.08	0.625
Secondary	22	900	Trapesium	2	1.5	0.8	0.003	2.50	0.023	1.4	6.31	0.222	0.87	1.222
Secondary	23	820	Trapesium	0.6	0.5	0.5	0.010	1.20	0.023	0.3	2.16	0.127	1.10	0.302
Secondary	24	2360	Trapesium	1.5	1.5	2	0.010	0.75	0.023	3.5	7.00	0.500	2.74	9.584
Primary	25	6700	Trapesium	4	3	2	0.010	2.00	0.023	7.0	12.94	0.541	2.89	20.197
Secondary	26	830	Trapesium	0.8	0.6	0.6	0.032	1.33	0.023	0.4	2.80	0.150	2.19	0.921
Secondary	27	510	Trapesium	2	1.2	1	0.031	2.00	0.023	1.6	6.47	0.247	3.00	4.799
Secondary	28	1460	Trapesium	1.8	0.9	1	0.029	1.80	0.023	1.4	5.92	0.228	2.76	3.723
Secondary	29	520	Trapesium	0.7	0.6	0.5	0.003	1.40	0.023	0.3	2.42	0.134	0.62	0.203

Channel orientation	Location	L	Shape Of Channel	Dimension b1	b2	h	S	m	n	A	P	R	V	Capacity
Secondary	30	840	Trapesium	0.6	0.5	0.8	0.024	0.75	0.023	0.4	2.60	0.169	2.07	0.911
Secondary	31	2500	Trapesium	1	0.5	0.5	0.024	2.00	0.023	0.4	3.24	0.116	1.61	0.603
Secondary	32	1670	Trapesium	2.5	1.5	1	0.009	2.50	0.023	2.0	7.89	0.254	1.67	3.339
Secondary	33	3940	Trapesium	3	4	2	0.024	1.50	0.023	7.0	10.21	0.686	5.27	36.858
Secondary	34	790	Trapesium	2	1.5	0.8	0.006	2.50	0.023	1.4	6.31	0.222	1.26	1.769
Secondary	35	3000	Trapesium	1.8	0.8	1	0.009	1.80	0.023	1.3	5.92	0.220	1.52	1.972
Secondary	36	2160	Trapesium	1	0.8	0.6	0.010	1.67	0.023	0.5	3.33	0.162	1.29	0.697
Secondary	37	1500	Trapesium	1	0.8	0.6	0.010	1.67	0.023	0.5	3.33	0.162	1.29	0.697
Secondary	38	140	Trapesium	0.5	0.4	0.3	0.010	1.67	0.023	0.1	1.67	0.081	0.81	0.110
Secondary	39	910	Trapesium	1	0.8	0.6	0.029	1.67	0.023	0.5	3.33	0.162	2.20	1.186

3.1.2. Condition of existing drainage

Based on previous calculations that the calculation of the existing channel capacity to discharge the design flood drainage channel conditions at this time can be evaluated. Design flood discharge is the sum of debits that go into the channel. The results of the evaluation will indicate that the channel is still able to function optimally, or is not able to function optimally. Based on the analysis evaluation in District Tamalanrea channel capacity of 41 channels that exist, there are 14 channel capacity is not able to accommodate the design discharge. So as to channel that does not meet the capacities, will be planned one of the technologies that is well catchment ecodrainase for handling.The technically meet the capacities but in the field occurred inundation for trash and sediment that much on the line, so it is necessary for handling channel normalization.

Table 2. The condition of existing drainage

Location	Flood Discharge	Capacity with Sediment	Inundation with Sediment	Requires	Capacity No Sediment	Inundation No Sediment	Requires	Area Of inundation	Height Of inundation	Period of Inundation
	m^3/sec	m^3/sec	m^3/sec	Capacity > Flood Disc.	m^3/sec	m^3/sec	Capacity > Flood Disc.	m^2	m	Hour
1	1.98	2.23	0.25	Qualify	2.97	1.00	Qualify	-	-	-
2	0.04	1.12	1.08	Qualify	1.37	1.34	Qualify	-	-	-
3	0.54	0.63	0.08	Qualify	0.72	0.18	Qualify	-	-	-
4	1.46	0.68	-0.77	Not Qualify	0.86	-0.60	Not Qualify	32,389.86	0,15-0,20	3
2 to 4	0.64	2.56	1.92	Qualify	3.20	2.57	Qualify	-	-	-
A	3.03	2.32	-0.71	Not Qualify	3.31	0.28	Qualify	-	-	-
5	1.40	1.52	0.11	Qualify	1.76	0.35	Qualify	-	-	-
6	0.77	0.37	-0.40	Not Qualify	0.56	-0.21	Not Qualify	11,579.70	0,15-0,20	3
7	1.37	0.26	-1.11	Not Qualify	0.27	-1.10	Not Qualify	59,226.46	0,15-0,20	3
8	0.36	0.17	-0.18	Not Qualify	0.35	-0.01	Qualify	-	-	-
9	0.39	0.25	-0.14	Not Qualify	0.32	-0.07	Not Qualify	3,773.56	0,15-0,20	3
10	0.45	0.19	-0.26	Not Qualify	0.21	-0.24	Not Qualify	12,761.65	0,15-0,20	3
11	0.10	0.50	0.40	Qualify	0.61	0.51	Qualify	-	-	-
12	0.63	0.66	0.03	Qualify	0.80	0.18	Qualify	-	-	-
13	1.01	0.09	-0.93	Not Qualify	0.13	-0.88	Not Qualify	47,750.19	0,15-0,20	3

Location	Flood Discharge	Capacity with Sediment	Inundation with Sediment	Requires Capacity > Flood Disc.	Capacity No Sediment	Inundation No Sediment	Requires Capacity > Flood Disc.	Area Of inundation	Height Of inundation	Period of Inundation
	m^3/sec	m^3/sec	m^3/sec		m^3/sec	m^3/sec		m^2	m	Hour
14	0.23	1.01	0.78	Qualify	1.35	1.12	Qualify	-	-	-
15	0.05	0.34	0.28	Qualify	0.41	0.35	Qualify	-	-	-
16	1.31	2.16	0.86	Qualify	2.89	1.58	Qualify	-	-	-
17	0.44	1.50	1.06	Qualify	2.25	1.81	Qualify	-	-	-
18	0.19	0.16	-0.03	Not Qualify	0.17	-0.03	Not Qualify	452.98	0,15-0,20	1
19	0.57	0.70	0.14	Qualify	1.12	0.56	Qualify	-	-	-
20	0.05	0.82	0.77	Qualify	0.85	0.81	Qualify	-	-	-
21	0.14	0.50	0.36	Qualify	0.63	0.48	Qualify	-	-	-
22	0.57	0.92	0.34	Qualify	1.22	0.65	Qualify	-	-	-
23	1.14	0.24	-0.90	Not Qualify	0.30	-0.84	Not Qualify	30,173.33	0,15-0,20	2
24	9.87	6.57	-3.29	Not Qualify	9.58	-0.28	Qualify	-	-	-
25	20.95	16.16	-4.79	Not Qualify	20.20	-0.75	Qualify	-	-	-
26	1.61	0.61	-0.99	Not Qualify	0.92	-0.68	Not Qualify	36,961.13	0,15-0,20	3
27	4.73	2.88	-1.85	Not Qualify	4.80	0.07	Qualify	-	-	-
28	1.36	2.98	1.62	Qualify	3.72	2.36	Qualify	-	-	-
29	0.91	0.16	-0.74	Not Qualify	0.20	-0.70	Not Qualify	37,929.79	0,15-0,20	3
30	0.50	0.85	0.35	Qualify	0.91	0.41	Qualify	-	-	-
31	2.25	0.51	-1.74	Not Qualify	0.60	-1.64	Not Qualify	59,128.81	0,25-0,30	3
32	5.94	3.00	-2.93	Not Qualify	3.34	-2.60	Not Qualify	140,217.95	0,35-0,4	6
33	30.56	25.80	-4.76	Not Qualify	36.86	6.29	Qualify	-	-	-
34	0.27	1.33	1.05	Qualify	1.77	1.50	Qualify	-	-	-
35	4.28	1.77	-2.50	Not Qualify	1.97	-2.31	Not Qualify	124,612.45	0,35-0,40	6
36	1.29	0.65	-0.64	Not Qualify	0.70	-0.60	Not Qualify	32,138.73	0,15-0,20	3
37	0.54	0.64	0.10	Qualify	0.70	0.16	Qualify	-	-	-
38	0.07	0.08	0.01	Qualify	0.11	0.04	Qualify	-	-	-
39	1.23	0.99	-0.24	Not Qualify	1.19	-0.04	Qualify	-	-	-

3.2 Environmental Aspect

Alternative inundation reduction referred to in this research is the need for change in the concept of conventional drainage systems become environmentally friendly drainage system in this study is the use of Infiltration wells.

IOP Conf. Series: Earth and Environmental Science **135** (2018) 012010 doi:10.1088/1755-1315/135/1/012010

Table 3. The condition of existing drainage

Channel	Location	Type Of Infiltration Well	Inundation (m3/sec)	A (km2)	I	C	Recharge (m3/sec)	Quantity Housing Infiltration wells	Quantity Road Infiltration wells
Secondary	4	Housing	-0.60	0.00048	223.59	0.65	0.01939	31	
Secondary	6	Road	-0.21	0.00050	224.69	0.56	0.01749		12
Secondary	7	Road	-1.10	0.00050	222.52	0.41	0.01268		86
Secondary	9	Housing	-0.07	0.00048	224.41	0.42	0.01258	6	
Secondary	10	Housing	-0.24	0.00048	225.24	0.45	0.01343	18	
Secondary	13	Housing	-0.88	0.00048	223.97	0.44	0.01315	67	
Secondary	18	Housing	-0.03	0.00048	224.35	0.62	0.01856	1	
Secondary	23	Housing	-0.84	0.00048	221.80	0.44	0.01302	64	
Secondary	26	Housing	-0.68	0.00048	223.34	0.55	0.01639	42	
Secondary	29	Housing	-0.70	0.00048	221.36	0.49	0.01447	49	
Secondary	31	Housing	-1.64	0.00048	218.72	0.37	0.01077	153	
Secondary	32	Road	-2.60	0.00050	219.51	0.54	0.01648		158
Secondary	35	Road	-2.31	0.00050	215.44	0.51	0.01527		151
Secondary	36	Housing	-0.60	0.00048	217.54	0.53	0.01550	38	
		Amount Of Infiltration Wells						468	407

3.3 Economic Aspect

A development activity is said to be economically viable if the cost of the investment (cost) required less than the benefits (benefits) obtained. Gains on drainage activity generally in the form of indirect revenue, for example, loss or reduction in loss due to waterlogging or flooding, environmental improvements, improved public health and aesthetic improvements.

Table 4. The Benefit, and Cost of infiltration wells

Channel	Location	Type Of Infiltration Well	Quantity Housing Infiltration wells	Quantity Road Infiltration wells	Cost Housing Infiltration Wells (Rupiah)	Cost Road Infiltration Wells (Rupiah)	Losses due to flood 5 years period (Rupiah)
Secondary	4	Housing	31		88,149,124.38		188,364,242.40
Secondary	6	Road		12		32,122,742.46	95,555,452.15
Secondary	7	Road		86		226,597,999.24	606,187,637.35
Secondary	9	Housing	6		15,835,193.80		16,041,598.05
Secondary	10	Housing	18		50,133,549.37		163,801,440.45
Secondary	13	Housing	67		191,642,432.65		495,236,725.25
Secondary	18	Housing	1		3,864,162.97		5,422,266.45
Secondary	23	Housing	64		183,428,266.79		184,273,040.45
Secondary	26	Housing	42		119,010,380.62		304,315,418.35
Secondary	29	Housing	49		138,311,010.28		206,394,904.30
Secondary	31	Housing	153		434,641,123.13		665,015,858.80
Secondary	32	Road		158		412,895,579.29	2,366,032,290.45
Secondary	35	Road		151		395,868,155.86	2,044,322,741.15
Secondary	36	Housing	38		109,424,652.98		272,533,562.95
	Amount		468	407	1,334,439,896.98	1,067,484,476.85	7,613,497,178.55

4. Conclusion

Urban street drainage need of improvement by infiltration wells viewed groundwater conditions of diminishing returns. Based on the analysis technical aspect found that there are 14 channel in Tamalanrea district was not qualify to accommodate wastewater and rainwater. Based on the analysis environmental aspect Infiltration wells absorb water into the soil and also reduce the inundation that occur in urban areas. Diameter 1,0 m, depth of water 4 m and soil permeability 0,00015 m/sec. Discharge of infiltration wells 0,0019 m^3/sec. Based on the analysis inundation treatment with the application of absorption wells in Sub Tamalanrea able to absorb 100% to the total amount of recharge wells as many as 875 pieces. Total costs required to manufacture 875 unit of infiltration wells is IDR 2,401,924,372.00. Based on analysis economic aspect (BCR) benefit cost ratio the result of rate is 3.19 and its mean that feasible to develop.

5. References

[1] Andayani S and Yuwono E B 2012 *Indikator Tingkat Layanan Drainase Perkotaan* (Jakarta: Universitas trisakti) vol 11 pp 148 -157

[2] Bachtiar S 1998 *Studi Penggunaan Sumur Resapan Untuk Mengurangi Masalah Genangan Di Dps Amprong Kecamatan Kedungkandang Kota Malang* (Malang: Jurusan Teknik Pengairan Fakultas Teknik Universitas Brawijaya) p 103

[3] Chow V T 1989 *Open Channel Hydraulics Alih Bahasa* (Jakarta: Erlangga) pp 16-28

[4] Suripin 2010 *Sistem Drainase Perkotaan yang Berkelnajutan* (Semarang: Andi Publishing) pp 4-19

[5] Mulyanto H R 2013 *Penataan Drainase Perkotaan* (Jakarta: Graha Ilmu) p 33

IOP Conf. Series: Earth and Environmental Science **135** (2018) 012011 doi:10.1088/1755-1315/135/1/012011

The Discrepancy between The Programs and Disaster Management Policy in Klapanunggal District, Bogor, West Java

D A Puspito Sari[1], I Listiyowati[2], T Nefianto[3] and Lasmono[3]

[1]Environmental Engineering Program Study, Faculty of Engineering and Computer Sciences, Universitas Bakrie, Indonesia
[2]Bogor City Public Health Office, Bogor, Indonesia
[3]Disaster Management for National Security Program Study, Faculty of Defense Management, Indonesia Defense University, Indonesia

Email : deffi.sari@bakrie.ac.id

Abstract. Bogor regency consists of 40 districts, 23 are prone tonatural disasters. Klapanunggal district is listed in 10 districts declared as most vulnerable to natural disasters. Natural disasters could lead to loss of property and infrastructure damage and will affect the food security in the region. Food shortages is one example of the condition which causes food insecurity. The aim of this research is to analyze the government's food security strategy in anticipation of disaster with a case study of food insecurity in Klapanunggal district. The analysis suggested that; 1) FSVA is an appropriate program to identify food shortage areas, 2) Food Shortage Relief Program (*Program Penanganan Daerah Rawan Pangan*-PDRP) is the optimal efforts in reducing food shortages in the region, 3)The mismatch between FSVA indicators and Food Shortage Relief Program makes Klapanunggal district difficult in achievingfree status food-shortage. Based on the analysis, it is suggested that the implementation of Food Shortages Relief Program could be carried out based on the priority issues and implemented with integrated coordination and assistance among stakeholders. Such priority issues, integrated coordination and assistance are fully analyzed in this study.

1. Introduction

West Java province is one of the provinces in Indonesia which has a high population density. Population wise, Bogor regency contained as many as 4,966,621 inhabitants in 2011 or 11.03% of the total population of West Java. The population of Bogor regency in the year 2012 has reached 5.077 million inhabitants with a population growth rate of 3.15% per year [1]. Bogor regency consists of 40 districts, among those districts 23 are prone to natural disasters. District Klapanunggal is listed in 10 districts declared most vulnerable to natural disasters [2]."In disaster-prone locations measures to improve disaster resilience should be an integral part of food security policies and strategies"[3]. Food shortage is one example of a condition which causes food insecurity. According to a report from Regional Disaster Relief Agency (*Badan Penanggulangan Bencana Daerah*-BPBD), Klapanunggal is a district that is prone to landslides and tornado. According to the Indonesia Disaster Risk Index (*Index Rawan bencana Indonesia*-IRBI), Bogor regency is on the top five most disaster-prone areas in Indonesia [4]. Agriculture is particularly vulnerable to disaster risk [5]. Frequently attacked by disasters is one of the triggers of food insecurity in Bogor. Disaster with its potential to disturb food supply can threaten food security. Although nationally Indonesia has enough food supply, food shortages can happen regionally, especially in disaster prone areas. Klapanunggal district is not only prone to disaster, it becomes a major area of focus with food-shortage status in Bogor regency.

The purpose of this study is to analyze the government's food security strategy in anticipating disaster with a case study of food insecurity in Klapanunggal district. The analysis includes the constraints and potential factors that exist and pose a threat to the implementation of the strategy and how to solve problems/obstacles in the implementation.

This study was conducted in Bogor regency, West Java, Indonesia in 2014. Primary data was collected by field observation and in-depth interview with district head, village leaders, head of regional planning agency and MWA training and consulting office leaders. The analysis was done qualitatively with purposive sampling and snowball methods. Secondary data for mapping were collected from regional planning agency that consist of data from 40 districts in Bogor regency.

2. Food Security Vulnerability Atlas (FSVA) Indicators

Food shortages should be a major concern for local government, in this case the local government is the district of Bogor. There are some indicators to identify food shortages, one of the indicators used by the government is Food Security Vulnerability Atlas (FSVA) that consists of nine indicators; one indicator of the aspect of food availability (ratio of normative consumption against the availability of staple food), three indicator aspects of access to food and livelihood (poverty, electricity and road access) as well as the five indicators of food utilization (life expectancy at birth, underweight birth, illiteracy, percentage of household without access to clean water and percentage of household with access to health facilities) [6]. The FSVA indicators are shown in figure 1.

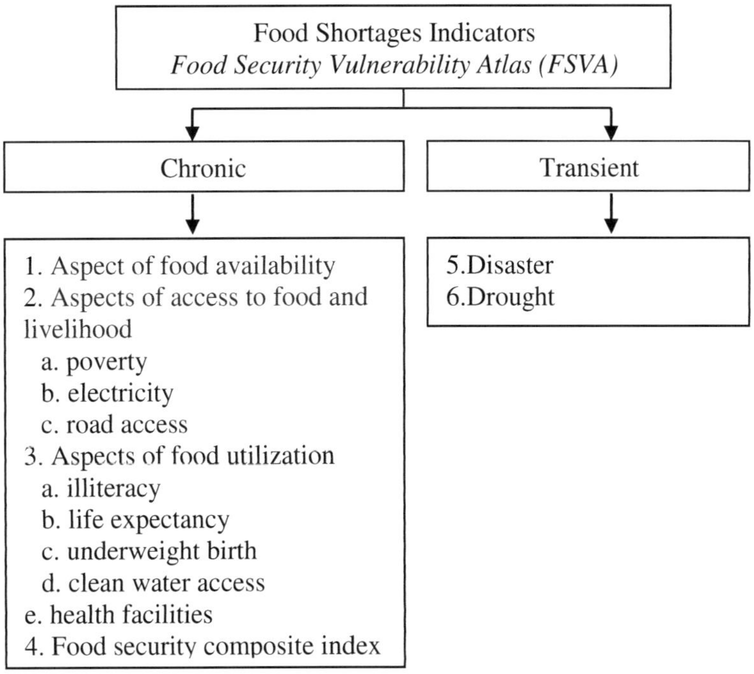

Figure 1. Chronic and Transient Food Shortages Indicator
Sources: BKP5, Bogor District. 2015 [7] and author.

Based on the combined indicators of Food Security Vulnerability Atlas (FSVA), Klapanunggal district aggregately falls into the district that faces food shortages as shown in table 1.

Table 1. FSVA Indicators of Klapanungal District

No.	Indicator	District Klapanunggal	Note
1	Food availability	High deficit	Priority 1
2	Food and Livelihood		
	a.Poor Household	13.71%(>10%)	Priority 5
	b.Electricity access (supply from state electrical company)	Without electricity access 76.5%	Priority 1
	c.Road access	30.68%	Priority 1
3	Food Utilization		
	a.Illiteracy	0.57%	Priority 6
	b.Life expectancy	71.03 years	Priority 6
	c.Underweight birth	15.11%	Priority 3
	d.Clean water access	44.98%	Priority 4
4	Food security composite index	0.82%	Priority 1
5	Vulnerability to transient food shortages due to disaster and drought	10.19%	Priority 1

Source :Bogor District Regional Planning Agency, 2012 [8].

From table 1 it appears that in terms of food availability Klapanunggal district is still in the category of low availability (high deficit), meaning that there is interference with the availability of food in Klapanunggaldistrict resulting in food shortages.There is not enough market to supply and stock food in Klapanunggal. However, government has secured the land for building new market and barn.

The percentage of poor households in the Klapanunggal district amounts to 7.721 (13.71%), a percentage that could not be underestimated even though only in priority number 5,the existence of large factories and industries in Klapanunggal district has given opportunity to work and increase the income so that poverty issue in Klapanunggal district is not a number 1 priority.Other districts in Bogor regency has more percentage of poor households such as Sukamakmur district with 53.27% poor household. Even though poverty is not a priority issue and people have high purchasing power, still overall indicator of FSVA shows that Klapanunggal district falls into the category of area with food shortage status.

Access to electricity supplied by State Electrical Company (*Perusahaan Listrik Negara-PLN)* inKlapanunggal districtis only 23.5%.Interestingly, with high purchasing power, people can provide their household with generator to produce electricity for their household.

Road access is only 30.68%, high industrial activities has brought trucks and heavy containers to use those limited roads in and out from the district, resulting in poor road condition with many holes and puddles after rain. Problems faced by Klapanunggaldistrict is that road access to the distribution centers of agricultural production were minimal. It can be seen in the village of Bojong and Ligarmukti, roads along the road to the area is in very poor condition. This is because of the present of large vehicles which take lime from limestone quarry, thus disrupting the distribution of agricultural production and crops.

Constraints arising from the road access indicator are: funding from the government to repair roads is minimal; monitoring of the Public Works Department is conducted once every6 (six) months; and although the road has been fixed, high mobility for the transport of raw materials to the cement factory is hampering Klapanunggal residents to use the roads optimally.

Both the rate of illiteracy and life expectancy are in priority number 6. Illiteracy is amounted to 0.57% and life expectancy is 71.03 years, underweight birth (15.11%, priority number 3) and clean water access (44.98% priority number 4), while all indicators in Food Utilization category are not in the first priority.

However, food security of Klapanunggal district has a composite index of 0.82% (priority number 1), so does vulnerability to transient food shortages due to disaster and drought (10.19%, priority number one).Dry region, large industrial areas, dense population have triggered fires in the Klapanunggal district. In the period 2009-2011 seven fires have occurred in the region. Klapanunggal district has the highest drought area in Bogor regency, that is 10.19%. This causes difficult access to clean water, both for food consumption and other household needs. The location of Klapanunggal district that is surrounded by industrial areas aggravates the availability of clean water. Drought covers the Lulut, Nambo and Bantar jati villages. The foregoing is a trigger tofood insecurity in the Klapanunggal district either directly or indirectly.

Having obtained the data necessary for the determination of the indicators, the analysis of food security and vulnerability was conducted using data for theyear 2012, with the final result of the composite score from those indicators. A combination of several composite index is processed using ArcView and Map Info 75 food security and vulnerability atlas. Shades of red indicated resulting in variations in the level of food insecurity and shades of green depicted food security conditions, the bolder color showed a higher level in terms of food security (bold green) or food insecurity (bold red). Klapanunggal subdistrict falls into the category of food-insecure as shown on the map (figure 2).

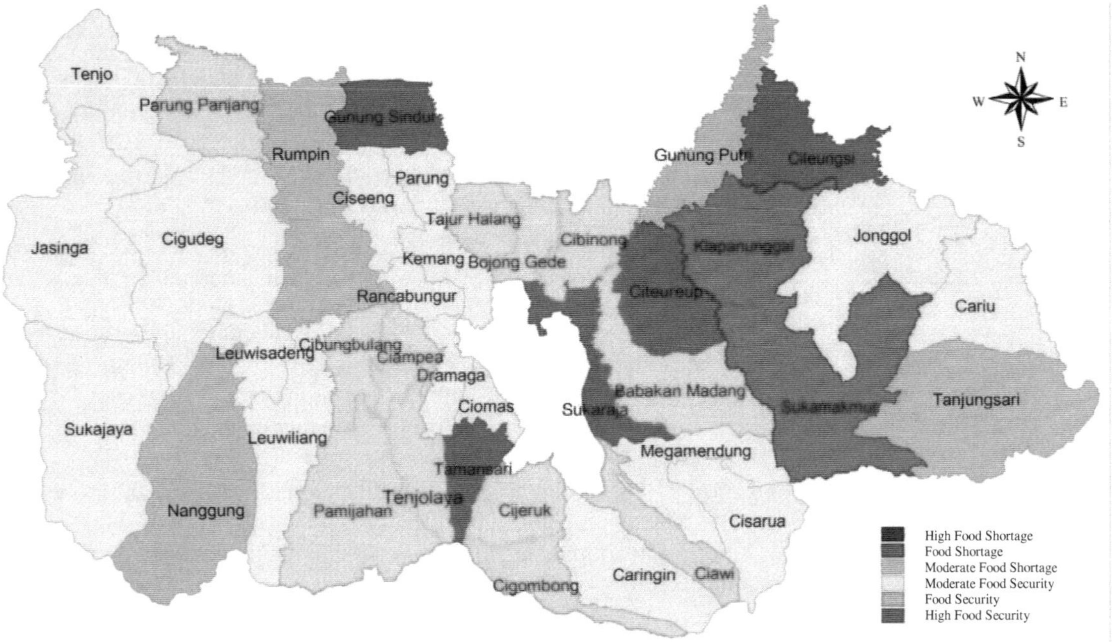

Figure 2. Food security and vulnerability atlas of Bogor district
Source: Author using data from Regional Planning Agency of Bogor District

3. Food Shortage Relief Program (*Program Penanganan Daerah Rawan Pangan*-PDRP Program)

Government has an important role and obligation to address food insecurity in the region. Hence the importance of the government is to issue a policy relating to Food Shortage Relief Program (*Program Penanganan Daerah Rawan Pangan*-PDRP) as a standard and management to improve the condition of areas facing food-insecurity issues. It resulted in the implementation of PDRP, the action plan and community based early warning systems. In practice, PDRP in Klapanunggal was implemented gradually from preparatory activities, socialization, the selection of villages and the location of the program, electoral assistance, implementation and monitoring and evaluation (monev).

Results of the activities that have been carried out as shown in table 2 shows that the activities are focused on the areas with minimum achievement of FSVA indicators. Low supply from PLN and poor road condition are the major indicators that need to be addressed. The Village of Cioray, Cibuntu and Leuwi karet have constrains in terms of electricity supply from State Electricity Company

IOP Conf. Series: Earth and Environmental Science **135** (2018) 012011 doi:10.1088/1755-1315/135/1/012011

(*Perusahaan Listrik Negara-PLN)*, but with the help from national and state budget funds, as many as 194 households that have not had access to electricity can obtain access to electricity. The village of Cioray has no road access to the village centre and takes approximately 4 hours to get to the centre of the village.

Table 2. The process of implementing Food Shortage Relief Program (*Program Penanganan Daerah Rawan Pangan*-PDRP) in Klapanunggal district

Process	Implementation
Preparation	To collect secondary data related to food-shortage areas in Klapanunggal district, to study the outcome for each indicator, to study the implementation strategy and goals to be achieved.
Socialization	Socialization to district and village heads regarding food shortage indicators through meeting and FGDs. Many of the participants were complaining of the food shortage status, especially those aware that the majority of Klapanunggal residents are not poor. By understanding the measurements and indicators, gradually the participant could accept and agree to set common goal to free Klapanunggal from food shortage area status.
Selection of village and location for program implementation	The villages are chosen based on the indicators. Almost all villages have food shortage issue, but the priority for each village depends on the situation of each village.
Selection of assistance	The head of district is chosen as the coordinator, but implementation technique is done by the district field staff (*Petugas Penyuluh Lapangan*-PPL) in coordination with village heads and village coordinators.
Implementation	For each village, focusing on indicator that have high priority: 1. Cioray, Cibuntu and Leuwikaret villages have constrains in terms of electricity supply from State Electricity Company (*Perusahaan Listrik Negara-PLN)*, but with the help from national and state budget funds, as many as 194 household that have not had access to electricity can get access to electricity. 2. Ciorayvillage has no access to the city centre, 208 households are using roads with poor condition. 3. Bojong and Klapanunggalvillages have difficult access to clean water. 4. Ligarmukti and Cikahuripan villages and some part of Bojong village are agricultural producing areas, but ownership of the land mostly belong to people from outside Klapanunggal district andthe harvest is brought and consumed outside Klapanunggal district. Each household now pursues planting sustainable food garden (*Kebun Pangan Lestari-KPL*) as well using vacant land in the area. 5. Ligarmukti village has water spring that has not been used optimally, now with the help from CSR, water supply management is start to supply the area without access to clean water. 6. Leuwikaret, Ligarmukti, Cikahuripan, Lulut, Bojong, Cioray and Cibuntu villages have limited health facilityand is now developing *Pos daya* with the help of the local midwife.
Monitoring and evaluation	Monitoring is done by district head in coordination with regency work force (*Satuan Kerja Perangkat Daerah*-SKPD).

Table 3 shows Food Shortage Relief Program (*Program Penanganan Daerah Rawan Pangan*-PDRP Program) in Klapanunggal district and these programs need to be evaluated, while in practice, some programs cannot be implemented. Of the 21 programs listed in the General Policy on Food Security, seven programs (33.3%) cannot be implemented properly.

Table 3. Food Shortage Relief Program (*Program Penanganan Daerah Rawan Pangan*-PDRP Program) in Klapanunggal District.

No	Existing Program	Implementation		Program Priority	
		Yes	No	Yes	No
1	Normal and Non formal Education Program		V	V	
2	Nutrition Improvement Program	V		V	
3	Community Health Program	V		V	
4	Procurement programs, enhancement and improvement of facilities and infrastructure sub-health centres and networks	V		V	
5	Improvement of maternal and child safety program	V		V	
6	Health for the poor program	V		V	
7	Bridge and database system improvement program		V	V	
8	Flood control program		V		V
9	Village infrastructure development program	V		V	
10	Poverty and community empowerment program	V		V	
11	Institution and social welfare program		V		V
12	Disaster preparedness program	V		V	
13	Electricity access development program	V		V	
14	Food security in agriculture, forestry program	V		V	
15	Food security in husbandry program	V		V	
16	Food security in fishery program	V		V	
17	Small and medium enterprise program		V	V	
18	Social and welfare planning program		V		V
19	Mother, child, and infant health program	V		V	
20	Women protection program		V		V
21	Family economy empowerment program	V		V	

According to the results from the interviews with Bogor's Regional Planning Agency, government and stakeholders in the Klapanunggal district, seven programs cannot be implemented because of obstacles in its implementation, namely; a) The limited funds allocated for the implementation of the program, b) Data reporting system that has not been integrated /not appropriate, b) The lack of participation of local communities, c) Limited human resources.

FSVA strategy undertaken by the Government of Bogor regency is appropriate for determining the area of food insecurity as well as for addressing plans to improve the condition in food-shortage areas. For Klapanunggal district, from nine indicators of FSVA, some indicators should be given priority, namely food availability, access to electricity and roads. These indicators should be the main program in the Region's Middle Term Development Plan (*Rencana Pembangunan Jangka Menengah Daerah*-RPJMD)in Bogor regency and implemented based on priority. When adjusted to six FSVA indicators, the Klapanunggal district's PDRP program that should be implemented is based on the priority shown in table 4.

Table 4. Klapanunggal PDRP Priority Scale Matrix Program Requirements

No.	FSVA (Based on Priorities)	Relevant PDRPProgram
1	Priority 1 of food availability	1. Food security in agriculture, forestry program 2. Food security in husbandry program 3. Food security in fisheryprogram 4. Family economy empowerment program
2	Priority 2 of electricity access	5. Electricity access development program 6. Village infrastructure development program
3	Priority 1 of road access	No program planned yet
4	Priority 3 of underweight children	7. Procurement programs, enhancement and improvement of facilities and infrastructure sub-health centres and networks 8. Nutrition improvement program 9. Mother, child, and infant health program 10. Health for the poorprogram 11. Community Health Program 12. Mother, child, and infant health program
5	Priority 1 of food security composite	No program planned yet
6	Priority 1 of vulnerability of transient food shortages	13. Disaster preparedness program

Sources : Bogor district Middle Term Development Plan and author's analysis.

From table 4, not all twenty one points of Klapanunggal's PDRP program should become priority. When adjusted to FSVA, thirteen PDRP program listed in table 4 should receive government attention and become core program to improve Klapanunggal's food-shortage status. Relevant PDRP programs are needed for road access and food security composite indicators which both have number one priority (according to FSVA indicator). Unfortunately, no relevant PDRP program is implemented yet to improve "road access" and improve food security composite indicators. The mismatch between FSVA indicators and food shortage relief program makes Klapanunggal district difficult in achieving food-shortage free status. Here we reveal the discrepancy between the program and disaster management policy in Klapanunggal district.

4. Conclusion

Obstacles encountered in the implementation of food security strategy in Klapanunggal area are: a) The implementation of the program to alleviate food shortages have not been based on FSVA priorities, b) The absence of village forum that specifically support the implementation of PDRP program, c) Lack of coordination and awareness of the community and local authorities about food-shortages in the region, d) Unavailability of district market to improve food availability in Klapanunggal district.

Government efforts in anticipating food insecurity in Klapanunggal district are : a) Implementation of the PDRP program, b) Socialization regarding food insecurity in the region ranging from policy makers at the level of Bogor regency, Klapanunggal district, villages in the sub district and community of Klapanunggal district, c) land preparation for the district market,d) development of barns in the Klapanunggal district.

FSVA is an appropriate program to identify food shortage areas. Using nine indicators of FSVA, the government can work to improve the condition in food-shortage areasand concentrate to work on improvement based on indicators with higher priority. Food Shortage Relief Program (*Program Penanganan Daerah RawanPangan*-PDRP program) is the optimal efforts in reducing food shortages in the region.However, the design of the program need to be synchronized with the priority area of improvement and tailored specifically based on the needs of the targeted areas. Thus government can avoid the discrepancy between the program and disaster management policy that could hamper the effort in achieving food security in disaster prone areas.

5 References

[1] BPBD 2013 Rencana Penanggulangan Bencana Daerah Tahun 2014-2019 Kabupaten Bogor BPBD Kabupaten Bogor 3
[2] Badan Penanggulangan Bencana Daerah 2013 Koran Tempo https://m.tempo.co/read/news/2013/12/22/214539448/cuaca-ekstrem-115-desa-di-bogor-rawan bencana
[3] De Haen H and Hemrich G 2007 The economics of natural disasters: implications and challenges for food security *Agricultural Economics* 37 31–45 doi:10.1111/j.1574-0862.2007.00233.x
[4] BPBD Kabupaten Bogor 2013 Data Indeks Rawan Bencana Indonesia menurut BNPB
[5] De Haen, H 2008 Food Security Strategies: Building Resilience Against Natural Disasters *EuroChoices* 7 26–33 doi:10.1111/j.1746-692X.2008.00108.x
[6] Bapeda Kabupaten Bogor and MWA Training and Consulting 2012Laporan Peta Kerentanan Terhadap Kerawanan Pangan di Kabupaten Bogor Tahun 2012
[7] BKP5 Kabupaten Bogor 2014 Laporan Sistem Analisis Kewaspadaan Pangan dan Gizi Kabupaten Bogor Tahun 2014 Bogor
[8] Bapeda Kabupaten Bogor 2012 Laporan Peta Kerentanan terhadap Kerawanan Pangan di Kabupaten Bogor Tahun 2012 Kerjasama Badan Perencanaan Daerah Kabupaten Bogor Tahun 2012 dengan MWA Konsulting, Lembaga Tata Kelola Ketahanan Pangan

IOP Conf. Series: Earth and Environmental Science **135** (2018) 012012 doi:10.1088/1755-1315/135/1/012012

Determining Domestic Container Shipping as an Enforcement of Indonesian International Hub Port

H I Nur[1] , S D Lazuardi[1], F Hadi[1] and M Hapis[1]

[1]Department of Marine Transportation Engineering, Faculty of Marine Technology, Sepuluh Nopember Institute of Technology, Indonesia

Email : hasaniqbaln@gmail.com

Abstract. According to Presidential Regulation Number 26 year of 2012 about the National Logistics System Development Blueprint, the Indonesian government proposed to build two international hub ports, which were in Port of Kuala Tanjung for the western region and Port of Bitung for eastern region. Therefore, the optimum routes and fleet size are required to support the enforcement of Indonesian International Hub Port. The optimization model is used to obtain the optimum route and fleet by minimizing the total shipping costs, while considering the container demand. The result of analysis obtained that the optimum route and fleet size for the western region of Indonesia were: (1) Kuala Tanjung-Belawan required 15 ships of 1,000 TEU; (2) Kuala Tanjung-Tanjung Priok required 73 ships of 2,500 TEU; (3) Kuala Tanjung-Tanjung Perak required 44 ships of 2,500 TEU. Meanwhile, the optimum route and fleet size for the eastern region of Indonesia consisted of: (1) Bitung-Sorong required 1 ship of 500 TEU; (2) Bitung-Banjarmasin required 3 ships of 500 TEU; and (3) Bitung-Makassar required 1 ship of 1,500 TEU.

1. Introduction

In the Masterplan for Acceleration and Expansion of Indonesian Economic Development 2011-2025 (MP3EI), Indonesian government consider to strengthen the National Connectivity in order to encourage inclusive and sustainable economic growth [1]. One of the implementations stated in MP3EI is to establish the Port of Kuala Tanjung as an International Hub Port for Western Region of Indonesia and Port of Bitung as an International Hub Port in the Eastern Region of Indonesia (Figure 1).The establishment of Indonesian international hub port is also supported by the Presidential Regulation Number 26 year of 2012 on National Logistics System Development Blueprint. In addition, the implementation of international hub port development programme is already in the priority of strategic project agenda on Presidential Regulation Number 3 year of 2016 in the Acceleration of National Strategic Project Implementation. Therefore, in order to obtain the additional national fleet capacity and to ensure the movement of goods, especially for export and import container cargoes, will require the capacity planning and its operational service on the main node of domestic container ports.

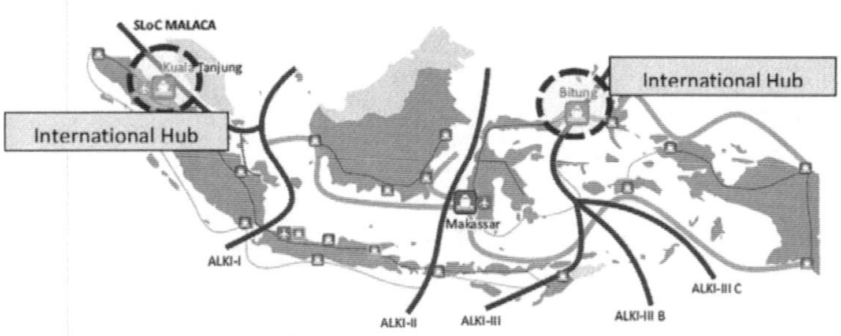

Figure 1. Theproposeof Indonesian international hub port [1]

2. Literature Review

2.1. Shipping Costs and Chartering Concept

In general, shipping costs are divided into four main categories: capital costs, operational costs, voyage costs, and cargo handling costs. Operating cost is sum of manning, stores, maintenance and repair, insurance and administration cost. While voyage cost is a function of fuel cost and port charges.

In the transport of cargoes or goods terms, generally used either owned or charter ship. In maritime transport, shipping chartering is divided into three types, such as bareboat charters, time charters and voyage charters [2]. And to approach total vessel operating costs, one of its components using the system time charter hire as s suitable for fixed cost, because fixed cost will be paid fixed in a certain period, and fixed cost does not depend on the ship operation. Then in the calculations of the total costs of vessel operation, by summing the time charter fees, port charges and the cost of cargo handling

2.2. Fleet Capacity and Shipping Routes Planning

In determining domestic container shipping, one of the main components are planning services or network, which focuses on building links from one to another nodes, in this case network planning can be defines as the movement of goods from point of origin to point of destination [3]. Some studies has used a variety of concepts in designing routes and fleet capacity in order to obtain optimal combination.

2.2.1. Liner Shipping Network Design Problem (LSNDP). Mulder and Dekker has conducted a study related designing the optimum routes in liner shipping they using combination of fleet-design, ship-scheduling and cargo routing with restrictions on availability of the number of ships operation. Whereas the formulation in this method used linear programming approach. [4]. In order to obtain optimum domestic shipping container, this research using combined approach in the relevant scope of LSNDP Principle, such as Feeder Network Design Problem and Multiple Commodities Problem.

2.2.2. Feeder Network Design Problem (FNDP). Previous research has using optimization on hub and spoke network concept to obtain optimum shipping routes problem. In the optimization plan they focus on fixed capacity slots services to meet demand satisfaction. The concept of hub and spoke network planning is an integration maritime network to delivering goods between two regions separated by sea water, one of them are selected as hub port and another being spoke ports. In the field operation, larger size of fleet provides services to the hub port, while smaller fleet used to provide between Hub to spoke or feeder ports [5]. The modelling of hub and spoke networks are consists of two type networks systems, namely direct and indirect feeder shipping [6]. As shown inFigure 2, direct feeder shipping are connecting between hub port and feeder port directly, while indirect feeder shipping connect hub port to more than one of feeder ports, this type also called shaped loops [7].

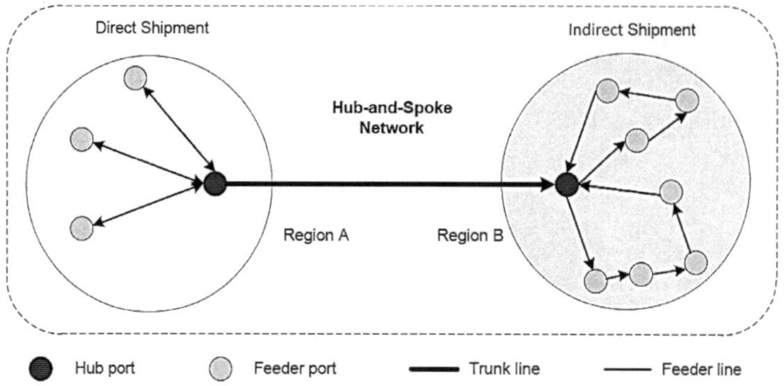

Figure 2. Feeder shipping as a part of Hub and Spoke Networks [9]

In other research also analyzing the optimum container shipping connectivity with using form of comparison scenarios for direct shipping and indirect shipping in achieving optimum allocation of demand in every port, and the optimum routes are provides extends to the needs of total capacity in each route with minimum total shipping cost [7]. But these studies have not completely reached the stage of identification on fleet sizes and its optimum operation routes of domestic container shipping. Then this research would continue to obtain the number of optimum fleet capacity and the selected optimum routes services for domestic container shipping in terms of enforcement Indonesian international hub port policy. Further related concept of optimum networks planning to meet demand satisfaction in this research, the concept of Multiple Commodities Problem also used to support Feeder Network Design Problem.

*2.2.3. Multiple Commodity Network Problem (MCNP).*Multi-Commodity Network Problem are part of important problem in several heuristics models, and it's become an appropriate method in determining shipping routes for container ships [8]. And recent condition on container shipping networks is strongly influenced by a multi-destination shipping purpose. Where affected double handling at each ports of call, definitely increasing the number of operation times and its costs. .The Multi-Commodity Networks Problem are working to define the correlation of fleet capacity to each of the main ports in this research.

3. Methodology

3.1. Research Scope
There are two scopes and limitations to obtain the main objective in this research, as follows:

- Only export and import container trades would be considered in the analysis. Meaning that container domestic trades are excluded.
- Shipping routes present the connecting of six main domestic ports, namely Belawan, Banjarmasin, Tanjung Priok, Tanjung Perak, Makassar, and Sorong. Moreover, the two purposed international hub ports, namely, Kuala Tanjung and Bitung International Hub Port are taken into account in the analysis as well.

3.2. Research Steps
a. Market Analysis
Firstly, we identify the market condition by analysing from both supply and demand sides. The supply side describes an overview about two purpose international hub ports and six main domestic ports as well as provides the ship specification including the number of ship and ship capacity. On the other hand, the demand side consists of the total international container volume per year and per region.
b. Transport Costs Analysis

In order to calculate the total costs, we firstly determine the operational costs, fuel costs, port dues and terminal handling charges. Moreover, we obtain the total costs by adding these costs.

c. Optimization Model
To obtain the optimal solution for selected routes and fleet sizing for the Indonesian international hub ports, we apply the Feeder Network Design Problem (FNDP) formulation by minimizing the total transport costs between main domestic ports and international hub port.

d. Domestic Container Shipping Connectivity Analysis
Dividing region scenario becomes two main Indonesia regions. Kuala Tanjung international hub port would be connecting to Belawan Port (BLW), Tanjung Priok Port (TPR) and Tanjung Perak Port (TPK) in the Western Region (WRI). Whereas Bitung International hub port concerns to connecting Banjarmasin Port (BJM), Makassar Port (MKS) and Sorong Port (SRG) in the Eastern Region of Indonesia (ERI). This grouping region is appropriate to the government plan in terms of cargo allocation belongs to these two international hub ports. Overall, we provide the conclusion and recommendation for optimum routes and number of fleet required for each region respectively.

3.3. Optimization Model
We used minimum total amount of marine transport cost for all vessel and routes alternatives on picking and delivery activities of import and export container between hub and domestic ports.

a. Objective Function.

$$Min \sum_{t\in T} \sum_{n\in N} \mathrm{X_{tn}} \times \mathrm{TC_{tn}} \quad (1)$$

Where TC_{tn} is total amount of marine transportation cost in Million-Rupiah/year, this formula has include fixed and variable cost components.

b. Network Constrains

$$\sum_{t\in T} \sum_{n\in N} \mathrm{X_{tn}} = 1 \quad (2)$$

WhereX_{tn}= *1*,if route and vessel sizes chosen, and X_{tn}= 0, if route and vessel sizes not selected.

c. Demand Transported Constrains

$$\sum_{t\in T} \sum_{n\in N} (\mathrm{X_{tn}} \times Fek_{tn} \times Q_{nt}) \geq Dex \qquad \sum_{t\in T} \sum_{n\in N} (\mathrm{X_{tn}} \times Fim_{tn} \times Q_{nt}) \geq Dim \quad (3)$$

Where Fek_{tn} as a number of frequency by export demand and Fim_{tn} as a number of frequency by import demand per year. To gain total cargoes transported, number of frequency times by Q_{nt} (ship capacity/frequency in TEUs). While Dex and Dim respectively are export and import container volumes in TEUs/year.

d. Port Compatibility Constraints

$$DK > DP + A = 0 \ and \ DK < DP + A = 1 \quad (4)$$

Where, *DK* as vessel draft, *DP* as a wharf depth and *A* as an allowance.

4. Domestic Container Shipping Analysis

4.1. Indonesian Export and Import Container Volumes
Overall export and import container trade in Indonesia continues to increase since 2010. Where there are 5.2 million TEUs trough Indonesian main port in year of 2008 and then decreased to 4.9 million

TEUs in 2009, while in 2013 has slowly grow significantly by average of 7.3% with total amount reached 7.3 million TEUs [6].

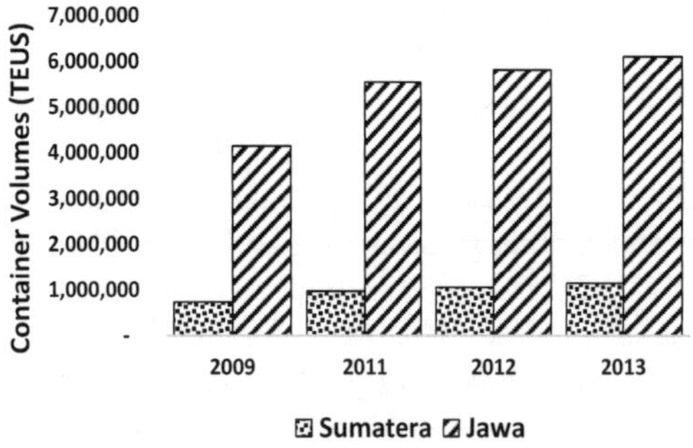

Figure 3. International Container Volume Growth in Java and Sumatera

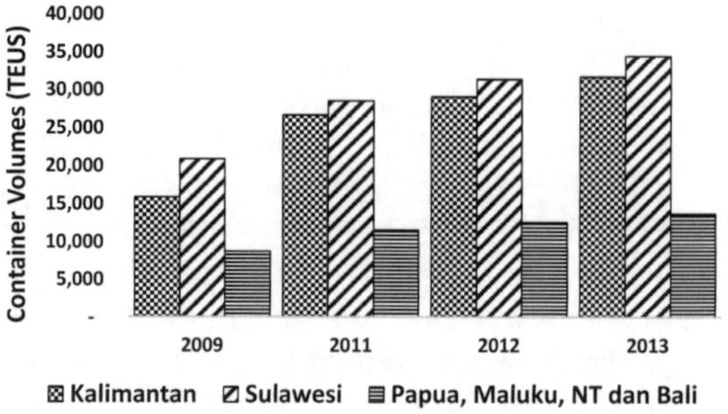

Figure 4. International Container Volumes Growth in Kalimantan, Sulawesi and Papua, NT, Bali

Total amount of future container volumes can be seen by applying the correlation approach among Indonesian economy growth rate as indicated of Indonesia Gross Domestic Product (GDP) on constant prices, against the historic data of international container volumes. Currently, number of export and import container proportion has reach respectively around 40% and 60% [8]. Tanjung Priok has 70% export container volumes and Tanjung Perak has 30% export container volumes based on total container volumes in Java region, meanwhile import container volumes proportion for Tanjung Priok and Tanjung Perak respectively has amount of 65% and 35% from total import containers in Java region [6]. Therefore in each region can be identified the container volume proportion both import and export as can be seen in Table 1below in the year of 2020 as convenient on establishment hub port international started.

Table 1. Potential International Container Volumes in 2020

Port	Codes	Export (TEUs)	Import (TEUs)
Belawan	BLW	713,694	1,070,540
Tanjung Priok	TPR	2,864,176	3,989,388
Tanjung Perak	TPK	1,227,504	2,148,132
Banjarmasin	BJM	19,592	29,388
Makassar	MKS	20,759	31,138
Sorong	SRG	8,491	12,737

The total amount of export and import container volumes above will be affected to the decision for optimum routes and number of fleets.

4.2. Indonesian Domestic Ports Depth and Container Vessel Proportion

Wharf depth and port channel were included as main part in port operation. Where both of this facility impactful in determining optimum fleet in optimization model result. Namely as compatibility variable for acceptable port capacity to serve specific means of vessels. For instance, according to Figure 5noticeable that Banjarmasin port depth has reach at 8 m LWS, when compare to maximum ability of entrance to the harbour, its limited to vessel within range of 300 TEUs to 800 TEUs with maximum draft up to 7 m.

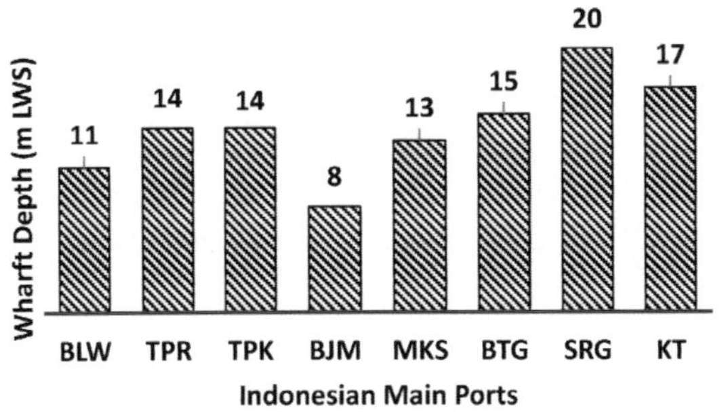

Figure 5. Indonesian existing ports depth within maximum 3,500 TEU vessel size to the deepest port

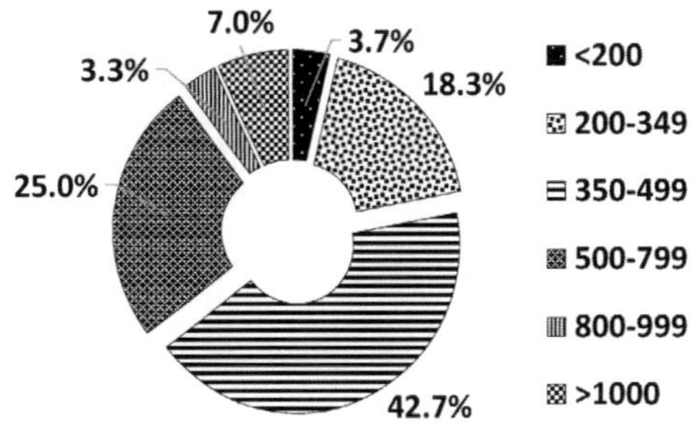

Figure 6. The vast majority within 300 to 500 TEUs container vessel at 42.7% of total in Indonesia

Meanwhile, according to distribution data of container freights by Indonesian Ministry of Transportation in 2012, domestic container vessel size has ranging from 200 TEUs capacity to 1,000 TEUs capacity. The proportion for existence number of container fleet operates in Indonesian sea water can be seen in Figure 6 above.

4.3. Regional Scenario and Vessel Alternatives for Optimization Model

According to the main purpose of two international hub port in Indonesia, Kuala Tanjung international hub port would be connecting to Belawan Port (BLW), TanjungPriok Port (TPR) and Tanjung Perak Port (TPK) in the Western Region (WRI). Whereas Bitung International hub port concern to connecting Banjarmasin Port (BJM), Makassar Port (MKS) and Sorong Port (SRG) in the Eastern Region of Indonesia (ERI) as shown in Figure 7.

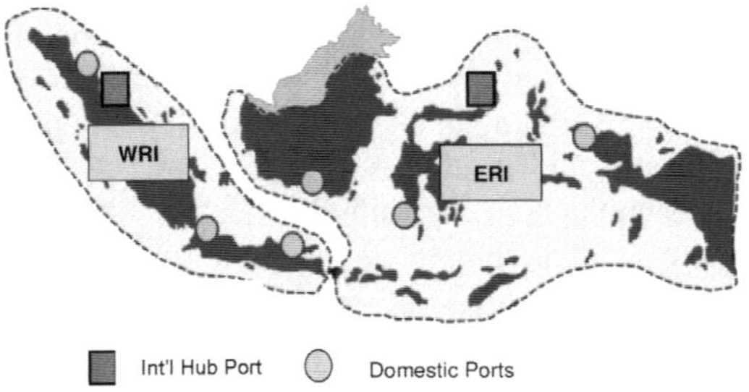

Figure 7. Port region allocation for eastern and western Indonesia.

Table 2. Container Vessel Size Alternatives

Alternatives	Vessel Size (TEUs)	Alternatives	Vessel Size (TEUs)
Vessel - 1	300	Vessel - 6	2,000
Vessel - 2	500	Vessel - 7	2,500
Vessel - 3	800	Vessel - 8	3,000
Vessel - 4	1,000	Vessel - 9	3,500
Vessel - 5	1,500		

Furthermore, to seek the optimal container vessel size in optimum routes, we define container vessel sizes alternatives in several number, specifically within 300 TEUs to 3,500 TEUs as convenient to the current average sizes of domestic container vessel in Indonesia (see in Table 2).

5. Result

The selected optimum routes are categorize has port to port service, which directly connects the international hub port to domestic hub port as can be seen in figure 8. Within this route, the total containers carried are 12,012 million-TEUs and 124 thousand-TEUs for western and eastern region respectively. As an impact on the total costs, it produces 29,310 Billion-Rupiah to operate 137 units of container vessel.

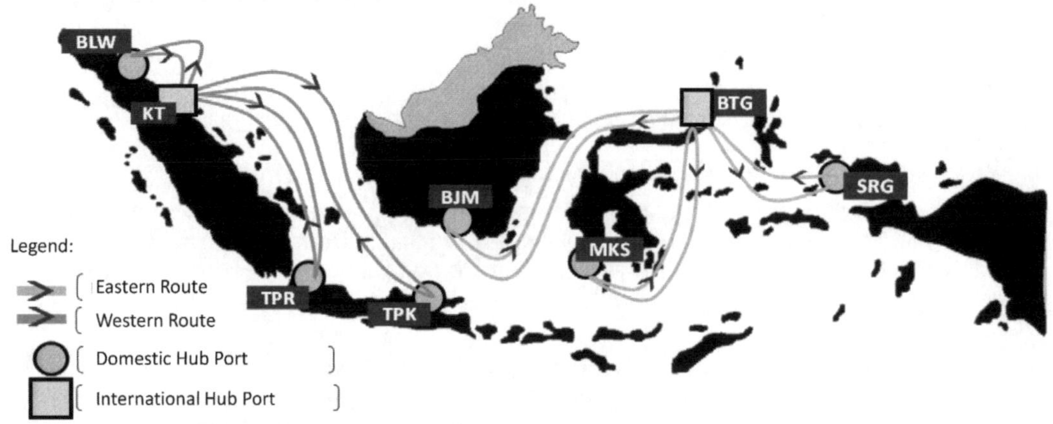

Figure 8. . Optimum routes selected as port to port categories

The optimum route for the western region of Indonesia consists of: (1) Kuala Tanjung-Belawan requires 15 ships of 1,000 TEU with total shipping costs of 2341.9 billion rupiah/year and the unit cost of 1.31 million rupiah/TEU; (2) Kuala Tanjung-TanjungPriokrequires73 ships of 2,500 TEU with total shipping costs of 16,664 billion rupiah/year and the unit cost of 2.43 million rupiah/TEU; (3) Kuala Tanjung-Tanjung Perak requires44 ships of 2,500 TEU with total shipping costs of 9.868 million rupiah/year and the unit cost of 2.922 million rupiah/TEU.

On the other hand, the optimum routes and fleets for eastern region of Indonesia are: (1) Bitung-Sorong requires 1 ship of 500 TEU with total shipping costs of 68.8 billion rupiah/year and the unit cost of 3.24 million rupiah/TEU; (2) Bitung-Banjarmasin requires3 ships of 500 TEU with total shipping costs of 233.71 billion rupiah/year and the unit cost of 4.56 million rupiah/TEU; and (3) Bitung-Makassar requires1 ship of 1,500 TEU with total shipping costs of 143.47 billion rupiah/year and the unit cost of 2.76 million rupiah/TEU.

Table 3. Optimum fleet sizes and routes

Routes	Fleet Size (TEU)	Quantity (Unit)	Total Frequency/y	Total Costs (Billion-Rupiah)	Roundtrip days
Kuala Tanjung-Belawan-Kuala Tanjung	1,000	15	1,260	2,342	4
Kuala Tanjung-Tanjung Priok-Kuala Tanjung	2,500	73	1,878	16,664	13
Kuala Tanjung-Tanjung Perak-Kuala Tanjung	2,500	44	1,011	9,868	15
Bitung-Sorong-Bitung	500	1	30	69	6
Bitung-Banjarmasin-Bitung	500	3	70	234	11
Bitung-Makassar-Bitung	1,500	1	25	143	10

Indonesian export and import container cargoes through international hub ports, Bitung and Kuala Tanjung, in the year of 2020 will produce the total potential revenue of 33.5 Trillion Rupiah per year based on selected routes and vessel capacity in the optimization result Table 3. In detail can be seen in Table 4 below:

IOP Conf. Series: Earth and Environmental Science **135** (2018) 012012 doi:10.1088/1755-1315/135/1/012012

Table 4. Potential revenue of Indonesian hub port enforcementin 2020

Routes	Quantity (Q) (TEUS/year)	Unit Cost (UC) (Million-Rp/TEUs)	Price (P) (Million-Rp/TEUs)	Revenue (R) (Billion-Rp/year)
Kuala Tanjung-Belawan-Kuala Tanjung	1,780,000	1,300,000	1,560,000	2,777
Kuala Tanjung-Tanjung Priok-Kuala Tanjung	6,800,000	2,399,999	2,760,000	18,768
Kuala Tanjung-Tanjung Perak-Kuala Tanjung	3,300,000	2,900,000	3,480,000	11,484
Bitung-Sorong-Bitung	48,000	4,500,000	5,400,000	259
Bitung-Banjarmasin-Bitung	51,000	2,700,000	3,240,000	165
Bitung-Makassar-Bitung	20,000	3,200,000	3,840,000	77

References

[1] Ministry of Indonesian Economic Affairs. (2011). *Masterplan Percepatan dan Perluasan Pembangunan Ekonomi Indonesia*. Jakarta: Kementerian Koordinator Bidang Perekonomian

[2] Wergeland, N.W. (1997) Shipping. Netherlands: Delf University Press

[3] Nur, H. I (2014). *Kajian Usulan Kebijakan Pendulum Nusantara: Tinjauan Sektor Pelayaran dan Kepelabuhanan.* Surabaya: Marine Transportation Engineering Department

[4] Mulder, J., & Dekker, R. (2014). Methods for Strategic Liner Shipping Network Design. *European Journal of Operational Research*, 367-377.

[5] Yunianto, I. T. (2014). *Model Penentuan Hub Port Sebagai Pusat Konsolidasi Angkutan Petikemas dan General Cargo.* Surabaya: Marine Transportation Engineering.

[6] Lazuardi, S. D. (2015). *Analyzing the National Logistics System Trough Integrated and Efficient Networks: a Case Study of Container Shipping Connectivity in Indonesia.* Netherlands: Maritime Economics and Logistics Erasmus University Rotterdam.

[7] Vad Karsten, C., Pisinger, D., Ropke, S., & Dangaard Brouer, B. (2015). The Time Constrained Multi-Commodity Network Flow Problem and its Application to Liner Shipping Network Design. *Transportatio Research Part E: Logistics and Transportation Review*, Vol 76, 122-138.

[8] Indonesian Statistics Center Department. (2016, April 5). *Produk Domestik Bruto Atas Dasar Harga Konstan 2000 Menurut Lapangan Usaha (Miliar Rupiah)*, 2000-2014. Retrieved from Badan Pusat Statistik: https://www.bps.go.id/linkTabelStatis/view/id/1200

[9] O. Polat, H. Gunther, O. Kulak (2014). *The Containership Feeder Network Design Problem: The New IzmirPort as Hub in the Black Sea, Maritime Economics & Logistics*

IOP Conf. Series: Earth and Environmental Science **135** (2018) 012013 doi:10.1088/1755-1315/135/1/012013

Freight Calculation Model: A Case Study of Coal Distribution

I T Yunianto[1], S D Lazuardi[1] and F Hadi[1]

[1] Department of Marine Transport Engineering, Faculty of Marine Technology, Sepuluh Nopember Institute of Technology, Indonesia

Email : dwilaz@gmail.com

Abstract. Coal has been known as one of energy alternatives that has been used as energy source for several power plants in Indonesia. During its transportation from coal sites to power plant locations is required the eligible shipping line services that are able to provide the best freight rate. Therefore, this study aims to obtain the standardized formulations for determining the ocean freight especially for coal distribution based on the theoritical concept. The freight calculation model considers three alternative transport modes commonly used in coal distribution: tug-barge, vessel and self-propelled barge. The result shows there are two cost components very dominant in determining the value of freight with the proportion reaching 90% or even more, namely: time charter hire and fuel cost. Moreover, there are three main factors that have significant impacts on the freight calculation, which are waiting time at ports, time charter rate and fuel oil price.

1. Introduction

One of the main problems in an effort to maintain the availability of national electricity supply is to preserve a continuous supply of primary energy. One of energy sources that has dominantly been used for maintaining the electricity supply in Indonesia is coal. However, there are limitations and challenges on coal distribution during its transportation from coal sites to power plant locations. They are including the spreading out of power plant locations from coal supply site, the variety of the number of coal required for each power plant, the difference of coal characteristics on every coal site, and so on [1].

In order to face these problems, it is required the eligible shipping line services that are able to provide the best freight rate in a certain route. In addition, there are three modes that can be used to serve the coal distribution in Indonesia, which are tug-barge, vessel and self-propelled barge (SPB).

Therefore, this study aims to obtain the standardized formulations for determining the freight rate based on the theoritical concept. The freight rate produced is the required freight rate (RFR), meaning the tariff that is able to cover all expenses incurred in the process of sea transport without adding a profit. The result shows the determination of required freight rate for coal distribution by using three alternative transport modes, namely tug-barge, vessel and self-propelled barge.

2. Literature Review

2.1. Dry-bulk shipping

Transport (in all modes) is the essential link between supplier and receiver, and the aim is to receive the goods in good condition, when and where they are needed with an affordable price. Since most of coal distributions are shipped by sea transportation mode, then understanding the type of ships commonly used, how they are operated and how the freight rate is determined are indispensable.

In general, types of ship can be classified into three groups: cargo ships, offshore mobile structures, and non-cargo ships. Furthermore, the group of cargo ships can be divided into four sectors, namely general cargo, bulk cargo, oil and chemicals and liquid gas. A dry bulk carrier is a ship specially designed to transport dry unpackaged bulk cargo, such as grains, coal, ore, etc in its cargo holds. Dry bulk carriers usually have several holds that are covered by hatches and equipment for loading and unloading of the cargo. In large part the design of dry bulk carriers depends on the density (stowage factor) of the cargo that will be transported [2].

2.2. Shipping costs

In general, the shipping costs can be classified into two main categories. First is the cost of operating the ship, and the second is costs of maintaining and financing the ship. These classifications can be broken down as follows [2]:

- Cost of operating the ship
    - Operating costs, which constitute the expenses involved in the dayto-day running of the ship - essentially those costs such as crew, stores, running repair, insurance and administration.
    - Vogaye costs are variable costs associated with a specific voyage and include such as items as fuel cost, port charges and canal dues
    - Cargo handling costs represent the expense of loading, stowing and discharging cargo.
- Cost of maintaining and financing the ship
    - Capital Costs, are the costs to finance the ship and depend on the way the ship has been financed. They may take the form of dividends to equity, which are discretionary, or interest and capital payments on debt finance, which are not.
    - Periodic Maintenance Costs are incurred when the ship is dry-docked for major repairs, usually at the time of its special survey. In older ships this may involve considerable expenditure, and it is not generally treated as a part of operating expenses.

3. Methodology

The keys to survival in the shipping market in which ship owners have to work with are the revenue received from operating the ship and cost of running and maintaining the ship. Revenue received from operating the ship is a multiplication of the freight (price) by volume of cargo transported. Although ship owners do not generally control the freight they receive per ton of cargo, there is a way to estimate whether the ship can generate positive cash flow or not. This can be done by calculating the Required Freight Rate (RFR).

Required Freight Rate (RFR) is the tariff that is able to cover all expenses incurred in the process of sea transport without adding a profit. Meanwhile, freight is the rate (selling price) charged by the ship owner to the tenant in which there is included ship owner's profit. Thus, the relationship between RFR and freight can be written in the following formula [3]:

$$Freight = RFR + Profit \tag{1}$$

To calculate the amount of RFR, firstly we determine all the costs, which are the ship owner's responsibility for the economic life of vessels by considering the increasing of each cost in a year. Afterwards, calculating the present value of all costs incurred during the economic life of vessels with the discount rate, which is applied by the method of Weighted Average Cost of Capital (WACC).

The purpose of calculating the total costs incurred is to calculate the cost burden uniformly over the economic life of vessels, then the method can be used to homogenize those expenses is the annuity method. This annuity calculation can be done by using the equivalence principle commonly used in theory of engineering economy [4]. The formula for calculating the value of an annuity on the present value can be formulated as follows:

$$A = P\left[\frac{i\,(1+i)^n}{(1+i)^n - 1}\right] \tag{2}$$

Where:
A = Annual equivalent value
P = Present equivalent value
i = interest rate (in this case is WACC)
n = economic life of vessels

By knowing the value of an annuity represents the annual fixed costs of vessel, which are including capital cost and periodic maintenance cost. Thus, if it is divided uniformly by the economic life of vessels, then it will be a basic calculation to determine the time charter hire.

Furthermore, total costs can be generated by adding the annual fixed costs and variable costs. Meanwhile, components of variable costs are voyage cost and cargo handling cost. The RFR can be known as unit cost, which can be calculated by dividing the total costs with the amount of cargo transported over one year. Thus, we obtain the unit cost of transporting the goods per unit of weight (USD/ton). In general, the cost calculation of unit cost can be formulated as follows [5]:

$$Unit\ cost = \frac{(LC+OPEX+CH)}{PS} \quad (3)$$

The unit cost of transporting a ton of cargo on a voyage is defined as the sum of the capital cost of the ship (LC), the cost of operating the ship (OPEX) and the cost of handling the cargo (CH), divided by the parcel size (PS), which for bulk vessel is the tonnage of cargo carried.

4. Result and Discussion

4.1. Transport modes alternatives

There are three alternatives of transport modes commonly used today for coal distribution in Indonesia, which are barge, vessel and Self-Propelled Barge (SPB).

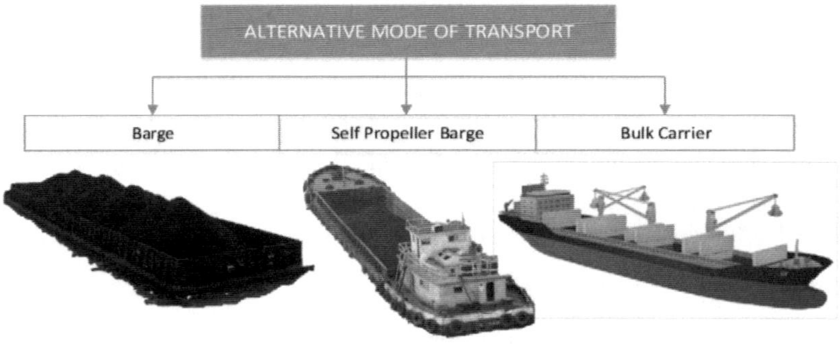

Figure 1. Transport modes alternatives

Before determining the required freight rate, we have to compute the total costs of coal distribution. Firstly we identify the specifications of each transport mode commonly used. Table 1 presents the specification of transport modes used on the freight calculation model.

Table 1. Transport modes specification

No.	Transport mode	Type	Payload (Ton)	Crew	Engine power (HP) Main engine	Engine power (HP) Auxiliary engine
1	Barge	BG 365'	12,500	10	3,000	300
2	Barge	BG 330'	9,500	10	2,400	300
3	Barge	BG 300'	7,400	10	2,000	200
4	Barge	BG 270'	5,000	10	1,600	200
5	Barge	BG 230'	3,000	10	1,200	200
6	Vessel	Panamax	65,000	24	15,400	2,400
7	Vessel	Handymax	45,000	22	11,500	1,800
8	Vessel	Small Handy	20,000	20	6,700	1,100
9	SPB	SPB 10.000	10,000	20	3,300	500
10	SPB	SPB 12.000	12,000	20	3,900	600

4.2. Time charter hire

The components of cost calculated on the time charter hire consists of two components namely capital cost and operating cost. In this case, the component costs repair and maintenance costs are also included in the operating cost component. By applying the formula on the Equation 2, thus we obtain the value of time charter hire for each type of transport mode as follows:

Table 2. Time charter hire

No.	Transport mode	Type	Payload (Ton)	Economic life	Ship's Price[a]	TCH[b]
1	Barge	BG 365'	12,500	15	37,000	733
2	Barge	BG 330'	9,500	15	32,000	648
3	Barge	BG 300'	7,400	15	29,000	596
4	Barge	BG 270'	5,000	15	25,000	528
5	Barge	BG 230'	3,000	15	21,000	460
6	Vessel	Panamax	65,000	25	320,000	5,419
7	Vessel	Handymax	45,000	25	250,000	4,278
8	Vessel	Small Handy	20,000	25	145,000	2,584
9	SPB	SPB 10.000	10,000	15	105,000	1,983
10	SPB	SPB 12.000	12,000	15	126,000	2,339

[a] Ship's price in million-rupiah per unit

[b] Time charter hire (TCH) in million-rupiah per month

4.3. Freight calculation

In order to calculate the required freight rate (RFR) for coal distribution in Indonesia, we select one of dedicated routes as an example to describe how the model calculation works by considering three types of transport modes.

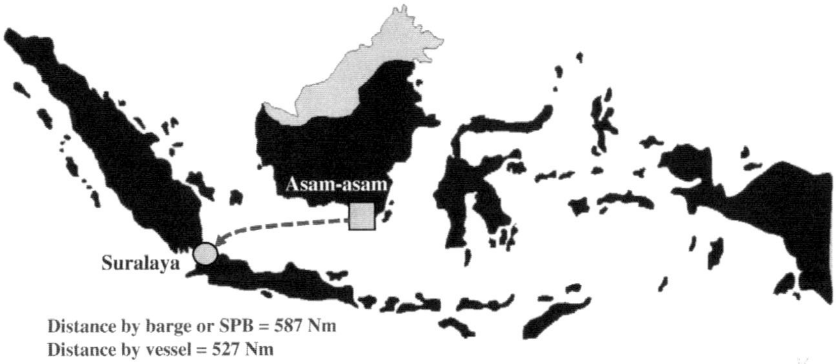

Figure 2. Coal distribution from Asam-asam to Suralaya

Asam-asam, one of the biggest coal suppliers in Kalimantan island, is selected as a coal site. On the other side, one of Asam-asam's consumers is Suralaya Power Plant in Banten – Java island. From Asam-asam to Suralaya has a distance of 527 Nm by vessel and 587 Nm by barge or SPB. Hence, in this calculation, Panamax vessel is selected to be analyse on this spesific route.

Table 3. Operational analysis by using Panamax from Asam-asam to Suralaya

Tranport mode operations		
Sea time	(hour/R.trip)	111.26
Port time		
Port of loading	(hour/call)	102.00
Port of discharging	(hour/call)	332.84
Total port time	(hour/call)	434.84
Total time	(hour/R.trip)	546.09
	(day/R.trip)	22.75

Transport mode consumptions		
Fuel consumption		
Main engine	(KL/R. Trip)	203.20
Auxiliary engine	(KL/R. Trip)	84.21
Total fuel cons.	(KL/R. Trip)	287.41
Fresh water cons.	(ton/trip)	109.22

Table 4. Freight calculation by using Panamax from Asam-asam to Suralaya

Total shipping cost		
Time charter rate	(Mln-Rp/month)	5,419.00
Fixed costs	(Mln-Rp/R.trip)	4,110.10
Fuel costs	(Mln-Rp/R.trip)	1,414.36
Port charges	(Mln-Rp/R.trip)	220.38
Port of loading	(Mln-Rp/R.trip)	110.19
Port of discharging	(Mln-Rp/R.trip)	110.19
Towing fee	(Mln-Rp/R.trip)	10.00
Fresh water cost	(Mln-Rp/R.trip)	8.74
Total cost	**(Mln-Rp/R.trip)**	**5,763.58**
Unit cost (RFR)	Rp/ton	88,670.40

Waiting time for both at Asam-asam and at Suralaya are six hours. Meanwhile, waiting time at Asam-asam and at Suralaya are three days and four days, simultenously. By applying the formula on Equation 3, as the result, the required freight rate on this route is IDR 88,670.4 per ton by using Panamax vessel, which has capacity of 65,000 ton.

If this route is served by tug-barge (330 feet) with the capacity of 9,500 ton and by SPB with the capacity of 10,000 ton, then the RFR produced are 126,555 per ton and 145,662 per ton, respectively. Overall, the other routes could be determined by the similar method by changing on the type of transport modes. The result for other routes by using three type of transport modes can be seen in the Figure 3 as below:

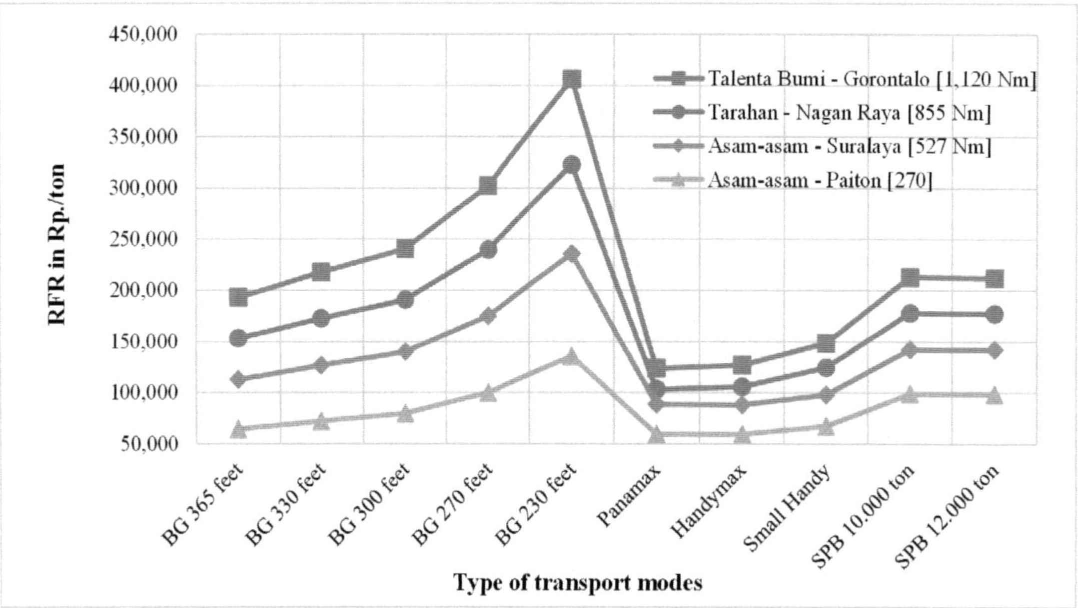

Figure 3. RFR for other routes by using three type of transport modes

From the figure 3 above, it can be concluded that the bigger capacity of transport modes used to transport the coal within the same length of distance, the cheaper unit cost produced. Kindly note that only several ports that could be served by bigger vessels. Thus, the decision for selecting the type of transport modes in the case of coal distribution should be considering about the number of coal distributed, the characteristics of ports and the length of distance between port of loading and port of discharging.

5. Conclusion

Based on the analysis, the type of transport modes used for distributing coal from coal sites to power plants depends on the volume of coal transported, the length of distance and the draught of ports. Meanwhile, selecting the type of transport mode for a certain route affects on the total cost during the coal distribution. Meaning that the bigger vessel used, the cheaper cost incurred.

There are two component costs are very dominant in determining the value of freight with the proportion reaching 90% or even more namely:

- Time charter hire, where the factors that determine the high and low of time charter rate are the type, the age and the capacity of transport modes.
- Fuel costs, where the factors that specify the high and low of fuel costs are engine power, fuel consumption and average speed.

In addition, there are three main factors that have significant impacts on the freight calculation, which are waiting time at ports, time charter rate and fuel oil price.

References

[1] PT PLN (Persero) 2016 *Annual Report PT PLN (Persero) of 2016* (Jakarta: PT PLN (Persero))
[2] Alan E B 2007 *Elements of Shipping, 8th Edition* (London and New York: Routledge Taylor & Francis Group)
[3] Stopford M 2009 *Maritime Economics 3rd Edition* (London and New York: Routledge Taylor & Francis Group)
[4] Buxton I L 1987 *Engineering Economics and Ship Design* (British Maritime Technology Limited)
[5] Blank P E L and Tarquin P E A 2012 *Engineering Economy* (McGraw Hill)

IOP Conf. Series: Earth and Environmental Science **135** (2018) 012014 doi:10.1088/1755-1315/135/1/012014

Fast Missile Boat Project Planning using CPM and What If Analysis Method

Silvianita[1], R Firmansyah[1], D M. Rosyid[1], Suntoyo[1] and D M Chamelia[1]

[1]Department of Ocean Engineering, Faculty of Marine Technology, Institut Teknologi Sepuluh Nopember, Surabaya, Indonesia

Email : silvianita.pc@gmail.com

Abstract. This paper discusses analysis of fast ship missile project planning with CPM and What If analysis.The poor performance can cause project delay and cost overruns, these two things often occur in a project management. Scheduling has an important role to control the project. Good scheduling is required because it will affect the product quality. Scheduling is useful to manage the activities so that the project completion time can be as short as possible and it is used to determine the costs of the project are in optimum condition. The alternatives that can be used to overcome the delay project namely increasing labor and equipment.Project planning can be done to determine whether the project is already in an optimum condition. Alternatives that can be used to overcome the delay in the project is to increase labor and equipment, the next is to look for the optimum solution taking into account the duration and gain of each alternative.

1. Introduction

On the construction of a project such as shipbuilding, construction management requires detailed work schedules to anticipate delays in project. Project delays can be anticipated with the acceleration in the implementation, but the cost factor need to be considered. Acceleration can be done by extra working hours, tools, and increasing the number of workers. The minimum added cost would be chosen for the best accelerations scenario [1].

The poor performance caused the delay time, poor quality, and cost overruns often occur in a project. Scheduling has an important role to control the project. Good scheduling required because it will affect the product quality. Scheduling activities are planned so that the project completion time can be as short as possible while taking into account the cost of the project are in optimum condition.

By using the Critical Path Method (CPM) planning process for all activities that must be completed in the project will be easier. CPM will be outlined in a whole list of activities required to complete the project, or the duration of time required for each activity, as well as human resource planning. From this process it will get critical path of anetwork diagram, which will include some activities that can be shortened to obtain the most efficient project execution [2].

This paper discusses project planning of fast ship missile PT.X. The objectives of this paper are to calculate the critical path of the project and to investigate the additional time and workers needed to accomplish the project . Project planning was conducted by the Critical Path Method (CPM) and What If Analysis. The project is starting to ship in early 2016 began the design, fabrication stage starting in September 2016 and expected to be completed by the end of 2017. The result of this paper can be used as a reference in order to avoid delays in the project. Fast Missile Boat has an overall length of 60m, width 8,1m, height 4,85m deck, and the total weight 460ton.

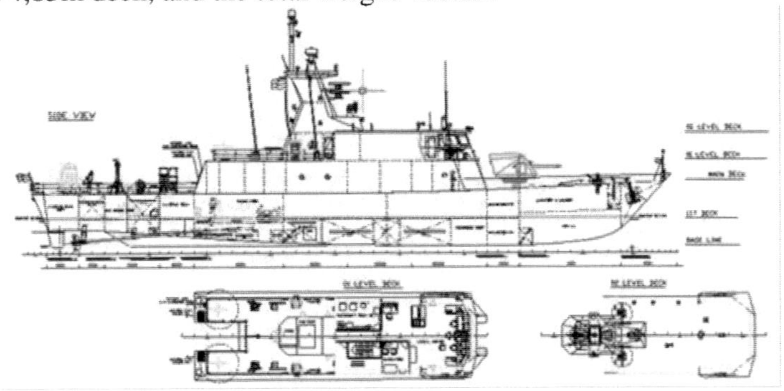

Figure 1. Fast Missile Boat Generral Arrangement[3]

2. Materials And Methods

The flow and procedure of the study is conducted by the stages as follow. Firstly, a literature review is performed by referring to the materials as contained in text books, journals, codes and standards, rules and regulations.

2.1 Data Collection

Collecting data in this research include the type of activities on development projects fast ship missiles, schedule and duration of the activity, the dependence between the activities with other activities, cost estimates and other support elements.

2.2 Compile Critical Path Method (CPM)

Making the project schedule using the Critical Path Method according to the sequence dependence of the project where all activities must be linked to each other in accordance with the logic of dependency that is common in ship projects. Then from the Critical Path Method will be known the critical path and critical activities that were likely to cause delays in Fast Ship Missile project.

2.3 Analysis of Projects Using What If Analysis

Create a project acceleration analysis using What If Analysis in order to determine the number of workers and extra work hours. This step is useful to accelerate the duration of the project so that the project completion time is not too late.

3. Results And Discussion

3.1 Earliest Event Time and Latest Event Time

In CPM (Critical Path Method) there are several activity namely EET (Earliest Event Time), LET (Latest Event Time), Total Float, Free Float, and Independent Float. In the CPM method will also get a critical path that connects the critical activities are activities that should not be delayed.

There are two methods of calculation that can be used to identify critical method. Firstly, the Forward pass which calculates the earliest times at which activity may start and the project completed. Secondly, Backward pass which calculate the latest start for activities [4].

3.2 Critical Path Method

CPM (Critical Path Method) is a project management technique that uses only one factor of time per activity. CPM is the fastest path to work on projects, where each project is included on the line was not given the time pause / break to process. If there is a delay in the critical path, it will affect the overall completion of the project. Completion of the overall project can be accelerated by the acceleration in activities that are on the critical path [5]. The critical path is the path that consists of activities that will result in a delay when the late completion of the project. In CPM (Critical Path Method) is known EET (Earliest Event Time), LET (Latest Event Time), Total Float, Free Float, and Independent Float. In the CPM method will also get a critical path that connects the critical activities are activities that should not be delayed [6].

Table 1. Earliest Event Time and Latest Event Time

ID Number	Activity	Duration (days)	Successor	EET	LET	Float
A	Preparation Plate & Profile	10	B	10	10	0
B	Steel Cutting & Bending Plate	23	C	33	33	0
C	Sub Assembly	109	U	142	142	0
D	Transport & Crane Area	60	E	70	72	2
E	Assembly Block	45	O	115	117	2
F	Fabrication Outfitting	10	I	10	53	43
G	Fabrication Outfitting Electricity	15	L	15	65	50
H	Fabrication Outfitting Machinery	20	J	20	50	30
I	Sub Assembly Outfitting	12	M	22	65	43
J	Sub Assembly Outfitting Machinery	15	K	35	65	30
K	Install Steelworks Machinery	5	N	40	70	30
L	Install Steelworks Electricity	5	O	20	70	50
M	Install Steelworks Outfitting	5	V	27	70	43
N	Pre Erection Outfitting	15	Q	55	147	92
O	Blasting & Painting Block	10	P	125	127	2
P	Erection Block	20	Q	145	147	2
Q	Fit Up	4	R,S	149	151	2
R	Welding	8	Z	157	159	2
S	Fairing Block	3	R	152	162	10
T	Install Out fittings & Piping	15	W	177	177	0
U	Install Electricity	10	V	152	152	0
V	Install Machinery	10	T	162	162	0
W	Repair Painting & Steering Gear	5	Y	182	182	0
X	Finishing Outfitting	22	Z	229	229	0
Y	Install Propeller	25	X	207	207	0
Z	Final Painting	10	–	239	239	0

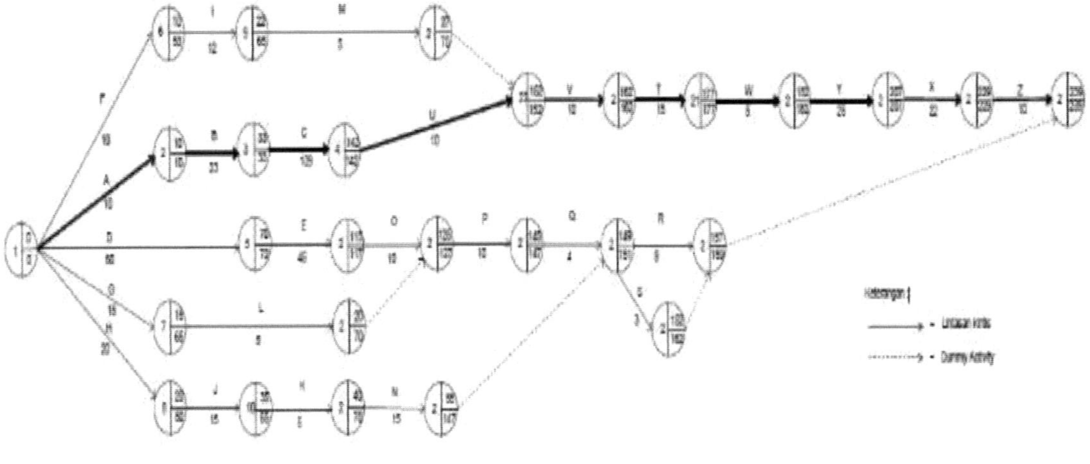

Figure 2. Critical Path Method

3.3 Project Delay

A project must have a schedule to keep on time. In a sense, the work should be completed within a predetermined time. However, due to constraints experienced, it cannot be met and the work is delayed. As a result, the entire schedule has been set originally to be chaotic because it impacts on the company's financial problems. This delay will prolong the duration of the project as well as the rising cost of construction can even affect both. As for the impact received by the owner and the client is a loss of resources to be placed on other projects and increased costs directly resulting in increased employee salaries, rent equipment, and other things that reduce profits [7]. Delays in a project can result in increasing the cost and time of execution of the project. This is often caused by design changes, the effects of weather, lack of labor, materials or equipment, and access to difficult locations. The delays are often caused by the owner of the project, which is due to an error in the process of planning, selection and specification of the influence of the things that are unexpected. So for all the parties concerned, project delays are often a source of friction because all parties will suffer the same loss of magnitude. This can be seen in terms of the owners, project delays will certainly bring losses since the owner will experience a reduction in income because of the delay operation of the facility that has been bookcd [8].

3.4 What If Analysis on CPM

Analysis of "What If" is a study which aims to monitor the project so as to avoid delays in project execution. Analysis of "what if" have to be done before the project is implemented, and can be used as a reference for the project manager to be able to quickly take the right decision and effective when there is a discrepancy with the actual schedule plan schedule. In fact the decision-making process is not separated from the factors of uncertainty and doubt. A good decision maker will consider all possibilities that would cause a mismatch to what has been planned. Results of the analysis presented in graphical form sensitivity is very communicative and easy to use, which graph shows the relationship between the type of activity which is accelerated by an additional number of workers or the number of additional hours of work per day. The construction project that is both flexible and complex is a high-risk work because the process out and have a lot of factors involved, so that the analysis of What If the CPM is necessary to apply. Analysis What If the CPM model of asking "What if there is a delay in one activity?", Hence the role of the float will look at non-critical activities [9]. Due to the events of late, it must be accelerated activity followers a project completed on time by

IOP Conf. Series: Earth and Environmental Science **135** (2018) 012014 doi:10.1088/1755-1315/135/1/012014

improving worker productivity in the activity concerned. Possible acceleration of activity can be done by increasing the number of working hours by the number of permanent workers, increasing the number of workers during normal working hours, and the last to create a new group of workers who work outside working hours [10].

Formula for calculating manpower addition is:

$$\Delta n = n' - n = \frac{\sum manhour}{d's \times H} - n \tag{1}$$

Formula for calculating working hour addition is:

$$\Delta H = H' - H = \frac{\sum manhour}{d's \times n} - H \tag{2}$$

$$\sum manhour = ds \times n \times H \tag{3}$$

Where:

Δn	= manpower addition
n	= normal total amount of manpower
Σmanhour	= amount of manhour
ΔH	= workinghour addition
H	= normal working hour
d's	= duration of acceleration

Based on the network diagram, activity C or Sub.Assembly will late until 7 days and it affects total duration of this project. Table 2 shows the calculation of the What If Analysis with scenarios 7 days delay.

Calculation for activity *Instal Outfitting & Piping* (T) :

Duration/ds = 15 hari
Float = 0
n/ normal amount of worker = 8 man
$\sum$ Manhour = n x H x ds = 960
H/ normal working hour = 8 hour
Duration of acceleration / d's = ds + float – delay = 8 day
Calculation requirement = d's<ds => yes (qualify)
1/2ds$\geq$delay=>yes (qualify)

Based on formula then the calculation are as shown below :
Additional workers

$$\Delta n = \frac{\sum manhour}{d's \times H} - n = 7 \text{ man}$$

Additional time

$$\Delta H = \frac{\sum manhour}{d's \times n} - H = 7 \text{ hour}$$

Table 2. What If Calculations

Activity	Duration / ds	∑man hour	Float (day)	Delay (day)	d's (day)	d's<ds	1/2 ds > delay	n (man)	H (hour)	Δn (man)	ΔH (hour)
T	15	960	0	7	8	yes	yes	8	8	7,00	7,00
U	10	480	0	7	3	yes	no	6	8	-	-
V	10	480	0	7	3	yes	no	6	8	-	-
W	5	160	0	7	-2	yes	no	4	8	-	-
X	22	1056	0	7	15	yes	yes	6	8	2,80	3,73
Y	25	1600	0	7	18	yes	yes	8	8	3,11	3,11
Z	10	240	0	7	3	yes	no	3	8	-	-

4. Conclusions

A study has been conducted for the case of Fast Ship Missile Project Planning. The findings of the study could be revealed as follows:

- Critical Path of this project is contained of Preparation Plat & Profile (A), Steel Cutting & Bending Plate (B), Sub. Assembly (C), Install Electricity Cable Connection (U), Install Machinery (V), Install Outfitting & Piping (T), Repair Painting & Steering Gear (W), Install Propeller (Y), Finishing Outfitting (X), dan Final Painting (Z).
- To prevent delays in Sub. Assembly activity (C) takes 13 additional workers and 14 hours of additional hours of work.

References

[1] Widyatmoko, Y. :*Analisa Percepatan Waktu Menggunakan Metode Crashing pada Kegiatan Pemancangan di Proyek Dermaga 115 Tanjung Priok dengan Aplikasi Program PERT Master*. Jakarta: Jurusan Teknik Sipil Universitas Indonesia.2008

[2] Soeharto, I. :*ManajemenProyek (Dari KonseptualSampaiOperasional)*. Jakarta: Erlangga.1995

[3] PT. X :*Naval Shipbuilding Project Management Document*.Surabaya.2016

[4] Ali, T.H :*Prinsip-prinsip Network Planning* (4 ed). Jakarta . 1-9, 2009

[5] Priowrjanto, H. G. :*KonsepDasarPerkapalan, UrutandanMetodePembuatanKapal: BagianProyekPengembanganKurikulumDirektoratJenderalPendidikanDasardanMenengah*. Jakarta.2003

[6] Frederika, A :*Analisis Percepatan pelaksanaan dengan Menambah Jam Kerja Optimum pada Proyek Konstruksi (Studi Kasus : Proyek Pembangunan Super Villa, Peti Tenget-Badung)*. Jurnal Ilmiah. 2010

[7] Mangitung,:*Analisis Dampak Percepatan Jadwal Proyek Terhadap Biaya Konstruksi dengan Teknik Statistika Non Parametrik*. Jurnal Ilmiah 2008.

[8] Nurhayati. *ManajemenProyek*. Yogyakarta: GrahaIlmu.2010

[9] Alifen, Setiawan, & Sunarto :*Analisa "What If" Sebagai Metode Antisipasi Keterlambatan Durasi Proyek*. DimensiTeknikSipil, 2(1).1999

[10] Syuhada, F. :*Analysis of Acceleration in Project Duration of Jetty Construction: Case Study at PT. Multi Baja Industri*. Surabaya: Jurusan Teknik Kelautan ITS.2015

IOP Conf. Series: Earth and Environmental Science **135** (2018) 012015 doi:10.1088/1755-1315/135/1/012015

Analysis Of Inland Waterway Transport For Container Shipping: Cikarang To Port Of TanjungPriok

T Achmadi[1],H I Nur[1], L R Rahmadhon[1]

[1] Department of Marine Transportation Engineering, Faculty of Marine Technology, Institut Teknologi Sepuluh Nopember, Indonesia

Email : triachmadi65@seatrans.its.ac.id

Abstract. Industry's development which is in the center of Cikarang's industrial estate causes a considerable increase from 7% to 13% of container's flow from and to Port of Tanjung Priok per year. Therefore, those obstacles rise the number of traffic congestion and transport cost. This research aims to analyze the potential alternative of transportation in order to transport containers at the route of Tanjung Priok to Cikarang utilizing Inland Waterways Transport through Cikarang Bekasi Laut (CBL) river. This research will be conducted by comparing component of total logistic cost that emerging caused by container trucks and vessels. Self Propelled Container Barge (SPCB) is a pointed alternative transportation in which it is used to transport containers through the waterways. The result of analysis obtained that the capacity of Cikarang Bekasi Laut river is 18,558 roundtrip per year. Furthermore, the collaboration of 3 SPCB operations, as well payload 32 TEUS can decrease the amount of road traffic congestion/density of Cikarang-Port of Tanjung Priok as much as 18.6%. The cost of containers shipping per unit transported by truck is IDR 2.2 Million per TEUs, whereas containers shipping transported by Inland Waterways cost only IDR 1.8 Million per TEUs.

1. Introduction

Development in the economic sector of Indonesia increased significantly in recent decades. Therefore, the trading activities is one of main elements that support economic development increased rapidly because it has been able to independently and actively operating. Cikarang Industrial Estate, West Java, is an area of major traders in the domestic and global market in Indonesia.

However, logistics activities have an impact on the trading activity in the area of Jakarta and West Java, especially in the Cikarang industrial estate. There are 11 industrial estate with total 3.115 units industry. Distance with land transport between Port of Tanjung Priok and Cikarang industrial estate are 60 kilometers[1].In 2015, the amount of container shipped from the port of Tanjung Priok and vice versa are 642.829 TEUs. Compared with the previous years, the average number of shipped containers increased by 10% each years[2]. So as a result of year-on-year traffic congestion in Jakarta continues to increase, especially at Port of Tanjung Priok lanes toward Cikarang. This traffic congestion due to growth of volume truck raised sharply accompanied by growth in cargo volumes. In addition to causing congestion, also cause other effects such as increased road maintenance costs.

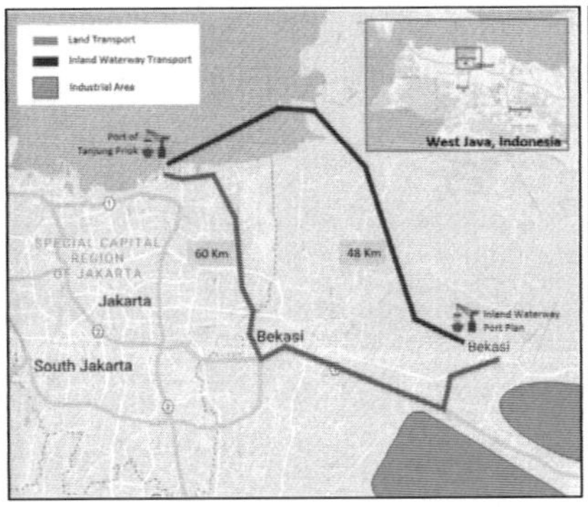

Figure 1. Distance comparison between land transport and inland waterway transport [1]

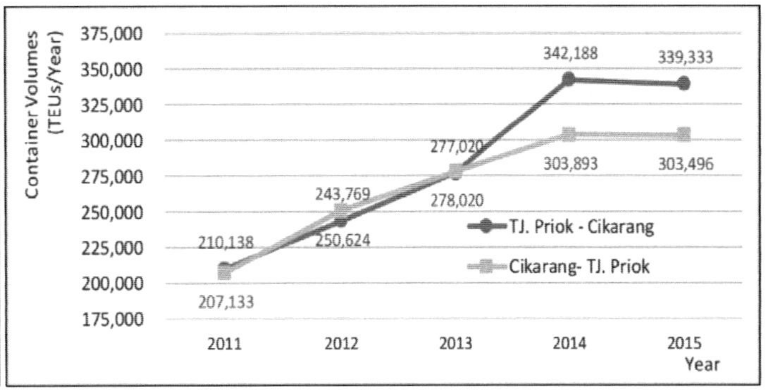

Figure 2. Comparison of total container shipment between Port of Tanjung Priok to Cikarang industrial estate and vice versa [2]

As a result of all is a high cost on the land transport. Logistics costs in Indonesia are the highest among the countries in Southeast Asia, which ranged between 25% - 30% of GDP [3]. Therefore, efforts are required which leads to the suppression of logistics costs. Nowadays, cargo from the port of Tanjung Priok to the Cikarang transported by container trucks. From this fact came up an idea to divert land transport modes using transport services through the river. It is necessary to do an analysis of the modes are most competitive in the service and the price paid by users for these services from the Port of Tanjung Priok to the Cikarang Industrial Estate and vice versa.

2. Literature Review

2.1. Inland Waterway Transport

Inland Waterway Transport (IWT) is a commercial transport that utilizes canal to move commercial cargoes from the main port to the hinterland that is connected with the canal. IWT is defined as a commercial transportation by ships that have not crossed the oceans [4]. IWT concept has been implemented in Europe, especially Northern Europe, as well as some countries in the Americas. The applications of the concept of IWT, in various countries have successfully overcome some of the problems caused by the implementation of freight transport. Port of IWT can be divided into several types, such as:

1. General Purpose Port, a combination of several users of IWT and various other modes of transportation such as trains and other modes of land.
2. Dedicated Cargo Terminal, made to transport the cargoes, sometimes combining the other modes but most do not occur modal transfer.
3. Industrial Port, which is generally the end of the path of IWT, and directly transporting raw materials and carry semi-finished goods from industrial estate.

IWT makes the distribution of cargoes to be more effective and efficient in the country which had applied. In addition, the implementation of IWT has succeeded in improving the movement of goods, lower levels of air pollution, lowering shipping costs and lower infrastructure costs that must be paid by the government.

2.2. Fleets Sizing

There are several approach method included in this research to determining the fleets capacity.

2.2.1. Canal Dimensions Approach. Dimensions of the river is used as a constrain in determining the ship's main dimensions. Ships draft should be sufficient to operate at the lowest water level with the fully loaded condition. The width of the river affects the the ship breadth that can cross the river, also including many lanes available to the width of the river[4].The curve's angle of the river affects to ships length which able cross the river[5].The bridge that cross the river to be a constrain for the ship height that can be through the river.

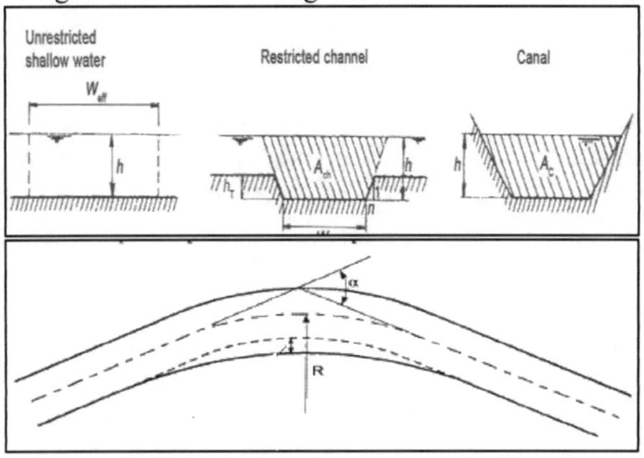

Figure 3. The width and curve angel of the river affects ship dimensions [5]

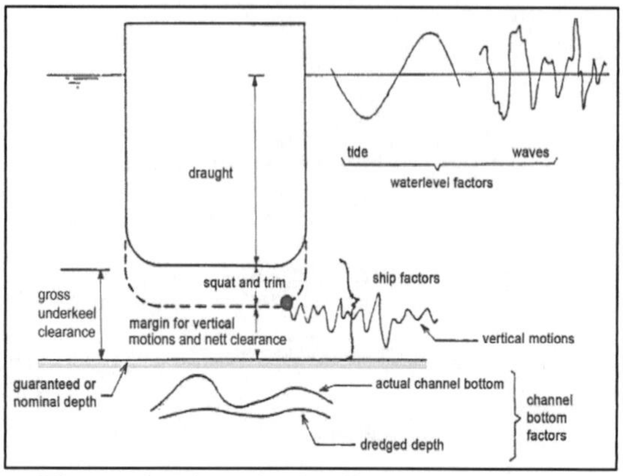

Figure 4. River depth affects to ship draft [4]

2.2.2. Parametric Design Approach. Every different country, even on a different river also has different ship characteristics of inland waterway transport. Ship design method to be operated on CBL river used spiral design method. Ship design process has a repeating cycle characteristics the most commonly described by spiral design [6].

2.3. Ship Investment

Long-term investment policy be considered as the issue of capital budgeting. Investment also means as an expense at the moment with the expected results of the spending will be received over the next 1 year. The main task in the investment policy is to make an estimate of expenditures and receipts of money to be received from these investments in the future. As these estimates include, Capital Cost, Operational Costs, Voyage Costs, Cargo Handling Costs, and Revenue [7]. Comparison of the value of the investment to the value of the future cash flow is used as the investment policy guidelines. The results of this comparison will be information to measure financial feasibility of an investment. In assessing the investment profit there are several criteria are used:

1. Criteria for investment based on the concept of profit / profit is the interest rate of return (IRR).
2. Criteria for investment based on the concept of cash flow does not take into account the time value of money or non-discounted cash flow which is the payback period method.The concept of the cash flow of the value of time and money or discounted cash flow, among other things, the Net Present Value (NPV), and Break Event Point (BEP).

2.4. What-if Analysis

Analysis about how if the assumptions used in these calculations change and how it affects feasibility conditions. With a what-if analysis, decision makers can determine the level of sensitivity of the decisions taken on the possibility of changes that occur in the used variables or assumptions [8]. If the sensitivity analysis performed on all the variables involved, the decision-making process will occur misleading precisely because too many what-if analysis that considered.

3. Methodology

3.1. Research Scope

There are two scopes and limitations to obtain the main objective in this research, as follows:

- This research steps analyze the supply side first, then known percentage of the total demand can be serviced with supply capacity.

- This research does not cover the ship design arrangement and port plan design.
- Dredging volume and dimensions of the river using data fromRegional Bureauof Ciliwung-Cisadane River Basin (BBWS Ciliwung-Cisadane).

4. Inland Waterway Transport Analysis

4.1. Ship Dimensions

Identify supply capacity by analyze ship dimensions. Ship dimensions describes about particular dimensions such as payload, draft, length, breadth, high, service speed, deadweight and cargo handling equipment.

Table 1. Ship particular SPCB 54 TEUs

Item	Value	Unit	Item	Value	Unit
L_{WL}	44.75	m	Power (P_B)	261	kW
L_{PP}	43.03	m	Total crew	7	Crew
B	10.53	m	Fuel capacity	2.38	ton
H	5.46	m	Payload	54	TEUs
T	3.64	m	Gross Tonnage	167.1	ton
Vs	5	knots	Net Tonnage	38.7	ton
C_B	0.828		DWT	1192.0	ton
LWT	356.97	ton	Trim	0.7	m

4.2. Voyage Calculation

In order to calculate the total costs, we firstly determine the operational costs, fuel costs, port dues and terminal handling charges. Moreover, we obtain the total costs by adding these costs.

Table 2. Voyage calculation and unit costs of inland waterway transport

Item	Unit	Total
Capital costs	Rp/year	2,170,144,715
Operational costs	Rp/year	6,460,498,179
Voyage costs	Rp/year	9,798,599,174
Cargo handling costs	Rp/year	12,209,400,000
Costs per ship	Rp/year	30,638,642,068
Capacity	TEUs/year	34,887
Trucking costs	Rp/TEUs	600,000
Stevedoring inland port	Rp/TEUs	427,000
Unit costs	Rp/TEUs	1,905,230

4.3. Feasibility of Ship Investment

As these estimates include, Capital Cost, Operational Costs, Voyage Costs, Cargo Handling Costs, and Revenue. Comparison of the value of the investment to the value of the future cash flow is used as the investment policy guidelines. The results of this comparison will be information to measure financial feasibility of an investment.

Table 3. Indicator of ship investment feasibility analysis

Item	Units	Value	Criteria	Min	Annotation
NPV	Million-Rp	32,413	OK	0	Positive Incr. Wealth
NPVI	Time	0.75	OK	0	Null
IRR	%	17%	OK	8%	MARR
IRRI	Time	2.09	OK	0	Null
BEP	years	7	OK	1	Construction Period
Cashflow on BEP	Million-Rp	7,365	OK	0	Positive Accum Cash

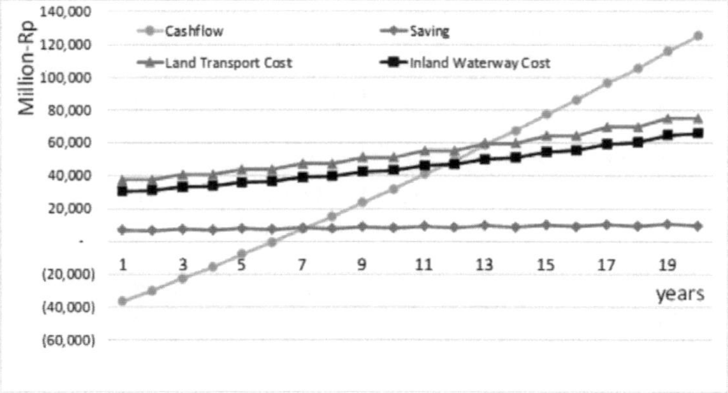

Figure 5. graph of ship investment feasibility analysis

4.4. Comparison with Land Transport

The results of operational analysis of the inland waterway transport compared with land transport by truck container. The components are compared include transportation costs, time, capacity, and fuel consumption.

Table 4. Table of comparison between inland waterway transport and land transport

Comparison	Land transport	Inland waterway transport	Units
Transport time	18-20	10	Hours / roundtrip
Dimension	256	51.6 (LoA)	Meter / 54 TEUs
Fuel consumption	5.184	2.377	Liter / 54 TEUs
Unit Cost	2.150.000	878.23	Rupiah / TEUs
Double Handling Cost	-	1.027.000	Rupiah / TEUs

4.5. Market Analysis

Firstly, we identify the market condition by analysing from supply sides. The supply side describes an overview about capacity of fleets and port occupancy, then known percentage of the total demand can be serviced with supply capacity.

Table 5. Channel capacity of Cikarang Bekasi Laut

Item	Unit	Value
Roundtrip days	Days/roundtrip	1
Commision days	Days/year	330
Roundtrip per annum	Roundtrip/year.ship	323
Channel capacity	Roundtrip/year	32,211
	TEUs/year	1,739,394

Table 6. Cargo transported with inland waterway transport

Item	Unit	Value
Port time Cikarang	Hours/trip	7.20
Voyage time	Hours/trip	5.06
Port time Priok	Hours/trip	7.20
Voyage time	Hours/trip	5.06
roundtrip hours	Hours/roundtrip	24.52
Berth occupancy	Ship/days	3
Berth	Units	3
Cargo transported	TEUs/year	313,981

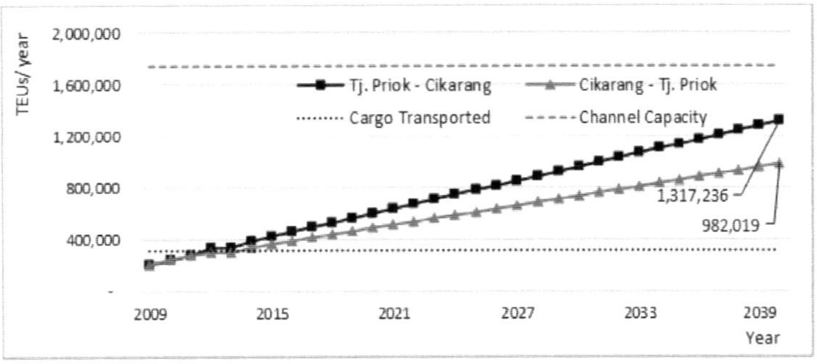

Figure 6. Graph of comparison between supply and demand

4.6. What-if Analysis

In this sensitivity analysis were considered significant variable is the value of service speed and the load factor of the container to be transported on NPV and the unit cost.

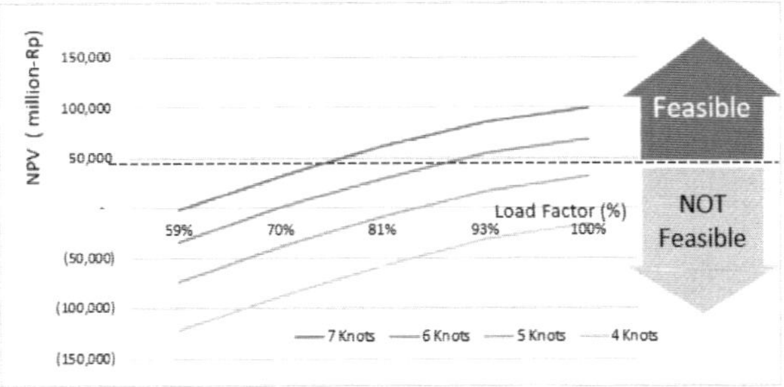

Figure 7. The effect of ship speed and the load factor variable of the NPV

5. Result

CBL channel capacity reached 32.211 roundtrip/ year or 1.739.394 TEUs/ year. Total cargo transported with 9 units SPCB 54 TEUs payload reached 313,981 TEUs/year. Based on the analysis of container shipping costs (port to port), unit cost through the inland waterway transport is by IDR 1,9 million/ TEUs lower when compared with land transport IDR 2,15 Million/ TEUs. With delivery time through the river to 10 hours/ trip and land transport 18-20 hours/ trip. The capacity of river channels reached 13.7% of the total cargo truck serviced. Inland waterway transport worthy to be selected as a mode of transportation that connects the Port of Tj. Priok - Cikarang. Known to the NPV of the investment SPCB are32 billion rupiah and BEP at seventh year with IRR value 17%.

IOP Conf. Series: Earth and Environmental Science **135** (2018) 012015 doi:10.1088/1755-1315/135/1/012015

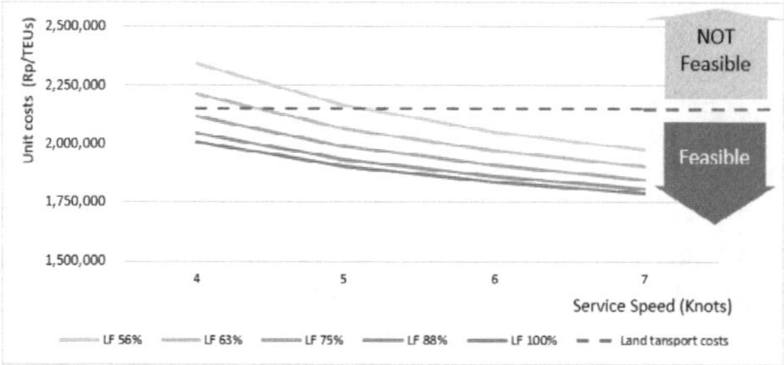

Figure 8. The effect of ship speed and the load factor variable of the unit costs and compared with unit costs of land transport

References

[1] Cikarang Dry Port. (2015). Annual Report. Creating New Cities. Bekasi. West Java. Retrieved from https://www.jababeka.com/menu-block-file/Annual%20Report%20KIJA%202015.pdf
[2] Jasa Marga Development Center. (2016). Traffic Report on Jakarta Outer Ring Road. Jakarta
[3] Indonesia Logistic and Forwarder Association. (2014). Logistic Reform for Indonesia Competitiveness. Jakarta.
[4] Velsink, H. (2016) Port and Terminal. TU Delft. Delft.
[5] Triadmodjo, B. (2012) Perencanaan Pelabuhan. Jogjakarta
[6] Manning, G.C.(1956) The Theory and Technique of Ship Design. MIT Press.
[7] Stopford, M. (1997) Maritime Economics.Taylor & Francis E-library. United Kingdom. Chapter 6 pp 159-172
[8] Law A.M. and Kelton W.D. (1999). Simulation Modeling and Analysis. McGraw-Hill Higher Education, Boston, MA.

IOP Conf. Series: Earth and Environmental Science **135** (2018) 012016 doi:10.1088/1755-1315/135/1/012016

Seasonal-Temporal Variability of Sea Surface Temperature Derived from Satellite and In Situ Data: Case Study of Indian Ocean Western Part of Sumatera Island

A Wahyudi[1], **S B Susilo**[1] and **J P Panjaitan**[1]

[1] Department of Marine Science and Technology, Faculty of Fisheries and Marine Sciences, Bogor Agricultural University, Kampus IPB Dramaga, Bogor 16680, Indonesia

Email : yudipro_91@yahoo.co.id

Abstract. It has been widely knowing that one of the important parameters to study about processes and phenomena in oceans is sea surface temperature (SST). This study is about the use of satellite and in situ data to find the seasonal and temporal SST value in Indian Ocean western part of Sumatera Island during 2010 to 2016. This result focuses on changes of SST and the accuracy of SST value derived from MODIS Satellite and ARGO Float. We divide the data into the dry and wet season based on a seasonal condition in Indonesia. The result shows the same pattern of SST value both from satellite and in situ data. The changes trend of SST value (Celsius) is increasing from 2010 to 2016 both in situ and satellite data. SST value increases 0.093 for in situ data and SST value increases 0.132 measured by MODIS satellite. The accuracy assessment (RMSE) value is 0,555. The coefficient of determination r^2 value shows a positive correlation between MODIS and ARGO data especially in dry season, while in the wet season the R^2 is smaller than in dry season.

1. Introduction

Dynamics and processes of biology, physic, and chemistry happen in oceans usually make a big influence on the living organism in it. We have to already know that oceans contain a high level of biodiversity. The coral reef is one of the ecosystems in oceans that have the complex ecosystem and high biodiversity. The other ecosystem like sea grass also have the same function as coral reef, many marine organisms use that for spawning ground, nursery ground and feeding ground and also protect from predators. Beside that coral reefs also protect coastal area from the high and big wave [1]. Parameters that exist in the oceans like physics, chemistry, and biology are an important parameter that must be studied to determine the biophysical condition of oceans. More deeply, finding out the mechanism and variability of marine ecosystem is required accurate and consistent measurement of the ocean condition and supporting data such as biological parameters in appropriate scale [2]. Classify the physical condition of oceans could help for explaining the patterns of the structure of the ocean. This classifies also have to provide and explain logically from different in situ measurement so in the result could be interpreted appropriately [3].

Oceanographic parameters such as seas surface temperature (SST), salinity, chlorophyll a concentration are forming a spatial pattern that is difficult to understand. Understanding these patterns is so important for understanding more process and dynamics of oceans. That processes and dynamics usually occur due to interactions among physical parameters in seawater. To understand the patterns of

SST value will help to see seawater process such as upwelling, front, mixing and also the other process that happen between air and seawater interaction. Then the result of that is valued to understand pattern and spatial structure in terms of time scaling to responses of changing in oceans process are important to be learned [4].

Sea surface temperature is one of the parameters that contributed to equivalence with dynamics and processes in oceans. Sea surface temperature has been widely used in many differences research to acquire an understanding of the ocean's dynamic especially to determine the physical and biogeochemical processes in ocean surface [5]. The changes of SST value could affect for instance the level stress on coral reefs. SST data could assist the management of coral reefs ecosystems to face the climate change phenomenon [6]. The availability of this SST data could really provide valuable information on coral bleaching events [7].

Remote sensing is a technique that being developed at this time to solve the problems both spatially and temporally. Remote sensing is expected to provide the information about oceans conditions. Many satellite data like Landsat, National Oceanic and Atmospheric Administration (NOAA), Moderate Resolution Imaging Spectro radiometer (MODIS) could help for manage marine ecosystem combine with in situ data to get a more accurate result [8]. Remote sensing is one of a method that could help the researcher to study and manage widely oceans ecosystems. Remote sensing data to monitor water quality in coastal region has many advantages compared with in situ or conventional measurement in the field [9]. One of the oceans parameters that could measure using satellite is sea surface temperature. The satellite which has a high resolution could generate SST data then detect the SST variability that occurs in that area [10]. The problem with using this kind of satellite imaginary that SST data is prevented infrared radiometer caused by the presence of clouds [11].

In the last few decade, it is has been knowing that Indian Oceans has a major influence on climate change that previously, though. Changes of environmental conditions on the Indian Ocean believed that would affect the distribution and the survival of organisms therein. It is known that the Indian Ocean is one of the fishing ground for tuna, around 20,4% [12]. The aims of this study are to look at the trend of increase the value of sea surface temperature and the impact to the surrounding area.

2. Materials and Methods

2.1. Study Site

The study took place in Indian Ocean western part of Sumatera Island. Located in $90^{O}00'00''$ – $106^{O}00'00''$ BT and $8^{O}00'00''$ LS – $6^{O}00'00''$ LU (Figure 1). Indian Ocean is a unique ocean that not only has a high mean value of SST but also the variability of SST value is plenty tremendous [13]. This fact has led many researchers to do the investigation in the Indian Ocean to determine the dynamics and processes that occur in Indian Ocean water body.

2.2. Satellite Data

The SST data mainly derived from MODIS on board Aqua satellites from 2010 to 2016. Level 3 of MODIS mapped data on night-time data with 4 km resolution is used in this study. The long wave SST algorithm of MODIS bands at 11μ is also used in this study.

2.3. In Situ Data

In situ data in this study derived from Array for Real-Time Geostrophic Oceanography (ARGO Float) deployed spread around the world within the period from 2002 till present. ARGO Float is globally coordinated project proposed with identify pressure, temperature and salinity structure of the mid and upper ocean through the distribution of autonomous profiling floats [14]. The objective of ARGO program is to perform and control a set of 3000 floats distributed in all oceans [15]. Within this study, we focused on sea surface temperature of ARGO Float data in the Indian Ocean within duration from 2010 to 2016. Sea surface temperature data derived from ARGO Float usually measure in < 5 m depth from the surface.

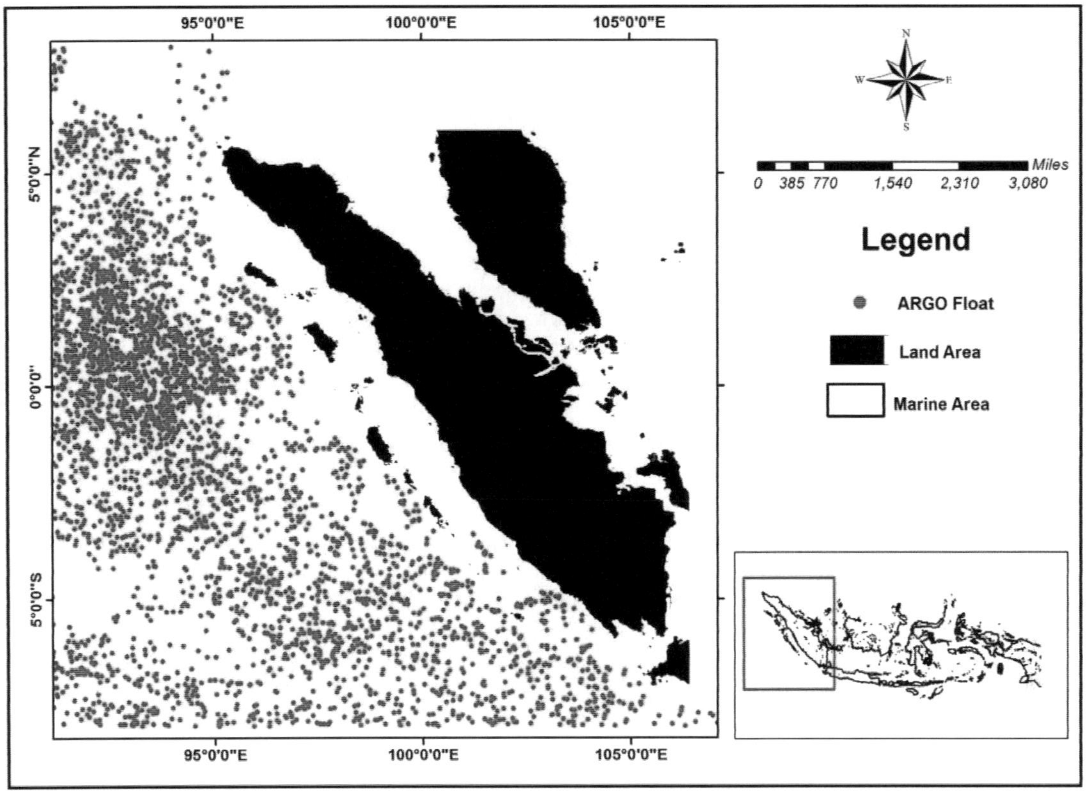

Figure 1. Study Site and ARGO Float Distribution (2005 – 2016)

2.4. Methodology

The extraction of MODIS level 3 data are using SEADAS 7.3.2 version to get SST value and distributions of SST in Indian Oceans. We calculated the average, maximum and minimum value of temperature seasonally (since of Indonesia season divided into the two seasons, wet and dry season, in order that we used November, January, and March for the wet season while May, July, and September for the dry season. This categorization also used in ARGO Float data). Reducing effect from the cloud on a sunny day, so in this research, we used MODIS satellite data acquisition in night data.

For the extraction of ARGO Float data, we used Ocean Data View 4.7.8 version and MS excel to get the SST data value. Although the ARGO Float data provide the vertical distribution sea surface data, but we only use SST data for this study.

2.5. Data Analysis

Data analysis in this research is focused on finding out SST trend using trend analysis models with linear type. The equation for calculate trend of SST values is:

$$Yt = \beta_o + (\beta_1 x\, t) + e_t \qquad (1)$$

While β_1 represents the average change of SST value in one period to next period.

In this research also calculate the Root Mean Square Error (RMSE). RMSE is a frequently used measure of the difference valued between MODIS satellite and the values obtained by ARGO Float, the environment that is being modeled. RSME used for predict the error value between SST value from MODIS satellite and ARGO Float. The equation for calculating the error value is:

IOP Conf. Series: Earth and Environmental Science **135** (2018) 012016 doi:10.1088/1755-1315/135/1/012016

$$RMSE = \sqrt{\frac{1}{N}\sum_{i=1}^{N}(xi - \hat{xi})} \qquad (2)$$

The RMSE value gives the deviation between SST in the field and SST value by MODIS satellite. High RMSE can be interpreted that SST of MODIS satellite is highly predicted or unpredicted from the actual SST value by ARGO Float.

This research also calculated the relationship between in situ and satellite data by used the coefficient of determination (r^2). r^2 value ranged between 0 – 1 which means that if the r^2 value is greater than 0,9 or approach 1 means that X has similarities to Y.

3. Result and Discussion

The result indicated that SST value is increasing during 2010 to 2016 both on satellite and in situ data (Figure 2). The graph shows the same pattern but SST value derived from ARGO Float has higher SST value then SST from MODIS satellite. It can explain that ARGO Float measures the SST in 5m depth or less but MODIS satellite only measure in the skin layer of surface water.

SST derived from ARGO data increases 0.093°C and SST data from MODIS satellite increases 0.132 °C. The SST data in 2011 show decreases from the year before both in situ dan satellite data, it can be explained because in 2011 there are La Nina phenomena occurred in that year. La Nina is phenomena that held in oceans with characteristic sea surface temperature value decreases from mean SST value in normal condition. This phenomenon also explains in 2011 that there are La Nina phenomena held in Pacific Oceans [16].

The average seasonal composite of SST value is calculated to see the variability of SST based on the different season in Indonesia (Tabel 1). The average SST in study site shows that dry season has higher SST value than wet season.

The determination coefficient (R^2) is also computed to see the accuracy of MODIS SST data in comparison to SST data from ARGO Float (Table 2). The R^2 value shows a positive correlation between MODIS and ARGO data especially in dry season, while in the wet season the R^2 is smaller than in dry season. Perhaps in the wet season, there is a lot of clouds spread around study site so produce noise contribute to some error in SST value. However, MODIS satellite categorized as passive satellite which makes MODIS satellite very sensitive to cloud cover and contribution another factor likes rainfall possibly make data error of SST.

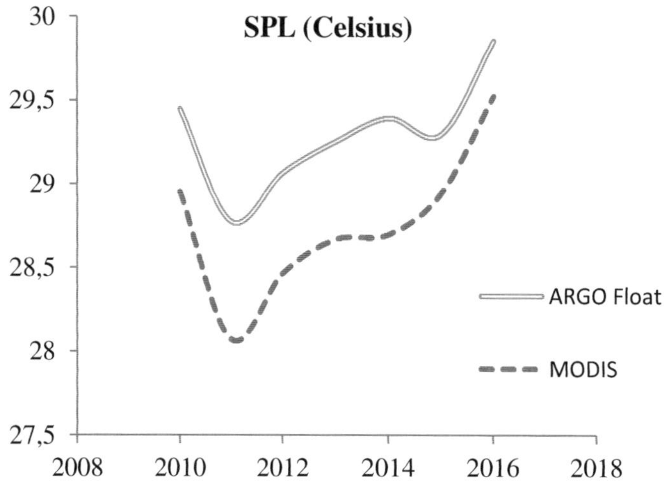

Figure 2. Sea surface temperature trends changes

Table 1. The average seasonal composite of SST value from 2010 to 2016

Data Base	Average Seasonal Composite													
	2010		2011		2012		2013		2014		2015		2016	
	Dry	Wet	Dry	Wet	Dry	Wet	Dry	Wet	Dry	Wet	Dry	Wet	Dry	Wet
ARGO	29.61	29.33	28.91	28.61	29.12	28.94	29.27	30.58	29.46	29.29	29.41	29.16	29.93	29.79
MODIS	28.88	28.99	28.20	27.81	28.49	28.33	28.60	28.59	28.80	28.35	29.16	28.65	29.65	29.44

Table 2. R^2 value for each season from 2010 to 2016

Year	Season	Match-up Numbers	R^2 Value
2010	Dry	121	0.391
	Wet	104	0.257
2011	Dry	150	0.555
	Wet	115	0.153
2012	Dry	180	0.625
	Wet	169	0.233
2013	Dry	139	0.372
	Wet	181	0.439
2014	Dry	142	0.301
	Wet	106	0.135
2015	Dry	157	0.833
	Wet	195	0.3
2016	Dry	149	0.517
	Wet	152	0.645

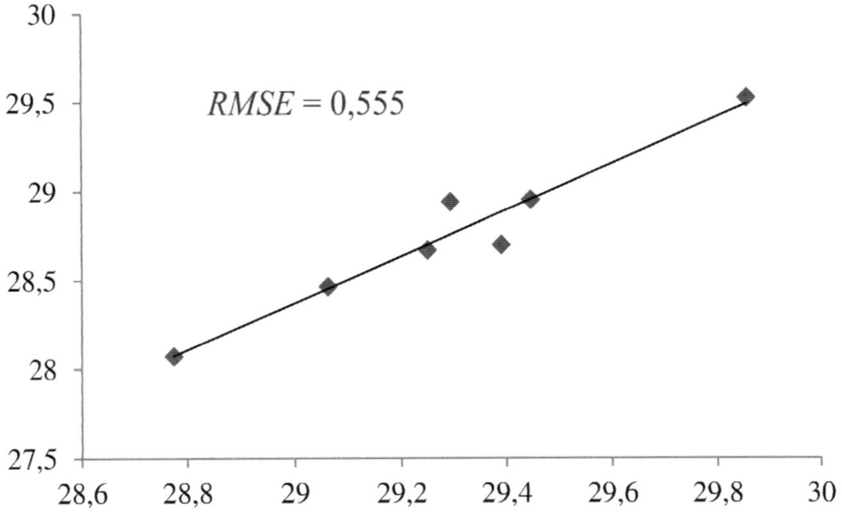

Figure 3. Comparisons SST value between *in situ* measured and MODIS satellite

The accuracy assessment of SST value from MODIS satellite have the same pattern as SST assessment derived from ARGO Float measurement. The value of RMSE is 0,555. RMSE value shows the accuracy of the data. While RMSE value is lower it means more accurate the data value measured by satellite to SST value measured by ARGO Float.

Indian Oceans western part of Sumatra possess complex dynamics and process on its seawater system. By using SST value could help the researcher to investigate another process in the areas namely upwelling, front, mass transport and other phenomena. The mapping that phenomena like upwelling are helpful because upwelling usually brings a lot of nutrient up. By knowing the upwelling event, it could assist the fishermen for catching fish by utilized the upwelling mapping phenomena site as their fishing ground.

4. Conclusion

Indian Ocean western part of Sumatera Island has trend changes of SST value from 2010 to 2016. Both measured by MODIS satellite and ARGO Float show the same pattern increasing of SST value. SST value measured by ARGO Float has increased 0,093^{O}Celsius and MODIS satellite has increased 0,132^{O}Celsius. RMSE value to predict how accurate SST value derived from MODIS satellite to ARGO Float is 0,555. The coefficient of determination r^2 value shows a positive correlation between MODIS and ARGO data especially in dry season, while in the wet season the R^2 is smaller than in dry season.

Acknowledgements

Authors would like to thank you for the International Argo Program and that made data freely available. (http://www.argo.ucsd.edu, http://argo.jcommops.org). The Argo Program is part of the Global Ocean Observing System: and also NASA Goddard Space Flight Center, Ocean Ecology Laboratory, Ocean Biology Processing Group for MODIS Aqua-Satellite data.

References

[1] Cinner J E, Cindy H, Emily S D, Austin T H, Nicholas A J G, Cristina C H, Nadine M, Tim R M 2013 Evaluating Social and Ecological Vulnerability of Coral Reef Fisheries to Climate Change. *PloS ONE.* **8(9)**

[2] Edwards M, Gregory B, Graeme C H, J A Koslow, Anthony J R 2010 Multi-Decadal Oceanic Ecological Datasets and Their Application in Marine Policy and Management. *Trends in Ecological and Evolution.* **25** p 602

[3] Chollet I, Peter J M, Frank E M-K, Chuanmin H 2012 Physical Environments of the Caribbean Sea. *Limnol. Oceanogr.* **57 (4)** p 1233

[4] Skakala J and Timothy J S 2016 Simple Heterogeneity Parametrization for Sea Surface Temperature and Chlorophyll. *J. JMARSYS.* **158** p 52

[5] Park, K-A, Lee, Eun-Y, Chang, Eunmi, Hong, Sungwook 2014 Spatial and Temporal Variability of Sea Surface Temperature andWarming Trends in the Yellow Sea, *Journal of Marine Systems.* **143** doi:10.1016/j.jmarsys.2014.10.013

[6] Mumby PJ, Ian A E, C M Eakin, William S, Claire B P, Helen J E, Susana E, Roberto I P, Laurent M C, Jamie R S 2011 Reserve Design for Uncertain Responses of Coral Reefs to Climate Change. *Ecology Letter.* **14** p 140

[7] McLoed E et al 2010 Warming Seas in The Coral Triangle: Coral Reef Vulnerability and Management Implications. *Coastal Management.* **38 (5)** p 518

[8] Rowlands G, Sam P, Bernhard R, Liisa M, Andrew B, Philip R 2012 Satellite Imaging Coral Reef Resilience at Regional Scale. A Case Study from Saudi Arabia. *Marine Pollution Bulletin.* **64** p 1222

[9] Madrinan, Max J M, Mohammad Z A-H, Douglas L R, Frank E M-K 2010 Using The Surface Reflectance MODIS Terra Product to Estimate Turbidity in Tampa Bay, Florida. *Remote Sens.* **2** p 2713

IOP Conf. Series: Earth and Environmental Science **135** (2018) 012016 doi:10.1088/1755-1315/135/1/012016

[10] Miyazawa Y, Hiroshi M, Toru M, Sergey M V, Xinyu G, Takuji W, Sourav S 2013 Data Assimilation of The High-Resolution Sea Surface Temperature Obtained from The Aqua-Terra Satellites (MODIS-SST) Using an Ensemble Kalman Filter. *Remote Sens.* **5** p 3123

[11] Hosoda K 2011 Algorithm for Estimating Sea Surface Temperature Based on Aqua/MODIS Global Ocean Data. 2. Automated Quality Check Process for Eliminating Cloud Contamination. *J. Oceanogr.* **67** p 791

[12] Pillai N G K and P Satheeshkumar 2013 Conservation and Management of Tuna Fisheries in the Indian Ocean and Indian EEZ. *International Journal of Marine Science.* **3** p 187

[13] Schott F A, Shang P X, Julian P M 2009 Indian Ocean Circulation and Climate Variability. *Rev Geophys.* **47** p 1

[14] Geetha G, T V S U Bhaskar, E P R Rao 2011 Argo Data and Product of Indian Ocean for Low Bandwith Users. *International Journal of Oceans and Oceanography.* **5** p 1

[15] Thierry C, et al 2012 Argo User's Manual, 2.4 Version., USA

[16] Tsuboi A, Tetsuya T, Kunio Y 2016 Seasonal Environmental Characteristics for the Tropical Cyclone Genesis in the Indian Ocean during the CINDY2011/DYNAMO Field Experiment. *Atmosphere.* **7** p 66

Application Side Casing on Open Deck RoRo to Improve Ship Stability

Hasanudin[1,2], I K.A.P Utama[1], Jeng-Horng Chen[2]

[1]Department of Naval Architecture and Shipbuilding Engineering, ITS, Surabaya-Indonesia
[2]Department of System Naval and Mechatronic Engineering, NCKU, Tainan-Taiwan

Email: hasanudinits@na.its.ac.id

Abstract. RoRo is a vessel that can transport passengers, cargo, container and cars. Open Car Deck is favourite RoRo Vessel in developing countries due to its small GT, small tax and spacious car deck, but it has poor survival of stability. Many accident involve Open Car Deck RoRo which cause fatalities and victim. In order to ensure the safety of the ship, IMO had applied intact stability criteria IS Code 2008 which adapted from Rahola's Research, but since 2008 IMO improved criteria become probabilistic damage stability SOLAS 2009. The RoRo type Open Car Deck has wide Breadth (B), small Draft (D) and small freeboard. It has difficulties to satisfy the ship's stability criteria. Side Casings which has been applied in some RoRo have be known reduce freeboard or improve ship's safety. In this paper investigated the effect side casings to survival of intact dan damage ship's stability. Calculation has been conducted for four ships without, existing and full side casings. The investigation results shows that defect stability of Open Deck RoRo can be reduce with fitting side casing.

1. Introduction

RoRo vessel is the most successful ship which able to carry passenger, goods and vehicles at same time. Besides the flexibility of the cargo, this type of vessel also able to operate in river, strait and even sea. This type of vessel also equipped with ramp door installed in after, side or forward of the hull as the entrance of the payload. The first RoRo was used in 1891 to carry train crossing strait which not connected with bridge. The first RoRo was equipped with rail which connected to a pier [1]. In Second World War, the principal of RoRo vessel was used to carry tank and other fighting vehicles as landing craft. The extreme popularity of RoRo was occurred during 1940 to early 1950s, RoRo used as short route ferry which pushes the development of land and sea transportation. Until 1994, there are 4,600 RoRo vessel to serve vehicles, mix of container and vehicles [2].

There are several problems of RoRo vessel, those are: lack of watertight, stability, low freeboard, safety of cargo, safety equipment and crew safety [2]. IMO has applied detail many rules such as SOLAS, MARPOL, MEPC, ILLC etc [3]. Nevertheless, there are still numerous accident occurred. Based on the accident data of ship with size >150GT from 1993-2013. The data shows averagely 150 ship lost each year. General cargo provides the highest contribution, it contributes 42% and RoRo vessel contributes 6% from total accident [4] [5]. however based on casualties data, the RoRo vessel accidents demand 1/3 of total loss of lives due to ship accidents [2]. The accident data also stated that more than 70% of RoRo accidents is collision case which followed by the capsized of the vessel less than 10 minutes [2].

Small size RoRo are mostly operated in archipelagic country to connect its spread islands [6]. Open deck RoRo vessel is the highest number of RoRo type which used by archipelagic country. The car deck in RoRo is designed open to reduce GT in order to decrease tax, also decrease usage of ventilation. There are numerous bad records of this vessel which taken loss of lives and goods due to lack proper application of safety standard [7]. Defect stability and damage compartments are the main factors of the sink of RoRo vessel. Also, the economical perspective of small GT become the contributor also. IMO has issued IS Code 2008 for intact stability and SOLAS 2009 which affect for cargo and passenger including RoRo vessel. If this regulation must be applied for developing country, there will be a lot of existing vessel must be grounded which may lead to huge economical loss[8]. This paper studies the solution for the existing vessel by modification of side casing by the point of view of ship stability.

2. Literature Review

The theory of buoyancy is believed introduced by Archimedes in 300 BC. The principle of floating object in water is developed as ship stability [9]. The research of stability criteria was introduced by Rahola in 1939 which studies asses' ship accident due to of flaw of stability. Based on the statistical data of capsized survived ship, an analysis are made and resulting criteria of safe stability. The basic of determination of dynamic stability is 40° roll. This Rahola criteria still not accounted the importance of hydrodynamic characteristic of vessel such as seakeeping. This criteria are based on coaster ship which sails in Baltic Sea. However, it is very popular among regulators. GL has adopted this Rahola criteria for its battle ship after World War 2. Even though this Rahola criteria was developed for special ship, but this criteria affects stability criteria widely [10] [11].

2.1. Intact Stability

The next step, IMO adopted the principal of Rahola criteria to measure the survival of intact stability [10] [12]. In 1948 IMO introduced regulation of intact stability which compiled in SOLAS'48 recommendation annex D, in 1960, SOLAS applied the first regulation of intact stability. In 1968 a regulation of stability are made based on righting arm and in 1985 the weather factors are adopted in intact stability criteria [13] [14]. The purpose of IS Code is to prevent ship capsized (from point of view MG and GZ area), the survivability of ship in bad weather, avoiding excessive free surface effect and guarantee the integrity of compartment watertight [15] .

IS Code 2009 is applied for passenger ships, oil tanker of 5000 dwt and above, cargo ships carrying timber on its deck, cargo ships carrying grain in bulk and high speed craft. Special IS code 2009 criteria can be applied to Passenger ships, Oil tankers of 5,000 dwt and above, Cargo ships carrying timber deck cargoes, Cargo ships carrying grain in bulk and High-speed craft. There are specific criteria recommended for Fishing vessels, Pontoons, Containerships greater than 100 m, Offshore supply vessels, Special purpose ships, and Mobile offshore drilling units (MODUs) [16].

For passenger ships/ RoRo must comply these criteria:

- Area under GZ curve for 0°- 30° ≥ 3.1513 meter.degree
- Area under GZ curve for 0° – 40° ≥ 5.1566 meter.degree
- Area under GZ curve for 30° – 40° ≥ 1.7189 meter.degree
- Maximum righting arm GZ at 30° or greater ≥ 0.20 meter
- Angle of maximum GZ righting arm ≥ 25°
- Initial GM_t at 0° measured at 1 radian ≥ 0.150 meter

RoRo vessel also must satisfies several special intact stability criteria:

- All crowding passenger at one side of the ship ≥ 10°
- Turn angle during maneuvering ≥ 10°
- Severe wind and rolling from side of the vessel, area a ≥ area b

2.2. Damage Stability

The regulation of damage stability generally using deterministic damage stability and recently has developed to probabilistic stability [17]. Deterministic damage stability is a survival criteria when one or more compartment is damage or flooded, where the damage determined during design stage. While probabilistic damage stability is a survival criteria when one or combination of group compartment experiences damage or flooded which calculated with probabilistic statistic. Subdivision of transverse bulkhead, longitudinal bulkhead and deck has significant influences to reduce probability of damage compartment. Less damaged compartment higher the chance of survivability. The sum of one and compartment combination or attained must be greater than required. The approach of probabilistic damage stability refers to the factual approach that one or group of compartment which causes ship capsize are occurred random. SOLAS 2009 part B-1 or SOLAS 1974 Revision 2009 has applied cargo and passenger ship to fulfil these criteria [18]. The calculation of probabilistic damage stability is complicated which requires software and high specification computer [19].

2.3. Side casing

Side casing first introduced as Glasgow Concept which adding the tanks at side of the hull ship. Low side casing positioned under the vehicle deck, the purpose is to reduce the speed of penetration during damage. Up side casing positioned left and right side of the main car deck, the purpose is adding freeboard and adding survival stability. Principle of up side casing the same with sponson that applied to improve ship stability. There are two type up side casing such as: inboard side casing which follow ship deck and outboard side casing which appendage out from hull. The study of correlation about side casing and intact stability showed up side casing can improve righting arm lever and peak of GZ curve more than 25°. it is difficult to satisfy because RoRo has wide deck [7]. Adding side casing on large RoRo also can improve survival of probabilistic damage stability, up side casing better than low side casing [18] .

3. Method

Indonesia is developing country which have thousands island is representation of Open Deck Small RoRo operation in the world. RoRo is favorite ship because can carry passenger, cargos and cars. In this paper the RoRo operation area is choice in the Bone Gulf, Celebes, Indonesia. The area has medium liner, worse weather at special months, crowded passengers, sea state 5 and many ships accident caused loss of lives, cargos and cars. This paper analyses four ship from the survey its area. Three ship is exist and one ship is capsize cause damage stability. The survey got some drawings and documents such as: stability booklet, lines plan, and general arrangement.

SOLAS 2009 Part B-1 give three loading conditions to calculate Probabilistic Damage Stability such as: Deepest Subdivision Draught (d_s), Light Service Draught (d_l), and Partial Subdivision Draught (d_p). Then combination one and group compartments can be made with triangle as shown in Figure 1. The combination depend on transverse bulkhead. Peak triangle show combination of group compartments. Tier one show single damage compartments, tier two show two compartments damage, tier three show third compartments damage etc.

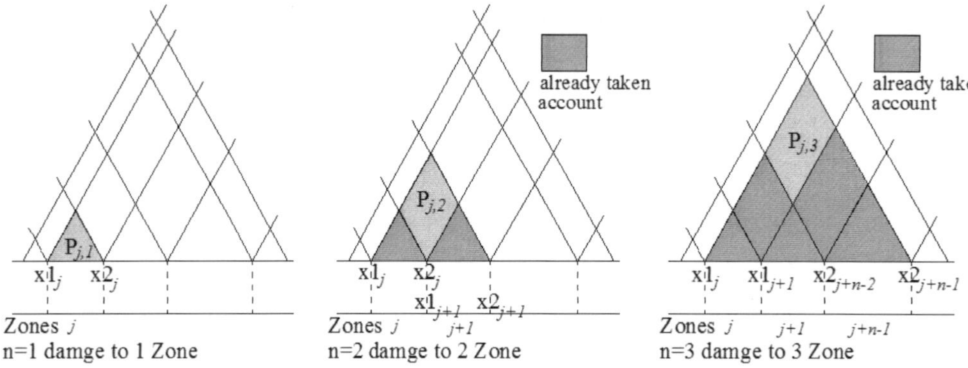

Figure 1. Single and combination damage group compartments

The calculation of probability flooded compartment (p_j) using distance function of after transverse bulkhead (x1) and the distance of the forward transverse bulkhead (x2) from AP. j in the floodable triangle is the number of transverse bulkhead in ship. The probability of damage compartment is based on Equation (1), for two flooded compartment uses equation (2), for three or more uses equation (3)

$$p_{j,1} = p(x1_j, x2_j) \quad (1)$$

$$p_{j,2} = p(x1_j, x2_{j+1}) - p(x1_j, x2_j) - p(x1_j, x1_{j-1}) \quad (2)$$

$$p_{j,n} = p(x1_j, x2_{j+n-1}) - p(x1_j, x2_{j+n-2}) - p(x1_{j+1}, x2_{j+n-1}) - p(x1_{j+1}, x2_{j+n-2}) \quad (3)$$

The calculation of survivability (s_i) from flooded compartment and group of compartment uses the function of GZ arm stability which experiences damage. For passenger ship, the GZ stability arm uses the minimum of intermediate survivability or the final survivability times moment as shown in equation (4).

$$s_i = minimum(s_{intermediate,i} \text{ or } s_{final,i} \cdot s_{moment,i}) \quad (4)$$

The calculation of Attained Index (A) is the indicator of ship survivability based on probabilistic damage stability. It calculated by summary of index probability of damage times survivability as shown in equation (5), due to the existing of transverse bulkhead and deck, index A is increased as stated in equation (6). The calculation of index A is must accounted three conditions Deepest Subdivision Draught (A_s), Light Service

Draught (A_l) and Partial Subdivision Draught (A_p). Where it will be times the constant value as written in equation (7).

$$A = \sum p_i \cdot s_i \quad (5)$$

$$A = \sum p_i \cdot (v_i \cdot r_i \cdot s_i) \quad (6)$$

$$A = 0.4 \cdot A_s + 0.4 \cdot A_p + 0.4 \cdot A_l \quad (7)$$

The calculation of Required Index (R) is the measurement of minimum index which required by SOLAS 2009. Part B-1. For cargo ships the R Index is the function of ship's length, while the passenger ships is the function of number of passenger carried by the ship. More passenger onboard the ship the R index is higher. Ship comply the probabilistic damage stability criteria when attained index is higher than required as written in equation (8).

$$A \geq R \quad (8)$$

4. Discussions

There are four RoRo open car deck which got from survey, the main dimension as shown in Table 1:

Table 1. Main Dimensions

Dimension	Unit	Ship A	Ship B	Ship C	Ship D
Loa	m	44.5	55.51	56.26	55.72
Lpp	m	41.83	51.07	52.50	49.66
B	m	11.30	14.00	13.10	16.2
H	m	3.70	3.70	2.70	3.80
Fb	mm	909	2.75	564	760
Gt		862	1112	1172	1376
Nt		259	1305	352	413
Passengers	persons	288	360	282	
Cars		17	25	30	
Crews	unit	18	19	22	

the lines plan are built in computer model using Maxsurf Modeler as shown inFigure 2. Ship A is inboard side casing and ship B, C, D are outboard side casing

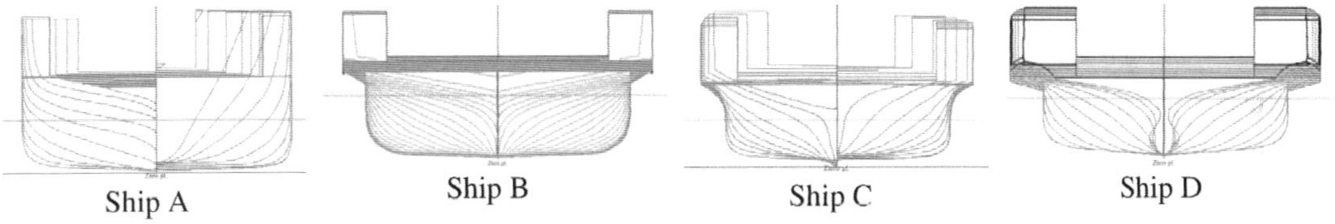

Figure 2. Lines plan

Based on lines plan, the 3D hull model are built with the existing side casing as shown in Figure 2. Ship A has the longest side casing, followed by Ship C, D and ship B. The nest modeling is to build three variation model of side casing. Without side casing, existing side casing and full side casing. For model without side casing, the model are built only until deck and for the full side casing are built continuous along the side of the ship.

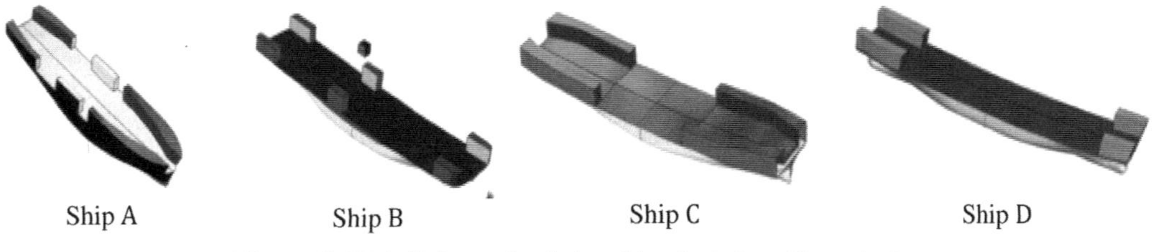

Figure 3. 3D hull form of existing ships (existing side casing)

The next step is to design the compartment and tanks based on the general arrangement. The permeability of the tank and compartment are determined by the fluid types inside the tank.

4.1. Intact Stability Analysis

This 4 ships are calculated with 3 variation of side casing: without side casing, existing side casing and full side casing. Each variation of side casing uses 3 load-cases: load-case deepest subdivision draught, light service draught, and partial subdivision draught. The result calculation can be showed at table 2-5

Table 2. Intact stability of ship without side casing

Ship	Intact Stability		
	d_s	d_p	d_l
A	Pass	Pass	Pass
B	Fail	Fail	Fail
C	Fail	Fail	Fail
D	Fail	Fail	Fail

Table 3. Intact stability of ship with existing side casing

Ship	Intact Stability		
	d_s	d_p	d_l
A	Pass	Pass	Pass
B	Fail	Pass	Pass
C	Pass	Pass	Pass
D	Pass	Pass	Pass

Table 2 Ship A without side casing satisfies all IS Code 2008 Criteria for all load cases. Ship B at deepest load case is not comply with IS Code 2008 due to the area under curve GZ 30°-40°, maximum GZ at 30° or greater and maximum angle GZ. Ship B at partial and light loading condition doesn't comply IS Code 2008 at maximum angle GZ. Ship C fails only for all loading conditions loading deepest, partial and light doesn't fulfil IS Code 2008 criteria maximum GZ. Ship D fails only for all loading conditions loading deepest, partial and light doesn't fulfil IS Code 2008 criteria maximum GZ. The area under curve GZ 30°-40° doesn't fulfil criteria at deepest and partial loadcase deepest and partial.

Table 3. Show ship A, C and D with existing side casing satisfies all IS Code 2008 Criteria for all load cases. Ship B at deepest load case is not comply with IS Code 2008 due to the area under curve GZ 30°-40° ≥ 1.7189 meter degree, maximum GZ at 30° or greater and maximum angle GZ ≥ 25.0 degree. It is caused L/B of ship B too large.

Table 4. Intact stability of ship with full side casing

Ship	Intact Stability		
	d_s	d_p	d_l
A	Pass	Pass	Pass
B	Pass	Pass	Pass
C	Pass	Pass	Pass
D	Pass	Pass	Pass

Table 5. Intact stability of all ship

Ship	Intact Stability		
	Without Side Casing	Existing	Full Side Casing
A	Pass	Pass	Pass
B	Fail	Fail	Pass
C	Fail	Pass	Pass
D	Fail	Pass	Pass

Table 4 show ship A, B, C and D with full side casing satisfies all IS Code 2008 Criteria for all load cases deepest subdivision draught, light service draught, and partial subdivision draught. Full side casing has positive effect to improve survival of intact stability.

Table 5 is summary from table 1 until table 3. Generally adding side casing can improve survival of intact stability. Ship A is the best ship which can satisfied IS Code 2008 Criteria From without side casing until side casing. For ship existing only ship B which not fulfil IS Code 2008 criteria. When all ships used full side casing all IS Code 2008 satisfied.

4.2. Probabilistic Damage Stability Analysis

Calculation of probabilistic damage stability was used Maxsurf Stability. Important thing before calculation arrange bulkhead position base on subdivision in general arrangement. Amount and position of bulkhead effect p_i value (probability damage compartments). Bulkhead after and foreword position are measured from AP. Combination of one compartment and group compartments four ships are showed by picture 8 as following:

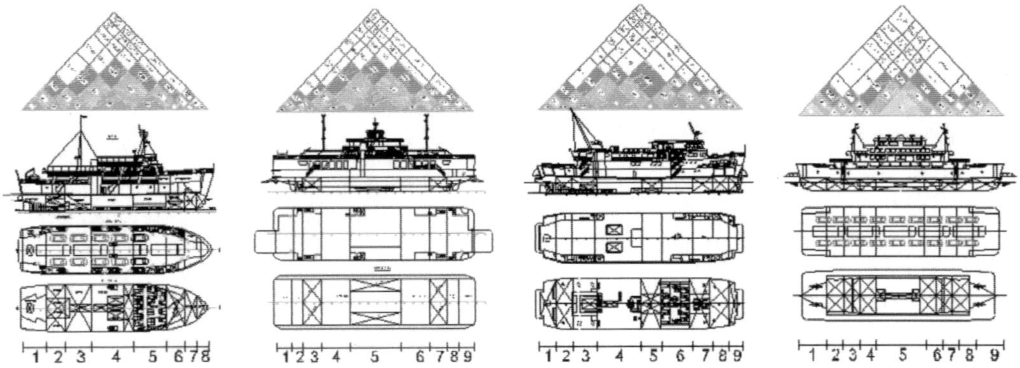

Ship A Ship B Ship C Ship D

Figure 4. Combination damage of group compartments and general arrangement

There are 8 compartments in Ship A, so the combination of the damage stability is 8 single damage compartment, 7 double group of damage compartment, 6 triple group compartment and so on. There are 9 compartment in Ship B, the combination of the damage stability are 9 single damage compartment, 8 double group damage compartments, 7 triple group compartments and so on. Ship C and D have the same damage compartments combination with ship B. The next step the probabilistic damage stability conducted where the results are shown in Table 6 to Table 9.

Table 6. Value A and R of ship without side casing

Ship	Probabilistic Damage Stability				
	A_s	A_p	A_l	A	R
A	0.588	0.892	0.854	0.763	0.706
B	0.424	0.538	0.571	0.499	0.714
C	0.636	0.819	0.555	0.693	0.706
D	0.510	0.502	0.502	0.505	0.688

Table 7. Value A and R of ship with existing side casing

Ship	Probabilistic Damage Stability				
	A_s	A_p	A_l	A	R
A	0.590	0.988	0.988	0.829	0.706
B	0.534	0.805	0.708	0.677	0.714
C	0.815	0.911	0.787	0.848	0.706
D	0.510	0.502	0.502	0.505	0.688

Table 6 shows the results of attained and required probabilistic damage stability for ship without casing. The results A compared to R value. It shows that ship A only A_s value which less than R, condition A_p>R, A_l>R, and A>R so this ship comply damage stability SOLAS 2009. All values of ship B (A_s, A_p, A_l and A) less than required, so this ship does not complies damage stability SOLAS 2009. Ship C only has A_p value which greater than R (A_p>R) and the other values are less than required. So it can be concluded that this ship does not complies damage stability SOLAS 2009. Ship D all the values (A_s, A_p, A_l and A) are less than R so it can be concluded than ship D does not complies damage stability SOLAS 2009. Generally for ship without side casing, only ship A which complies damage stability SOLAS 2009.

Table 7 shows the results of attained and required probabilistic damage stability for ship existing casing. The results A compared to R value. It shows that ship A only A_s value which less than R, condition A_p>R, A_l>R, and A>R so this ship complies damage stability SOLAS. Only values A_p of ship B more than required, so this ship does not complies damage stability SOLAS. All values of ship C (A_s, A_p, A_l and A) more than required, so this ship can be concluded complies damage stability SOLAS 2009. Ship D, all the values (A_s, A_p, A_l and A) are less than R so it can be concluded than ship D does not complies damage stability SOLAS 2009. Generally for ship with existing side casing, half of them complies damage stability SOLAS 2009.

Table 8. Value A and R of ship with full side casing

Ship	Probabilistic Damage Stability				
	A_s	A_p	A_l	A	R
A	0.656	0.934	0.930	0.822	0.706
B	0.677	0.986	0.910	0.847	0.714
C	0.829	0.937	0.982	0.903	0.706
D	0.975	0.940	0.938	0.953	0.688

Table 9. Summary of probabilistic damage stability calculation

Ship	Probabilistic Damage Stability		
	Without side casing	Existing	Full Side Casing
A	Pass	Pass	Pass
B	Fail	Fail	Pass
C	Fail	Pass	Pass
D	Fail	Fail	Pass

Table 8 indicates that all A values are greater than R so it is concluded that all ships with full side casing satisfy probabilistic damage stability. Ship A has As value less than R, however the A value is greater than R so the ship passes criteria of damage stability SOLAS 2009 so as the ship B.

Table 9 summaries the results of probabilistic damage stability from table 2 until table 4 for comparison A and R. If the value of A is greater than R it passes, contrary if A less than R it is fail. From Table 9, all variation of ship A passes criteria of the probabilistic damage stability SOLAS 2009. From several existing ships only two ship that satisfy the probabilistic damage stability SOLAS 2009, those are ship A and ship C. When existing side casing of ship is extend become full side cashing, all ships satisfy the probabilistic damage stability SOLAS 2009.

5. Summary

Based on the analyses and discussion previously, it can be concluded that the installation of side casing in open car deck RoRo vessel able to increase survivability of the intact stability and damage probabilistic stability, it can to be solution for the ship which have defect stability. There are several detail conclusions which can be drawn, those are:

1. One of four the ships existing which operation in Bone Gulf, Celebes, Indonesia does not comply the IS Code 2008 criteria. The ship fails in maximum GZ angle > 25°.
2. Based on the intact stability, almost all ship without side casing do not comply IS Code 2008, when added existing casing, three of fourth of them are comply the criteria and it comply all criteria when full side casing are added. The full side casing installation has significant influence in increasing survival of intact stability.
3. Ship which satisfies IS Code 2008 not always fulfil damage stability criteria. There is only two ship with existing side casing which fulfil the damage stability criteria.
4. The probabilistic Damage Stability analysis stated that all ship without side casing not comply SOLAS 2009, however when it added side casing, only one ship comply IS Code 2008. All ships comply SOLAS 2009 when they added full side casing. It can be concluded that additional side casing increases survivability and Probabilistic Damage Stability.
5. Ship A with all variation side casing and load cases satisfy IS Code 2008 Criteria and Damage Stability SOLAS 2009 Criteria. The best ship to fulfil stability is ship A followed by ship C, ship D and ship B.

Based on the conclusions, there are several recommendations in order to increase survivability of ship, those are:

1. For existing open deck RoRo vessel which doesn't comply intact stability and damage stability can be modified with the installation of side casing in order to comply the criteria of stability. The new designed open car deck RoRo is recommended use side casing.
2. The analysis conducted in this paper neglected the water trap on car deck. It assumes that the ingresses water in car deck directly wiped to freeing port or by scupper which equipped with pumps so the water dumped to sea.

Acknowledgments

The authors would like to thank the Ministry of Research, Technology and Higher Education of Republic Indonesia for supporting this work under research schema of INSINAS DIKTI 2016 with contract number 01888/IT2.11/PN.08/2016. This paper was calculated using Maxsurf Education Version 15. The author would like to address gratitude to Suryo Anggoro, Habibi Si'qon, Bagus Jamilludin and Ahmad Baidowi for their assistance during the completion of this paper.

References

[1] J. Marshall, *The Guinness Railway Book*. Guinness Books, 1989.

[2] A. Embankment, "IMO and RoRo safety," London SE1 7SR, United Kingdom, 1997.

[3] Bereau Varitas, *Safety of RoRo passenger & cruise ships*. 2016.

[4] A. Papanikolaou, K. Bitha, E. Eliopoulou, and N. P. Ventikos, "Statistical analysis of ship accidents occurred in the period 1990-2012 and assessment of safety level of ship types," in *Proceedings of the Maritime Technology and Engineering Conference (MARTECH)*, 2014, pp. 227–233.

[5] B. B. Primorac and J. Parunov, "Review of statistical data on ship accidents," in *Maritime Technology and Engineering*, 2016.

[6] JICA, "The Master Plan and Feasibility Study on the Establishment of an Asian Roll-On/Roll-Off (RoRo) Shipping Network and Short Sea Shipping," Japan Marine Science Inc., Japan, EI JR 13-069, 2013.

[7] S. Anggoro, "Analysis of the Intact Stability of Indonesian Small Open-deck Roll-on/Roll-off Passenger Ferries," UNIVERSITY OF NEW SOUTH WALES, 2008.

[8] R. Tagg, "Comparison of survivability between SOLAS 90/95 and SOLAS 2009 ships—A retrospective view 10 years on from project HARDER," in *Proceedings of the 14th International Ship Stability Workshop, Brest, France*, 2014, vol. 29.

[9] A. Francescutto and A. D. Papanikolaou, "Buoyancy, stability, and subdivision: from Archimedes to SOLAS 2009 and the way ahead," *Proc. Inst. Mech. Eng. Part M J. Eng. Marit. Environ.*, vol. 225, no. 1, pp. 17–32, 2011.

[10] V. Bertram and H. Schneekluth, *Ship design for efficiency and economy*. Butterworth-Heinemann, 1998.

[11] J. Arjava and R. Jalonen, "The Rahola Criterion-The life and work of Professor Jaakko Rahola," 2015.

[12] L. Marinel-Danut, A. Cristian, and H.-P. Radu, "The Analysis of Intact Ship Stability Regulations," *Univ. Marit. Constanta Analele*, vol. 13, no. 18, p. 45, 2012.

[13] A. Francescutto, "The intact ship stability code: Present status and future developments," in *Proceedings of the 2nd International Conference on Marine Research and Transportation, Naples, Italy, Session A*, 2007, pp. 199–208.

[14] A. Francescutto and N. Umeda, "Current status of new generation intact stability criteria development," in *Proceedings of the 11th International Ship Stability Workshop, Wageningen, The Netherlands*, 2010, pp. 21–23.

[15] IMO, "Stability and Subdivision," 2016. [Online]. Available: http://www.imo.org/en/OurWork/Safety/StabilityAndSubdivision/Pages/Default.aspx. [Accessed: 27-Aug-2016].

[16] IMO, *International Code on Intact Stability (IS Code)*. 2008.

[17] A. Papanikolaou, *Ship Design: Methodologies of Preliminary Design*. Springer, 2014.

[18] E. S. Ravn and J. J. Jensen, "Probabilistic Damage Stability of RoRo Ships," Technical University of DenmarkDanmarks Tekniske Universitet, Department of Naval Architecture and Offshore EngineeringInstitut for Skibs-og Havteknik, 2003.

[19] D. Vassalos and A. Jasionowski, "SOLAS 2009–raising the alarm," in *Contemporary Ideas on Ship Stability and Capsizing in Waves*, Springer, 2011, pp. 103–115.

Physicochemical Characteristic of Municipal Wastewater in Tropical Area: Case Study of Surabaya City, Indonesia

I M W Wijaya, E S Soedjono

Environmental Engineering Department, Faculty of Civil Engineering and Planning, Institut Teknologi Sepuluh Nopember, Surabaya, Indonesia

Email : wijaya16@mhs.enviro.its.ac.id

Abstract. Municipal wastewater is the main contributor to diverse water pollution problems. In order to prevent the pollution risks, wastewater have to be treated before discharged to the main water. Selection of appropriated treatment process need the characteristic information of wastewater as design consideration. This study aims to analyse the physicochemical characteristic of municipal wastewater from inlet and outlet of ABR unit around Surabaya City. Medokan Semampir and Genteng Candi Rejo has been selected as wastewater sampling point. The samples were analysed in laboratory with parameters, such as pH, TSS, COD, BOD, NH_4^+, NO_3^-, NO_2^-, P, and detergent. The results showed that all parameters in both locations are under the national standard of discharged water quality. In other words, the treated water is securely discharged to the river

1. Introduction

Surabaya is the second largest city in Indonesia, and become a business centre in east part of Java Island. It is a reason of high population living in Surabaya City. Increasing of population in Surabaya City is in line with the water demand and impacting wastewater generation. In addition to industries, municipal wastewater which is discharged from the households or domestic activities is also the main contributor of wastewater. Domestic wastewater effluent is the main contributor to diverse water pollution problems. Domestic wastewater management is a big problem particularly with improvement in living standard of community [1]. Grey water is generated as result of living habits, used product, and artificial installation [2]. Even today, domestic wastewater is disposed to drainage, river, or lake without proper treatment before [3].Some of problems caused by wastewater, such as eutrophication, increasing water treatment cost, decreasing recreational value of water, health risks to humans and livestock, loss of oxygen and undesirable changes in aquatic ecosystem. In order to prevent the pollution risks, wastewater have to be treated before discharged to the water body. Selection of appropriated treatment process need the properties of wastewater as the references.

The option of reusing treated wastewater is becoming necessary and possible. It is as the result of increased climate change, thus leading to droughts and water scarcity, and the fact that wastewater effluent discharge regulations have become stricter leading to a better water quality [4]. Wastewater treatment unit which is commonly used in Surabaya City is anaerobic baffled reactor (ABR). ABR is a series up-flow *anaerobic sludge blanket* (UASB) which has simple design and low cost in operation and maintenance [5]. It is one of treatment unit with communal system and serves 50-100 households. The wastewater flows up and down in the ABR pass through the baffles and causes the flow of influent have more intensive contact with an anaerobic biomass. It improves processing performance for biodegradable matter [6]. This communal system has been applied in several areas in Surabaya City. For example, in Medokan Semampir and Genteng Candi Rejo Sub-District. Domestic wastewater from the households in both area is treated in ABR unit and the effluent is reused for watering and discharging to water body. It is important to determine the characteristic of the pre-treated and post-treated wastewater in explaining its impact to the main water.

Chemical compositions in the domestic wastewater are highly diverse substances from simple compounds to complex polymers. Types and amount of substances show the characteristic of domestic wastewater. Characterization at the overall substances is important to expand the knowledge in selecting appropriate wastewater treatment processes or models [7,8]. Determination of characteristic of domestic wastewater is very important in order to evaluate the existing treatment plants and selection of appropriate treatment plant. Besides that, it also necessary to determine the utilization of treated or untreated wastewater based on their contents [8]. The concentrations and ratios between various parameters in wastewater influent can influence the selection and function of treatment processes. The quality of treated wastewater are important for evaluating WWTP performance and subsequent impacts or risks on human health, surrounding environment and design of advanced wastewater treatment and/or reclamation processes. Consequently, it is critical to conduct systematically analyses on wastewater quality parameters of municipal WWTPs [9]. Wastewater characteristic is related to water quality standard that is aimed to protect the designated use of water body [10].

Generally, the characteristics of domestic wastewater are specifically represented by some physicochemical parameters, such as pH, chemical oxygen demand (COD), biochemical oxygen demand (BOD), total dissolved solid (TDS), total suspended solid (TSS), dissolved oxygen (DO), total nitrogen (TN), total phosphate (TP), and potassium. The other minority substances such as metal, toxic material, detergent, and bacteria (4,9,11–14). Huang [7] has investigated the characteristic of organic matters in domestic wastewater showed that fibre (20,64%) was the largest group of organic matters and followed by proteins (12,38%) and sugar (10,65%) were next largest. The sources of wastewater influent are mainly from municipal wastewater and industrial streams, and also possible mixture of storm water and surface water. In other word, wastewater influent quality can also be affected by the discharge of industrial streams, storm water, surface runoff, etc. (9,15). Grey wastewater is defined as wastewater without any input from toilets, which means that it corresponds to wastewater produced from bathtubs, showers, hand basins, laundry machines and kitchen sinks, in households, office buildings, schools, etc. It is less polluted than black wastewater which is contained wastewater from toilet, such as faces and urine [8].

Water will become the most strategic resource in many parts of the world within the next decades. The identification of critical limit in water quality parameters and their concentrations provide an opportunities for improving sustainable water utilization in the future [9]. According to the Indonesia discharge quality standard of pollutant for municipal wastewater, the quality standard for pH, BOD, TSS and oil and grease corresponding are 6-9, 100 mg/L, 100 mg/l, and 10 mg/L, respectively. The treated wastewater from WWTP will be discharged into the water body for reutilization. Indonesia government has been classified the water quality level to determine the appropriate utilization for each water classes. There are four water classes and each has water quality index involved some water quality parameters. The aim of present study was to investigate the physicochemical characteristic of municipal wastewater in urban area in case of Surabaya City. Those physicochemical characteristic would be presented as pH, COD, BOD, DO, TSS, Nitrogen, Phosphorus, and detergent.

2. Materials and Methods

2.1. Domestic Wastewater Sampling Point

The objective of this study is analysing the characteristic of municipal wastewater from communal wastewater treatment plant in Surabaya City. Domestic wastewater was taken from inlet and outlet of 2 communal ABR in Surabaya City. Location of the ABR was in Medokan Semampir (MS) and Genteng Candi Rejo (GCR) Sub-District. ABR in MS with 2 compartments is serving about 175 households while in GCR with 5 compartments is serving more. Santosa [22] has been classified all areas in Surabaya based on the health risk level. Both selected locations of study are belong to low risk level of wastewater service area. Those samples were analysed in Laboraturium Pemurnian Air, Department of Environmental Engineering, Faculty of Civil Engineering and Planning, Institut Teknologi Sepuluh Nopember.

2.2. *Physical parameters*

Physical parameters that measured in this study were TSS and pH. TSS was measured based on the gravimetric method. The samples were vapored on temperature 105^{0}C in the oven then substances were weighed on digital scale. pH was measured by pH meter.

2.3. *Chemical parameters*

The chemical characteristics of domestic wastewater in this study were presented by DO, COD, and BOD. The number of BOD and COD indicates the level of oxygen depletion due to degradation of organic matter [6]. COD was measured based on the closed reflux method. Chemical oxygen demand (COD) determination was carried out by refluxing Using potassium dichromate K2Cr2O7 as the oxidant in the presence of HgSO4 and H reagent and then titrating excess K2Cr2O7 against Fe(NH4)2·(SO4)2·6HO using ferroin indicator [16,17]. BOD was measured the BOD_5 method according to standard method.

2.4. *Nutrient content and detergent*

Nutrient concentration is also important in determining wastewater characteristic. Wastewater with high nutrient content increase the risk of eutrophication in water body. Concentration of nutrient should be reduce to a level protective of the receiving stream [17]. Nitrogen and phosphorus become two kind of nutrient that were measured in this study. Form of nitrogen, such as ammonium, nitrite, and nitrate was measured according to standard method. Phosphorus in this study was measured as orthophosphate and determined using a spectrophotometric method. LAS was observed in all sample as the representative of detergent substance in wastewater. LAS concentration measured by using MBAS method. MBAS will respond to any molecule with an anionic centre capable of forming a stable ion pair with methylene blue and a hydrophobic group to allow the dye complex extraction into the chloroform phase [18].

3. Results and Discussion

3.1. *Physical characteristic*

A pH less than 7 in wastewater influent is an indication of septic conditions while values less than 5 and greater than 10 indicating the presence of industrial wastes and do not appropriate with biological treatment process. The pH range for suitable biological life is quite narrow (typically 6-9) [4]. Low pH level may influenced by the methanogen bacteria activity. High level of carbonates influenced calcium and magnesium ions become insoluble minerals leaving sodium as the dominant ion in solution. High alkalinity water can intensive soil conditions, which will have implications for agriculture [12]. Alkalinity is an important indicator when selecting grey water treatment in term of anaerobic system. Lowering pH is dangerous for microorganism in organic substance [3]. The pH analysis for both influent samples were 6,85 and effluent were 7,15 and 6,95, respectively. The value was following the pH range similar with result [4].

TSS potentially increase the sedimentation in water body and cause depletion of oxygen. TSS is essential for controlling the treatment process and compliance with effluent regulations [16]. The results of the present study show that the TSS ranged from 18 mg/L for both influents wastewater. After passing through the treatment process, the treated water has TSS about 15 mg/L for MS and 12 mg/L for GCR area. The difference of TSS concentration between MS and GCR was not significant, even though ABR in MS has 2 compartments while GCR has 5. It shows that ABR in GCR has longer hydraulic retention time (HRT) than ABR in MS. However the TSS concentration in treated water did not exceed the general standard of 50 mg/L stipulated by the rule of water quality control. The low of

TSS concentration in influent and effluent indicates that it has been reduced in the sewer stream. The TSS in influent were extremely lower than TSS in previous research conducted by Hahn and Figueroa, with 510 mg/L of TSS [5].

3.2. Chemical characteristic

Domestic waste water from kitchen sinks and dishwashers contributes most of the biodegradable organic substances and particulate nitrogen. The presence of large biodegradable organic substances, kitchen waste water is more contaminated by the thermal tolerant coliforms than other sources [2]. Oxygen demand may be in the form of BOD or COD, is the amount of oxygen used by microorganisms in degrading organic material in wastewater. According to Hudson [16] BOD may cause oxygen depletion in water body and lead to nuisance odor and fish kills. It is involved the measurement of dissolved oxygen used by microorganisms in the biochemical oxidation of organic matter. COD is number of oxygen which is consumed by oxidisable matter in the wastewater [19].

The COD will always be higher than the BOD. This is because the COD measures substances that are both chemically and biologically oxidized. The ratio of COD:BOD figures the proportion of organic material contain in wastewaters. Some polysaccharides, such as cellulose, can only be degraded anaerobically, so they will not be involved in the BOD [4]. COD is more resistance to treatment and may accumulate include humic residues, detergents, phenols, cyanide, residual hormones, pharmaceuticals and pesticides [16]. High organic matter can lead the high oxygen consumption by the aerobic bacteria and competing with the aquatic organism [15].

BOD is amount of organic content in the wastewater which is biologically degradable with oxygen consumption. BOD usually indicate as 5 days oxidation of biodegradable organic matter at 20^0C by microorganism [19]. BOD in the influent and effluent in MS measured of 9 mg/L and 5 mg/L, respectively. According to COD measurement, the BOD/COD ratio for influent and effluent was 0,6 and 0,55, respectively. It indicates that the wastewater contained more biodegradable organic matter. It was similar with organic matter concentration in Genteng since the result showed that 8,00 mg/L and 3,00 mg/L of BOD for inlet and outlet, respectively. COD concentration in influent and effluent was 15 mg/L and 6 mg/L, respectively. In this case, decreasing of COD concentration needs to investigate because ABR classified as biological treatment process. The value of BOD and COD for both location is still under the rule of water quality control for discharge of effluent on water stream.

Table 1. Physico-chemical characteristically wastewater in Medokan Semamir and Genteng

Parameter	Medokan Semampir		Genteng	
	Inlet (mg/L)	Outlet (mg/L)	Inlet (mg/L)	Outlet (mg/L)
pH	6.85	7.15	6.85	6.95
TSS	18	15	18	12
Nitrit	0	4.30	2.18	0.24
Nitrat	0	2.45	0.92	0.20
Ammonium	12.82	5.09	38.91	0.32
Orthophosphate	1.08	0.25	0.44	1.16
COD	15.00	9.00	15	6
BOD	9.00	5.00	8	3
Detergent (LAS)	0.46	0.15	0.34	0.25

3.3. Nutrient and detergent

Municipal activities, agriculture, and rapid urbanization led to increased nitrogen and phosphorus discharge to water system. More often, the effluents from municipal wastewater treatment plant fail to meet the national standard for effluent quality [20]. Excess nutrients, mostly N and P is main cause of eutrophication which is result in oxygen depletion, biodiversity reduction, fish kills, odor, and increased toxicity [16]. Bathroom grey water discharge contain both of nitrogen and phosphorus due to exclusion of urine and feces [2]. Nitrogen appears in municipal wastewater as ammonia (NH_3), nitrate (NO_3^-) and nitrite (NO_2). Organic nitrogen is decomposed to ammonia. Urea and protein are the major source of nitrogen in municipal wastewater [21].

In the septic tank process, organic nitrogen is converted to total ammonium nitrogen in anaerobic condition and due to low oxygen present, amount of it is converted to nitrate. The concentrations were much higher than the safe limit of 1.0 mg/L for long term exposure of aquatic macro organisms [15]. Organic and inorganic forms of nitrogen can generate eutrophication impact in lake, river, estuarine and
coastal waters. Nitrogen sources in domestic wastewater are food waste, soap, and fertilizer. In complete nitrogen cycle, ammonia is oxidized to nitrate, creating an oxygen demand and low dissolved oxygen in surface waters. Nitrogen in the form of ammonia is toxic to fish and exerts an oxygen demand on receiving water by nitrifiers [4]. Ammonium results of influent and effluent water in MS is lower than in GCR with concentration of 12,82 mg/L and 38,91 mg/L, respectively. Generally, it caused by the service area in GCR is larger than in MS.

Phosphorus appears in wastewater as orthophosphate, polyphosphate and organically bound phosphorus. The last two components counted usually for up to 70% of the influent phosphorus. Microbes utilize phosphorus during cell synthesis and energy transport [21]. Phosphorus was investigated in the present study as orthophosphate. This phosphate is dissolved the water as dominant phosphorus substance in anaerobic condition. In the present study, orthophosphate in the inlet and outlet from MS is 1,08 and 0,25 mg/L, respectively. Different result showed from Genteng area which is the outlet concentration is higher than in the inlet. It indicates that there was another phosphorus source which is contribute in the higher phosphorus, such as detergent. Surfactants are the mayor ingredients of detergent which is commonly used in laundering activity. The presence of surfactants that is belong to synthetic compounds in natural water body leads an aesthetic loss caused by foam. Surfactants may be toxic to the organism in soil or water [3]. Study from Braga and Varesche [3] showed that LAS concentration in specific laundry wastewater is 163,65 mg/L. This result is very higher than LAS concentration in this research that is 0,46 mg/L in MS and 0,15 mg/L in GCR. LAS concentration is contribute to the presence of phosphorus. Detergent is also the source of phosphorus due to the presence of phosphorus-five in detergents [2].

4. Conclusion

Municipal wastewater in Medokan Semampir and Genteng Candirejo characteristically has low concentration of physico-chemical. All test results are showed still under the national standard of waste water quality. Treated wastewater which is discharged to the stream has been controlled by using communal wastewater treatment plant.

Acknowledgments
We are grateful for the Ministry of Research and Higher Education of Indonesia for the PMDSU scholarship which includes fund this part of research.

References

[1] Rani Devi, Dahiya RP. COD and BOD removal from domestic wastewater generated in decentralised sectors. Bioresour Technol. 2008; **99(2)** :344–9.
[2] Li F, Wichmann K, Otterpohl R. Review of the technological approaches for grey water treatment and reuses. Sci Total Environ [Internet]. 2009;**407(11)**:3439–49.
[3] Braga JK, Varesche MB a. Commercial Laundry Water Characterisation. Am J Anal Chem. 2014;2014(January):8–16.
[4] Akpor, O. B., Muchie M. Environmental and public health implications of wastewater quality. African J Biotechnol. 2011;10(13):2379–87.

[5] Hahn MJ, Figueroa L a. Pilot scale application of anaerobic baffled reactor for biologically enhanced primary treatment of raw municipal wastewater. Water Res [Internet]. 2015;87:494–502.
[6] Hendriarianti E, Karnaningroem N. Evaluation of Communal Wastewater Treatment Plant Operating Anaerobic Baffled Reactor and Biofilter. 2016 p:7–12.
[7] Huang M, Li Y, Gu G. Chemical composition of organic matters in domestic wastewater. Desalination [Internet]. 2010;262(1-3):36–42.
[8] Eriksson E, Auffarth K, Henze M, Ledin A. Characteristics of grey wastewater. Urban Water. 2002;**4(1)**:85–104.
[9] Sun Y, Chen Z, Wu G, Wu Q, Zhang F, Niu Z, et al. Characteristics of water quality of municipal wastewater treatment plants in China: Implications for resources utilization and management. J Clean Prod [Internet]. 2016;**131** pp 1–9.
[10] Miller-Pierce MR, Rhoads N a. The influence of wastewater discharge on water quality in Hawai'i: A comparative study for Lahaina and Kihei, Maui. Mar Pollut Bull [Internet]. 2016;**103**(1-2). pp 54–62
[11] Broaddus A. eScholarship provides open access, scholarly publishing services to the University of California and delivers a dynamic research platform to scholars worldwide. Berkeley Plan J. 2013;26(1) pp 217–20.
[12] Bai S, Srikantaswamy S, Shivakumar D. Urban Wastewater Characteristic and its Management in Urban Areas—A Case Study of Mysore City, Karnataka, India. J Water Resour Prot [Internet]. 2010;02(08) pp 717–26
[13] Mojiri A. Effects of municipal wastewater on physical and chemical properties of saline soil. J Biol Env Sci [Internet]. 2011;5(14) pp 71–6.
[14] Yang L, Shin HS, Hur J. Estimating the concentration and biodegradability of organic matter in 22 wastewater treatment plants using fluorescence excitation emission matrices and parallel factor analysis. Sensors (Basel). 2013;14(1) pp 1771–86.
[15] Ling T. Domestic Wastewater Quality and Pollutant Loadings from Urban Housing Areas. Iran J Energy Environ [Internet]. 2012;3(2) pp 129–33
[16] Hudson K. *Operational Performance of the Anaerobic Baffled Reactor Used to Treat Wastewater from a Peri-Urban Community*. 2010;
[17] Apha/Awwa/Wef. *Standard Methods for the Examination of Water and Wastewater. Stand Methods*. 2012. pp 541.
[18] Shon HK, Vigneswaran S, Kandasamy J, Cho J. *Characteristics of Effluent Organic Matter in Wastewater*. UNESCO - Encycl life Support Syst Water wastewater Treat Technol. 2007 pp 1–17.
[19] Islam B, Musa a, Ibrahim E, Sharafa S, Elfaki B. *Evaluation and Characterization of Tannery Wastewater*. J For Prod Ind [Internet]. 2014 **3(3)** pp 141–50.
[20] Wang C, Yu X, Lv H, Yang J. *Nitrogen and phosphorus removal from municipal wastewater by the green alga Chlorella sp*. J Environ Biol ISSN. 2013; **34** pp 421–5.
[21] Kladitis G, Diamantis N, Grigoropoulou H. Ammonia and Phosphorus Removal in Municipal Wastewater Treatment Plant With Extended Aeration. Int J [Internet]. 1999 **1** pp 47–53.

IOP Conf. Series: Earth and Environmental Science **135** (2018) 012019 doi:10.1088/1755-1315/135/1/012019

Boundaries Delineation of Marine Management Sharing According to Local Government Law No. 23/2014 (Case Study: Surabaya, Sidoarjo, Bangkalan and Sampang)

Khomsin[1] and S Intan Ary Prayogi[2]
[1,2] Geomatics Engineering Department, Faculty of Civil Engineering and Planning, Institut Teknologi Sepuluh Nopember Surabaya, Indonesia

Email : khomsin@geodesy.its.ac.id

Abstract. Regional autonomy is the right, the authority, and the obligation of autonomous region to set up and manage their own affairs and interests of the community in accordance with the potential and peculiarities of each area. To implement regional autonomy, the autonomous region must be clear where the location of its borders. Boundary area is divided into two, namely boundaries in the land and boundaries in the sea. Based on the authority of region government that regulated in Law of Republic Indonesia Number 23 in 2014, the regional maritime boundary consist of maritime management boundary for the province and maritime income sharing boundary for the district/city. This study aimed to determine the maritime income sharing boundary between Surabaya City, Sidoarjo, Bangkalan and Sampang District related to the presence of tanah oloran. Tanah oloran is located in the border of Surabaya City and Sidoarjo district which is currently being disputed border and seizure of property by the two districts/cities. The results of research represent that the claim ownership of Tanah Oloran can impact on maritime income sharing boundaries of Surabaya City and Sidoarjo District with region maritime overlapping is 2,258 ha and will benefit for the region maritime income sharing Sidoarjo District Government.

1. Introduction

According to Local Goverment Law, Indonesian Government delegate to Local Government (Provinces and Cities/Districts) to manage their region. Therefore, Local Government has rigth, authority and obligation to set up and manage their own affairs and interest of local communities. According this law, Local Governments have authorized to organize and to manage governmental affairs and public interests at its own initiative included in region boundary. Region boundary is dividing the territory of the one authority area to another [1][2]. Boundary area is divided into two, namely boundaries in the land and boundaries in the sea.

Referring to Law No. 32/2004 [1] and has been amended by Law No. 23/2014 [2] about Local Government in Indonesia, there ara some fundamental differences, especially on regarding the management of marine areas in their regions. According to Law No. 32/2004 the provincial government has the authority to manage the sea up to 12 nautical miles from shoreline and district/city has the authority to manage the sea until 1/3 of the provincial authority [1]. Meanwhile on Law No. 23/2014, district/city government has not authority to manage marine areas but they just get the sharing income marine managed by province as far as 4 nautical miles [2]. Based on the authority of region government that regulated in this law, the regional maritime boundary consist of maritime management boundary for the province and maritime income sharing boundary for the district/city. Province government has right to manage maritime zone until 12 nautical miles from hig water line coastline. Althoughh distrcit/city does not have right to manage maritime zone, they will get sharing income boundary from province until 4 nautical miles.

In addition, another fundamental difference is the referenced that is used to delineate the maritime boundary area. Based on UNCLOS 1982 Article 5 Section II [3], the baseline of the territorial sea is the low water line along the coast and the baseline used in the Law No. 32/2004 [1] and the Regulation of Minister of Home Affair No. 76/2012 [4] is also refer to the low water line (LWL)) as the same as UNCLOS 1982 reference. On the other hand, Law No. 23/2014 uses high water line (HWL) shoreline as reference to delineate maritime boundary [2]. Therefore, there was a significant impact on the maritime boundary into the regional authority to manage their maritime zones.

This paper will asses the sea boundary impact of 'Tanah Oloran' (land accretion) in Sedati, Sidoarjo which direct to north or to the sea of Surabaya. Tanah Oloran is a new land that is formed naturally by river or beach sediment around the estuary through a long process. Tanah Oloran begins with the sea water abrasion. Due to abrasion, mud and sand were exposed to abrasion drift away towards the river mouth and settles there. From time to time, silt and sand increasing more widely so it appears a new land. The new land lead coastline change due convertion of sea dimention become land dimention. Until now, Tanah Oloran still a disputed territory between Surabaya City and Sidoarjo District [5].

The claim ownership of Tanah Oloran can impact maritime boundary between Surabaya City and Sidoarjo District, and the region in front of it, namely Bangkalan and Sampang district. In order to get a solution solving existing problems, it is necessary to study handling of a maritime boundary between Surabaya, Sidoarjo, Bangkalan and Sampang. In this case refers to Government Law No. 23/2014, since this law was the most recent regulations concerning maritime boundary area. Delimitation of maritime income sharing boundary related solutions for disputes Tanah Oloran by carthometric method, namely the determination of boundary on a map or satellite image and set in the form of a boundary map with a list of coordinate points of boundary.

2. Method

In this paper, we will focus on Tanah Oloran case in Sedati Sidoarjo which impact to maritime boundary between Sidoarjo and Surabaya (adjacent administration) and Sidoarjo- Surabaya and Bangkalan – Sampang as opposite administration. This case has geographical position on 07°11 'LS - 07°22' LS and 112°48 'BT - 113°2' BT (Figure 1)

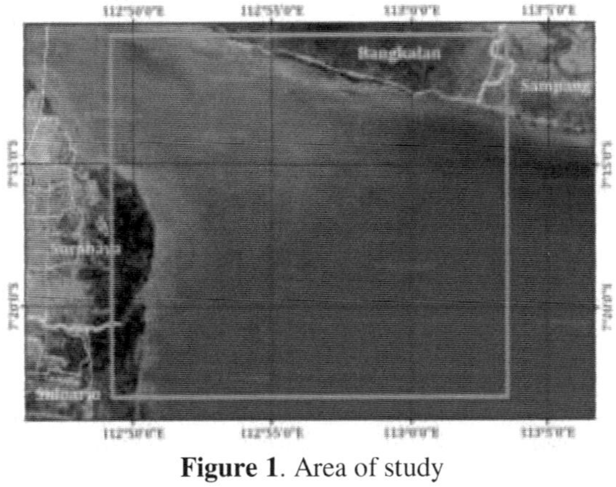

Figure 1. Area of study

Remote sensing technology is growing up until this time. Resolution also continously increased. The result was also getting help in everyday human activities, one of which is to deliminate the shoreline so as to minimize the cost required for the survey / measurements in the field.

Generally, research methodology which used in this paper can be seen at Figure 2 as follow. The data used in this study is high-resolution satellite images SPOT 7 in 2015. SPOT 7 has high resolution to distinguish information that is spatially or spectraly adjacent [6]. So, this image is getting help in everyday human activities. One of the application is to high water line coastline delineation to minimize the cost required for field survey. After that buffering to delineate as far as 4 nautical miles from the coastline [4]. For areas that are overlapping sea territorial claims, conducted border demarcation with equidistance method for adjacent area and median line for adjacent each other [7].

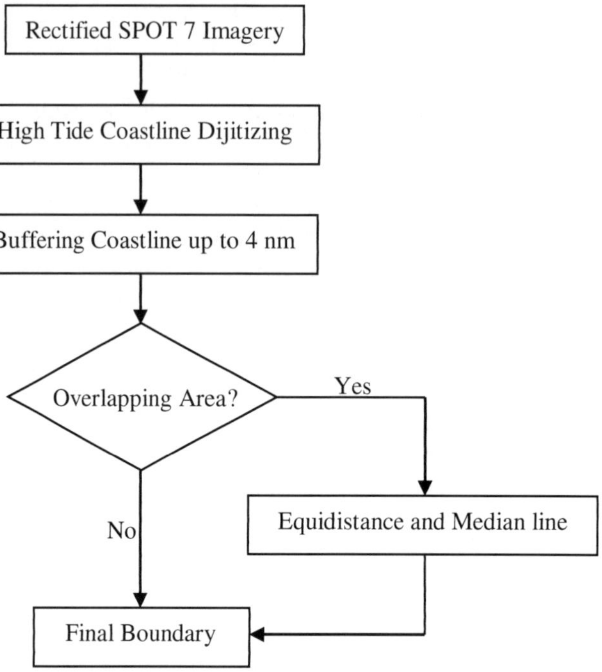

Figure 2. Research Methodology

3. Results And Analysis

3.1 Coastline Digitizing

According to Local Government Law No. 23/2014, the coastline used to determine maritime boundary is meeting boundary between sea and land parts in the event of the highest tides. Coastline at high tide position can be obtained easily through a variety of ways, both directly (field measurement) and Carthometric (digitizing high resolution satellite images). In this case, high water level of costline is conducted with the interpretation of the visual appearance of objects natural color composite or actually because of firm boundaries between sea and land (Figure 3) [8].

IOP Conf. Series: Earth and Environmental Science **135** (2018) 012019 doi:10.1088/1755-1315/135/1/012019

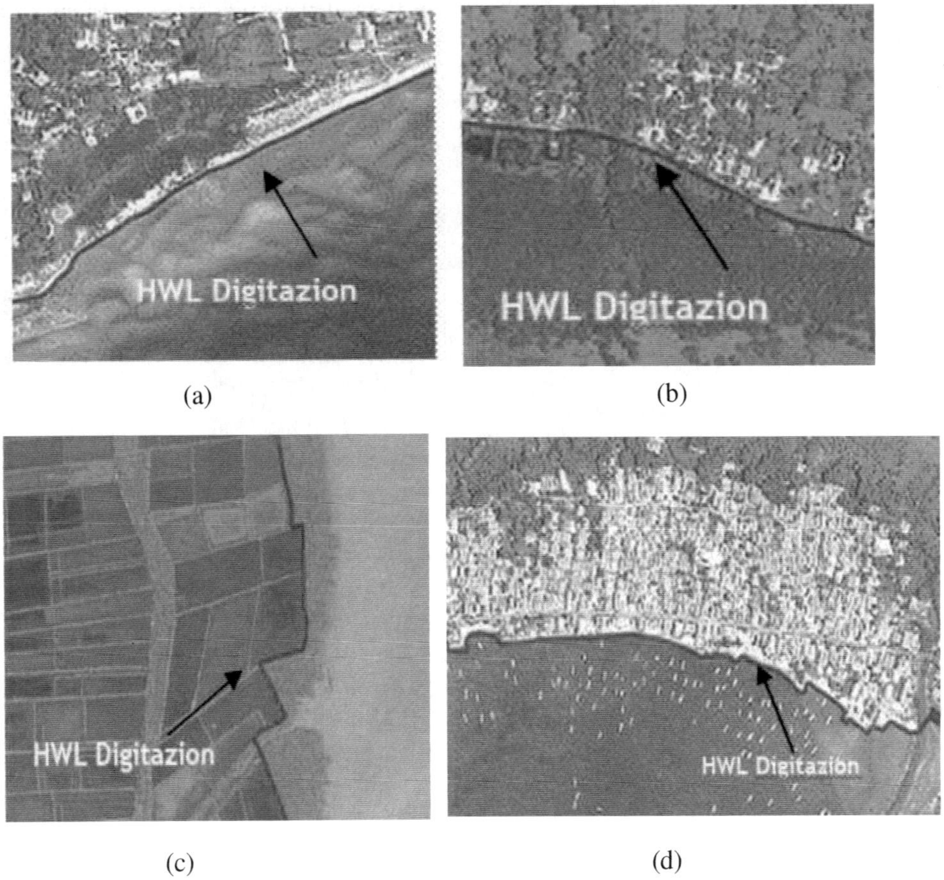

Figure 3. Digitizing of High-Water Level of Coastline at Bangkalan District

On the sand beach area, digitization is done following the trace or former sea on the sand when tide that look from the color of sand which still contains water and does not contain water (Figure 3a). For the area trees beach (mangrove) digitization is done following the outermost trees closest to the land. For the area, the beach trees (mangrove) done digitization in the outermost trees closest to the land. Due to the mangrove is plant species that live in the intertidal area that boundary area between land and sea where tidal influence is still occur. Then mangrove can be flooded during high tide and at low tide inundation free (Figure 3b) [9]. Moreover, the existence of mangrove capable of inhibiting the ocean waves coming from the ocean so it does not directly hit the land (Figure 3b) [10]. An artificial beach areas such as ponds and buildings, done digitization by the line of the outer boundary of a pond or permanent building which is located on the coast or the nearest land. To determine the high-water coastline in ponds area, digitization done following the outer boundary or pond dikes. Because the embankment of a pond created by humans which is designed so that sea water does not overflow to the mainland when the tide (Figure 3c) [8]. For a beach area that many permanent buildings standing man-made, then the digitization done by following the outer limit of the building (Figure 3d).

3.2 Buffering 4 Nautical Miles

District/city get sharing income boundary until 4 nautical miles from high water level coastline. After HWL delineation from SPOT 7 in figure 1, the next step is to buffer HWL shoreline up to 4 nautical miles. Superimposed is marine area for the results showed that the distance between Surabaya and Bangkalan less than 8 nautical miles (Figure 4). So delimitation maritime boundaries for the results on the site is done by the median line (median line) [4]. For adjacent areas such as between Surabaya and

Sidoarjo and between Bangkalan and Sampang, determining maritime boundaries performed with the principle of equal distance (equidistance) [4].

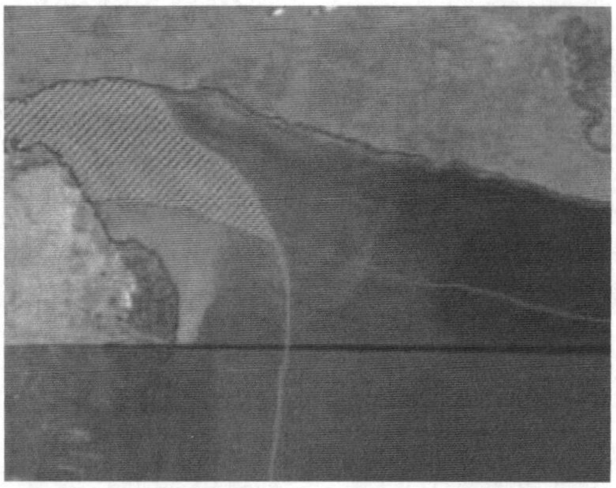

Figure 4. Superimposed maritime boundary from Surabaya-Sidoarjo to Bangkalan-Sampang

In this paper, we focus on Tanah Oloran case in Sedati, Sidoarjo with 2 alternatives. First, if Tanah Oloran is in Surabaya with assuming no land accretion (boundary between Surabaya and Sidoarjo is straight forward from river to sea). Second, if land accretion is in Sidoarjo teritory because this land was certified by land agency Sidoarjo. [5]. Both of these alternatives has different impact to maritim bondaries between Surabaya, Sidoajo, Bangkalan and Sampang (Figure 5). Provincial maritime boundary can be seen at Figure 4 with the dark blue. This area indicated that the distance from HWL coastline more than 4 nautical mile from Surabaya and Sidoarjo and also more than 4 nautical miles from HWL coatline Bangkalan daan Sampang. District/city maritime boundary can also be seen at Figure 4 with light blue and this indicated that the distance less than 4 nautical miles from HWL coastline.

High Water Coastline — Maritime Income Sharing Boundary

Figure 5. Maritime income sharing boundary (Left) If tanah oloran include area of Surabaya (Right) If tanah oloran include area of Sidoarjo District

Tanah Oloran has no impact on maritime management boundary between Bangkalan and Sampang because of the distance more than 8 nautical miles from Tanah Oloran shoreline. If Tanah Oloran is Sidorajo territory, so the impact of the area differences of maritime income sharing between Surabaya City and Sidoarjo District is 2,258 ha (Figure 6).

IOP Conf. Series: Earth and Environmental Science **135** (2018) 012019 doi:10.1088/1755-1315/135/1/012019

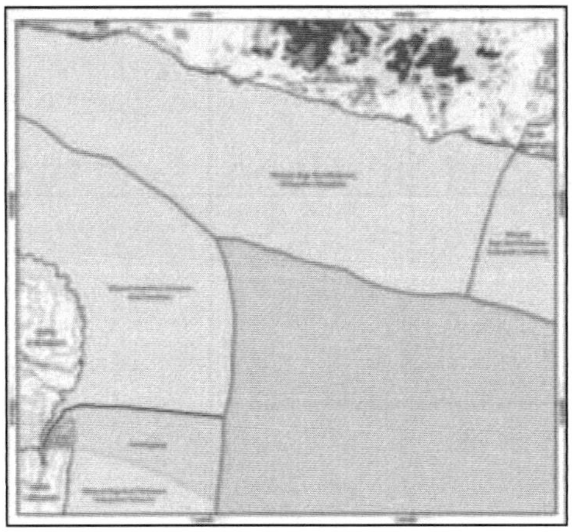

Figure 6. Overlapping of Maritime Income Sharing Boundary

4. Conclusion

There are two alternative solutions the disputes of Tanah Oloran in demarcation of maritime income sharing boundary of Surabaya City, Sidoarjo, Bangkalan and Sampang District, that if land accretion is Sidoarjo territory, so the maritime income sharing boundary for Sidoarjo increased 2,258 ha and otherwise Surabaya decreased as wide as Sidoarjo.

References

[1] Republik Indonesia. 2004. Undang-Undang Republik Indonesia Nomor 32 Tahun 2014 Tentang Pemerintahan Daerah. Lembaran Negara Republik Indonesia Tahun 2004 Nomor 125. Jakarta.

[2] Republik Indonesia. 2014. Undang-Undang Republik Indonesia Nomor 23 Tahun 2014 Tentang Pemerintahan Daerah. Lembaran Negara Republik Indonesia Tahun 2004 Nomor 244. Jakarta.

[3] Windari, R. 2009. Hukum Laut, Zona-zona Maritim Sesuai UNCLOS 1982 dan Konvensi-Konvensi Bidang Maritim. Badan Koordinasi Keamanan Laut Republik Indonesia. Jakarta

[4] Kementerian Dalam Negeri. 2012. Peraturan Menteri Dalam Negeri Republik Indonesia Nomor 76 Tahun 2012 Tentang Pedoman Penegasan Batas Daerah. Lembaran Negara Republik Indonesia Tahun 2004 Nomor 1252. Jakarta.

[5] Harwinda, Z.B. 2015. "Implementasi Peraturan Menteri Dalam Negeri No. 76 Tahun 2012 pada Daerah Kecamatan Sedati Kabupaten Sidoarjo dan Kecamatan Gunung Anyar Kota Surabaya". Surabaya : Jurusan Teknik Geomatika FTSP ITS.

[6] http://www.satimagingcorp.com/satellite-sensors/spot-7 : SPOT-7 Satellite Sensor.

[7] Khomsin, Widiastuti, R., Alif, T F., Artanto, E. 2015. A Determination Analysis of Regional Maritime Boundary Based on Regulation of Home Ministry Affair Number 76 in 2012 (Case Study: Dispute of Galang Island Border between Surabaya and Gresik). Elsevier: Procedia Earth and Planetary Science p 83 -93.

[8] Khomsin, 2017. Analysis of East Surabaya Shoreline Determination Using Tidal Data, Image Satellite and RTK-GNSS. Applied Mechanics and Materials. ISSN. ISSN: 1662-7482, Vol. 862, pp 41-45. Trans Tech Publications, Switzerland

[9] Farhaeni, Mutria. 2016. "Komodifikasi Buah Mangrove Untuk Pemberdayaan Masyarakat Pesisir di Desa Tuban, Kecamatan Kuta, Kabupaten Badung, Bali". STIE BIITM Kuta Badung.

[10] Soerianegara. 1987. "Masalah Penentuan Batas Lebar Jalur Hijau Hutan Mangrove". Jakarta.

Potential of Fishing Port Development in the East Java

N Rosana[1], V D Prasita[2]

[1] Department of Fisheries, Faculty of Engineering and Marine Sciences, Hang Tuah University, Indonesia
[2] Department of Oceanography, Faculty of Engineering and Marine Sciences, Hang Tuah University, Indonesia

Email : nurul.rosana@hangtuah.ac.id

Abstract. The development of fishing ports in East Java is one of efforts to increase the province revenue, develop the fishing industry and to improve the welfare of fishermen. Profile of capture fisheries in East Java should be provide information that can be developed based on priorities and types of resources. The purpose of this study was to determine several districts in East Java which is a priority for the development of fishing ports based on the potential that exists. The method used is descriptive and spatial analysis in order to obtain an overview of the districts in East Java that has the potential to be developed. Results of the study is the districts in East Java are priorities for the development of fishing ports based on aspects related to the fishery is Banyuwangi, Trenggalek and Jember.

1. Introduction

East Java is one of the provinces that have the potential fishery very large and diverse. In managing this potential required facilities and infrastructure that supports the fishery increasing and the welfare of the community, especially fishermen. The level of development of the fishery sector of the southern region of East Java is relatively lower compared to the northern and central regions. The lag is due to the utilization of the existing fishing is not optimal, which are caused by a lack of support infrastructure, lack of data and information on the fishery potential and the lack of fish processing technology [1].

Fishery Port [2] is an area of water that covered / protected and safe enough from the effects of wind and waves, equipped with various facilities such as logistics, fuel, workshops and means of transportation of goodsThe development of fishing ports in East Java is one of efforts to increase the province revenue, develop the fishing industry and to improve the welfare of fishermen. East Java is one of the provinces that have a high potential and diversified fisheries, mainly marine fisheries. Profile of marine fisheries in the northern and southern regions of East Java should be examined in order to provide information about the development of fishing ports in East Java, so it can be developed based on priorities and types of resources. The purpose of this study was to determine several districts in East Java which is a priority for the development of fishing ports based on the potential that exists.

2. Research Methods

This study was conducted in January-August 2016. The method used in this research is descriptive method, the analyzed data is the statistical data of fisheries in East Java. Existing data were analyzed using the average value and percentage. For the value of the utilization rate of the fish used Schaefer analysis. From the analysis conducted spatial analysis using software Arc View by digitization spatially to compare the potential of fisheries in the south and north of East Java, in order to obtain some of the maps are: map coastal districts in East Java, the map center for fish landing, maps fishery production fishing, fisheries production value map, a map number of fishermen, fish utilization rate map, a map of the relative value of production index and map the number of fishing gear. Data fisheries based on the Department of Fisheries of East Java Province [3].

3. Result and Discussion

East Java Province has 19 (nineteen) coastal counties and 1 (one) municipalities bordering the sea, which are spread throughout the north and south. In the northern region there are 11 coastal counties comprising: Tuban, Lamongan, Gresik, Probolinggo, Pasuruan, Situbondo, Sidoarjo, Bangkalan, Sumenep, Sampang, Pamekasan and 1 municipality of Surabaya. In the south there are eight coastal districts consisting of Pacitan, Blitar, Malang, Tulungagung, Trenggalek, Jember, Banyuwangi, Lumajang (figure 1).

Fish landing in Java can be divided into 4 groups: PPN (Nusantara Fishery Port), PPP (Fishery Port Beach), PPI (Fish Landing Centres) and TPI (Fish Auction Place). GIS analysis derived from the distribution of fish landing (PPN, PPP, PPI and TPI) of 27 units in the southern region and 35 units in the northern region. Percentage distribution of PPN, PPP, PPI and TPI in the northern region by 56%, while in the south by 44% (Figure 2).

The development of fishing ports in East Java-based spatial analysis is based on the potential of fisheries in the northern and southern regions of East Java. Fishery parameters that analyzed were : Total Unit of fishing gear, the number of fishermen, the number of fish landings, the number of fishing production, the value of fisheries production, the utilization rate of the fish and the relative index value of production of fishing ports. Analysis was used to obtain images of 1-8 using descriptive statistical average value and percentage, then do digitized using Arc View program and spatially processed.

From the parameters of fisheries above, it can be seen that the conditions in the southern region of East Java lower than in the northern region (Figure 2-8). This condition can be used as a basis for the future development of fishing ports in East Java. The southern region of East Java are left in all relevant aspects of fisheries, should receive priority in the development of infrastructure related to increasing the quantity and quality of the fishing port with a different status (TPI, PPI, PPP and PPN).

When viewed from the geographical position of the southern region of East Java is more steep and hilly, the road infrastructure is very important developed in quantity and quality, so that the distribution of catches of fish landings to consumers more quickly so that consumers get good quality fish. Instead also will facilitate the distribution of related equipment in fishing operations and fish processing companies. Judging from fish potential in southern Java dominant large pelagic fish, would be very open in the development of export opportunities, which will generate local revenues and local fishermen.

Having in mind that the southern region of East Java are the priority areas in the development of fishing ports, we then need to be conducted to determine the three districts in southern Java that have more value to serve as the District prioritized the development of port fishery seen from five aspects of fishing is related to the number of fishermen , the total number of fishing gears, the total production of the catch, the production value of the catch and the amount of fish landings.

Data of capture fisheries in the southern region of East Java is a source of information that can be used to analyze the profile of capture fisheries that can later be used as a reference in developing the fishing port in the south of East Java. The research objective was to determine the profile of capture fisheries in the southern region of East Java as the basis for the development of Fishery Port. The study results were obtained based on the number of fishing unit, production and value of the catch and the number of fishermen, the northern regions still dominate higher value than the southern region of East Java [4].

IOP Conf. Series: Earth and Environmental Science **135** (2018) 012020 doi:10.1088/1755-1315/135/1/012020

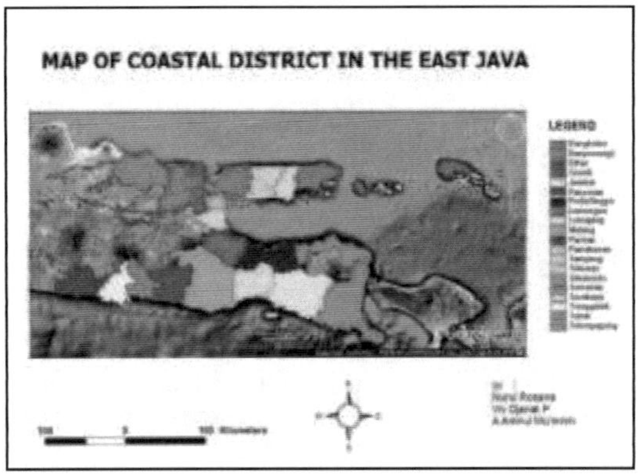

Figure 1. Map of coastal district in the east

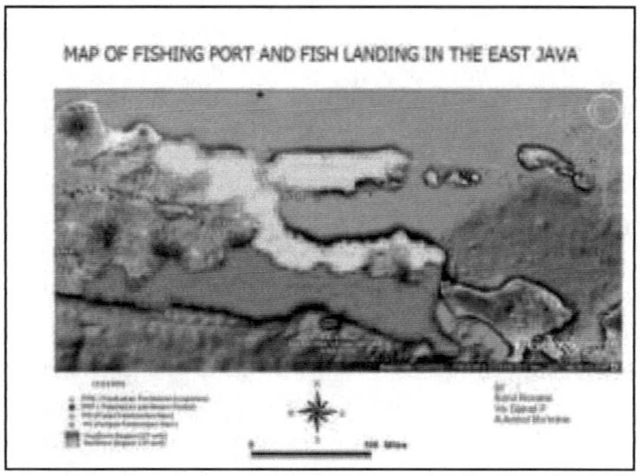

Figure 2. Map of fish landing in east java

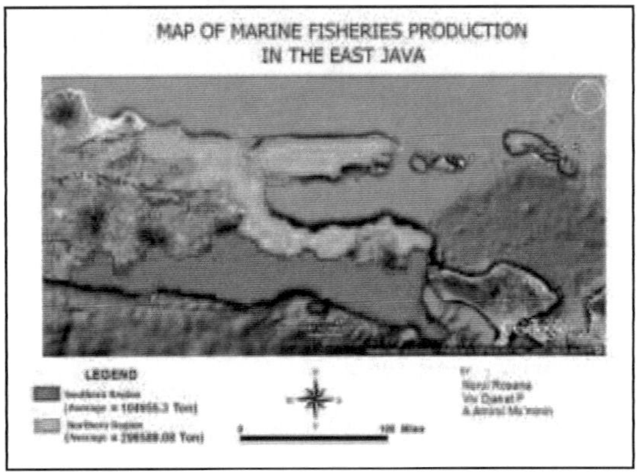

Figure 3. Map of Fisheries Production in East Java

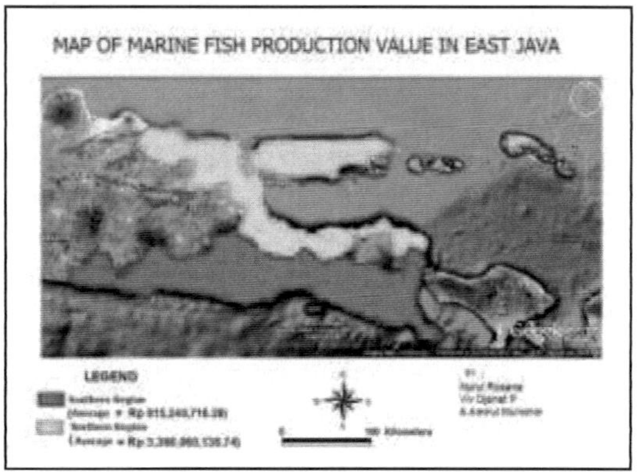

Figure 4. Map of fisheries production value in east iava

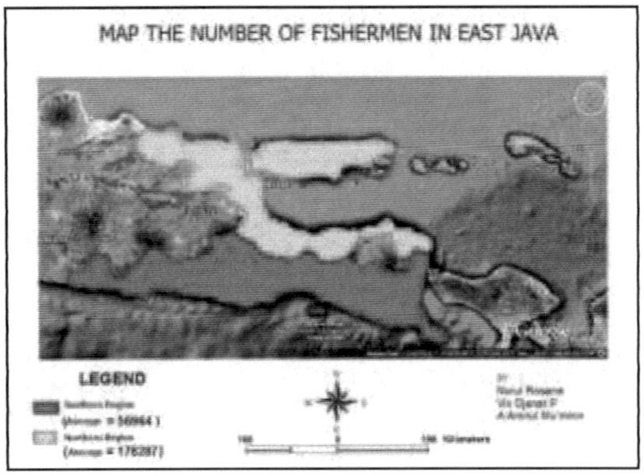

Figure 5. Map of the number of fishermen in east iava

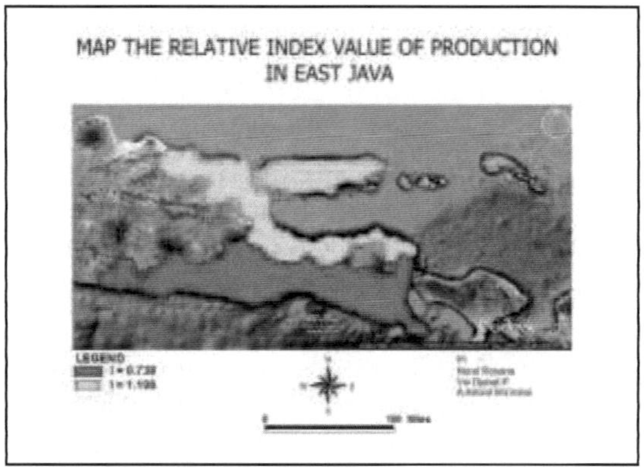

Figure 6. Map of the relative index value of production in east java

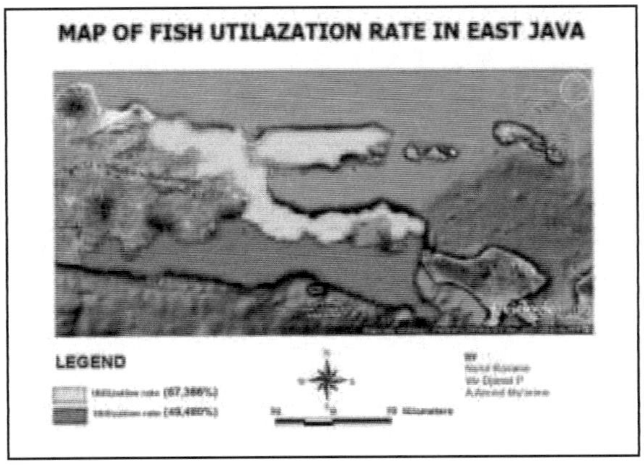

Figure 7. Map of the utilization rate of the fish in

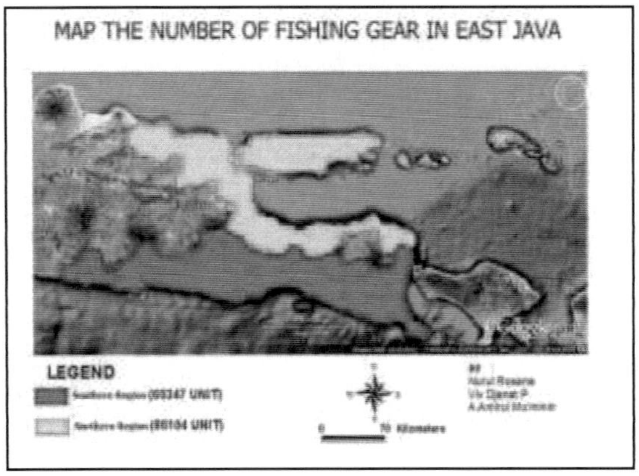

Figure 8. Map of number of fishing gears in east java

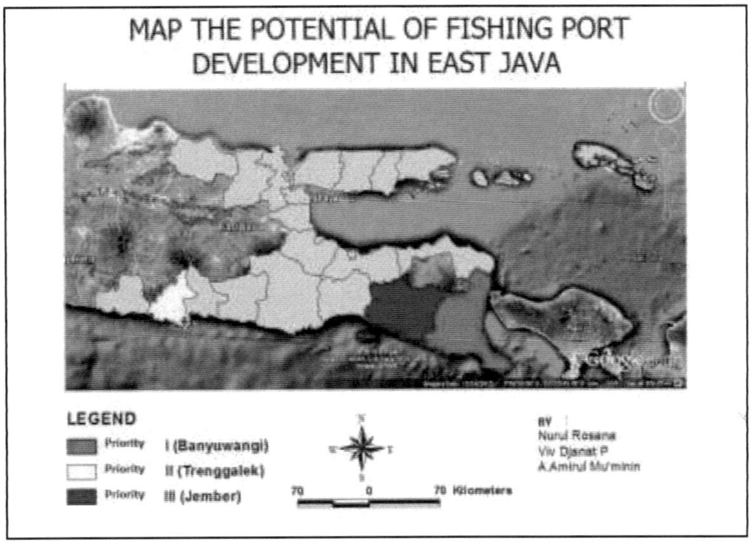

Figure 9. Map the potential district of fishing port development in east java

The Districts in East Java that has the potential for development of fishing ports is Banyuwangi, Trenggalek and Jember. The first rank is Banyuwangi, followed by Trenggalek and last is Jember. These districts have higher potential value in the aspect of capture fisheries compared to other coastal districts in East Java (Figure 9).

4. Conclusion

This research concludes of the study is that the districts in East Java are priorities for the development of fishing ports based on aspects related to the fishery is Banyuwangi, Trenggalek and Jember.

References

[1] Rosana N and Prasita. *Analysis of Potential And Fisheries Resource Utilization Rate In the South, East Java*. Neptune; University of Marine Science Magazine Hang Tuah Surabaya, No. 2, Vol 10, (2004).

[2] Lubis E. *Introduction to Fishing Port. Material Fishery Port*. Laboratory of Fishery Port, Resource Utilization Department of Fisheries, Faculty of Fisheries and Marine Sciences, Bogor Agricultural University, (2000).

[3] Anonymous. *Fisheries Statistics Book of East Java Province*. Department of Fisheries and Marine Resources of East Java Province, (2009-2013)

[4] Rosana N, Prasita and Radinal. *Profile Fishing in North and South East Java as the Basis for Development of Fishery Port*. Proceedings: National Seminar on Marine University Trunojoyo. Pages 104-113. Madura, (2015).

IOP Conf. Series: Earth and Environmental Science **135** (2018) 012021 doi:10.1088/1755-1315/135/1/012021

Longitudinal Pipeline Scour Propagation Induced by Wave-Current Interaction For the South Sumatra-West Java Submarine Pipeline

Suntoyo[1,a], B. Perkasa[1], T.J. Atikasari[1] and A. Wisudawan[1]

[1]Ocean Engineering Department, Faculty of Marine Technology, Institut Teknologi Sepuluh Nopember (ITS), Surabaya, Indonesia (60111)

[a]Email : suntoyo@oe.its.ac.id

Abstract. Scouring process around subsea pipelines could reduce the soil bearing capacity which affected to the pipe stability. Scouring initial process against time should be known to discover scouring propagation. This paper aims to analyze the time scale calculation of 32 inch diameter in-trench pipe, until meet the maximum-scouring-depth stage. Embedment (e/D) variation is given to know the impact to the scour propagation. Wave and current condition presented to meet the real condition. Wave orbital particle velocity (U_w) is calculated to obtain the non-dimensional factors ($U_c/(U_c+U_w)$) and KC. The results showed according to the deeper pipe embedment, it takes longer time to reach the maximum scouring depth.

Keywords: Pipeline Scour Propagation, wave-current interaction, embedment ratio

1. Introduction

One of the important aspects to be considered in designing and planning of submarine pipelines is scouring process phenomenon. Pipeline existence would affect the flow pattern and characteristic, thereby inducing sediment scouring around pipe structure. Scouring process towards the longitudinal (along the pipe) on long run could form free spans. Based on DNV OS-F101 [1], formed free span has a certain safety standard level to avoid the failures such as local buckling, fatigue, and fracture.

The interaction between pipe, seabed, and flow (wave and current) is a complex phenomenon that results in local scour as shown in [2]. The occurrence of scours begin with a small span occurred by constant wave and current force or pipe is held on an uneven surface. Scour occurs when current flows and wave accelerate below pipe so it widen and deepen the free span. Some scouring development researches against current and wave condition has been studied, e.g. [3, 4, 5, 6, 7, 8, 9].

Scouring depth analysis over time needed to monitor the development of scouring depth. There are a few researches which study the time scale scouring propagation. [6] showed that the shield parameter has a role on the scouring process and it shows how long a scouring process occurs where the greater Shields parameter value would affect the smaller time scale. The larger the Shields parameter cause the larger the sediment transport which influence the higher scour propagation. The sediment transport studies under various wave motions has been done by [10, 11, 12, 13, 14]. While [9] mentioned that current gives increasing time scale scouring time scale significant impact when it combines with wave. In this paper, the time scale of longitudinal scouring propagation will be discovered until it reaches the maximum scouring depth. This research conducted on in-trench pipe condition and given three different embedment depths, so it can be known the influence of different embedment depth (e/D) to the time scale of scour propagation on in-trench condition.

IOP Conf. Series: Earth and Environmental Science **135** (2018) 012021 doi:10.1088/1755-1315/135/1/012021

2. Theoretical Background

2.1. Governing Parameters

In this case, pipe obtains force from two sides caused by the bottom flow due to wave and unsteady current flow due to tides. Parameters in scouring calculations which important due to wave and current influences are particle velocity ratio, KC number, and the Shields parameters. Particle velocity ratio could be defined by Equation (1)

$$m = \frac{Uc}{Uc+Uw} \tag{1}$$

where U_c is the current velocity in 100% depth (m/s) and U_w is the velocity of the wavelength particle (m/s).

Wake-pattern formation and extension is determined by KC parameter (Keulegan-Carpenter number). When KC number is small then the orbital motion also small against the total pipe width, if it comes to greater KC number, then affected water particle is farther than the total pipe width. KC is defined by

$$KC = \frac{U_w T_w}{D} \tag{2}$$

where T_w is period of wave (s) and D is the pipe diameter (m).
To express the Shield parameter due to the wave-current interaction given in Equation (3) proposed by [15], as follows

$$\theta cw = (\theta c^2 + \theta w^2 + 2\,\theta w\,\theta c\cos\varphi)^{1/2} \tag{3}$$

where, φ is the angle between current and wave. Where shield parameters of current use the following formula

$$\theta c = \frac{U_f}{g(s-1)d}^{2}, \quad Ufc = \sqrt{\frac{\tau_\infty}{\rho}} \tag{4}$$

U_{fc} is the current friction velocity, and $\tau\infty$ is the shear stress for undisturbed flow. Shield parameters of wave use the following formula

$$\theta_w = \frac{\tau_w}{\rho g\left({\rho_s}/{\rho} - 1\right) d_{50}}^{2} \tag{5}$$

where $\tau_w = \frac{1}{2}\rho f_w U_w{}^2$, $f_w = 0.237 r^{-0.52}$, $r = \frac{U_w T}{2\pi k_s}$, τ_w represents shear stress flow undisturbed under wave action, ks is the nikuradse equivalent sand grain roughness, and f_w is wave friction factor.

2.2. Time Scale Scouring

At a certain period, scour depth will reach the equilibrium condition. The time scale is taken into calculation in this study. The formula used to calculate the time scale is as follows

$$S_t = S\left(1 - \exp\left(-\frac{t}{T}\right)\right) \tag{6}$$

S_t is the equilibrium scouring depth; T is the predicted time of the scour depth and the time it takes (usually calculated with a slash drawn from t = 0). Time scale on steady currents calculated by Equation (7) and time scale on wave conditions defined by Equation (8)

IOP Conf. Series: Earth and Environmental Science **135** (2018) 012021 doi:10.1088/1755-1315/135/1/012021

$$T^* = \frac{(g(s-1)d^3)^{1/2}}{D^2} T \tag{7}$$

$$T^* = \frac{1}{50}\theta^{-5/3}. \tag{8}$$

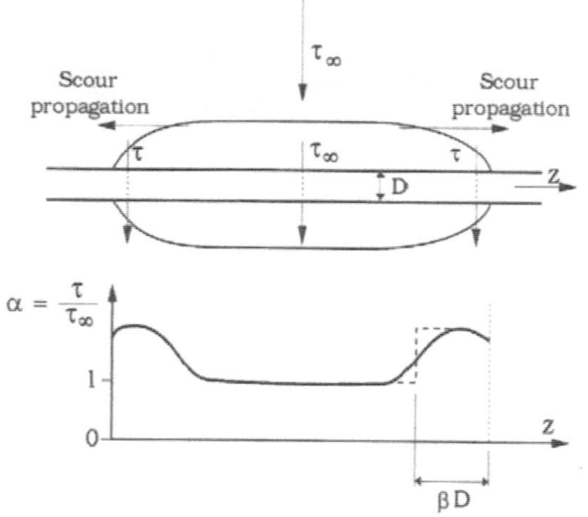

Figure 1. Skecth of the scour propagates along the length pipeline [6]

2.3. Scouring Propagation Rate along the Pipe

In two-dimensional modeling, scoured part is centered until it reaches the equilibrium. So that bed shear stress is equal to the undisturbed value τ_∞. However, bed shear stress increase as it moves at the end of span which occurs along βD as shown in Figure 1. Occurred erosion on the span angle is determined by the difference of sediment transport that come out from the corner and which come in to the corner area. The erosion rate determined by sediment properties, bed shear stress, pipe embedment, and α - β coefficients. However, by assuming β=1 then erosion volume rate can be calculated by this following formula

$$c = \frac{q_{co} - q_o}{e(1-n)} \tag{9}$$

Where q_{co} as sediment transport in support free span area (corner), q_o is transport sediment in flat bed area, e is embedment, and n defined as porosity. While q_o can be determined using Equation (10), as follows

$$qb = 8\sqrt{(s-1)gd_{50}}(\theta - 0.047)^{3/2}. \tag{10}$$

2.4. Study Area and Environmental Data

The location of South Sumatera – West Java Gas Pipeline (SSWJ) offshore pipeline used as a case study in the present paper given in Figure 2. The SSWJ pipeline gas transmission transport natural gas from the Grissik gas production field in South Sumatra Province to Rawa Maju in West Java Province to meet the rising energy demand in West Java. The capacity of the Project pipeline is 460 million standard cubic feet per day [16]. The water depth, soil type, wave and current data for Zone 17 and 18 with Kilometer Points (KP) from 139 up to 162 were used to evaluate the longitudinal pipeline scour propagation given in Table 1, Table 2 and Table 3, respectively. In which, it has the pipeline diameter (D) 32 inch lyaing on seabed and in-trench condition. While the embedment depth set on 15% of pipe diameter (e=-0.15D).

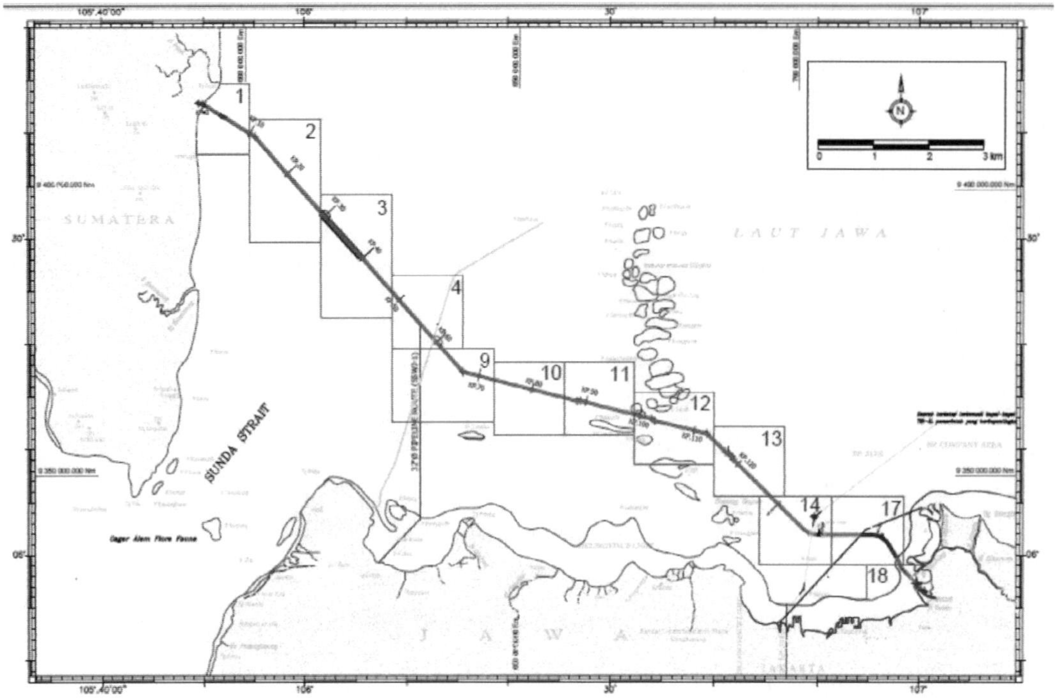

Figure 2. Offshore Pipeline Route of South Sumatra-West Java (SSWJ) [16]

Table 1. The Water Depth in Zones 17 and 18 for KP 139.6 to 161.3

Zones	KP from	KP to	min depth (m)	max depth (m)
17	139.6	155.7	5.80	21.80
18	155.7	161.3	0.00	5.80

Table 2. Soil Type for KP 149.6 to 161.3

KP Range		Soil Type	Bulk Density (kg/m3)	Zone
From	To			
149.6	161.3	Sand	1258	17-18

Table 3. Wave and Current Data for Zone 17 and Zone 18

Zone	Z17	Z18
Wave		
Hs (m)	3.8	3.46
Ts (sec)	7.88	7.52
Current		
0% of depth	1.28	1.42
50% of depth	0.69	0.83
100% of depth	0.38	0.52

3. Result And Disscussion

3.1. Time Scale Propagation of Pipeline Scour

The modelling results of time scale propagation pipeline scour at zone 17 and 18 for the pipeline diameter(D = 32 inch) with variation of the embedment i.e. e=-0.05 D, e=-0.15 and e=-0.25 given in Table 4 and Figure 3. It can be shown that the maximum scour depth (S_{max}) and the time (t_{max}) needed to reach the maximum scour depth increase with the increase the water depth (d) and the embedment. For all the variation of the embedment e=-0.05 D, e=-0.15 and e=-0.25 when the water depth (d) = 15.30 m was obtained t_{max} and S_{max} higher than others water depth as shown in Figure 3. The result shows SmaxTmax for e=-0.25D slightly longer than Tmax for e= -0.15D & e= -0.05D.

Table 4. Time scale propagtion of pipeline scour for Zone 17 and Zone 18 with the pipeline diameter(D) = 32 inch in variation of the embedment value (e=-0.05 D, e=-0.15 and e=-0.25)

No	d (m)	$U_c/(U_c+U_w)$	e = -0.05D		e = -0.15D		e = -0.25D	
			T_{max} (h)	S_{Max} (m)	T_{max} (h)	S_{Max} (m)	T_{max} (h)	S_{Max} (m)
1	7	0.15	5.44	0.208	5.49	0.22	5.55	0.234
2	8.66	0.183	4.86	0.216	4.9	0.229	4.95	0.243
3	10.32	0.2	4.9	0.212	4.95	0.226	5	0.239
4	11.98	0.211	5.3	0.205	5.35	0.217	5.41	0.231
5	13.64	0.219	5.93	0.195	6.01	0.208	6.07	0.22
6	15.3	0.225	6.85	0.186	6.92	0.197	6.98	0.209

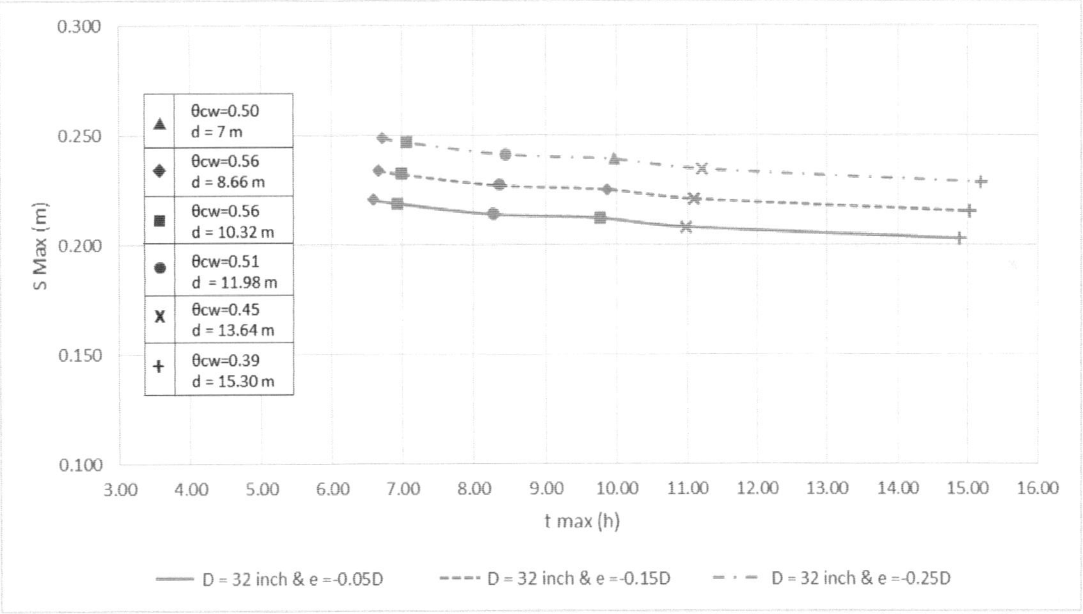

Figure 3. Time scale propagtion of pipeline scour for Zone 17 and Zone 18 with the pipeline diameter(D) = 32 inch in variation of the embedment value (e=-0.05 D, e=-0.15 and e=-0.25)

Figure 3 showed how the scouring propagation process against time starts from the initial time until it reaches the maximum scouring time. It can also be seen that the steep slope of the graph is a time scale value (T) which is a progressive scouring process (63.14% of maximum scouring depth). The deepest depth (15.30 m) has the longest time of maximum scouring as well as the larger the embedment (e) values lead to a slightly longer scouring time.

3.2. Longitudinal Scour Propagation

The longitudinal scour propagation was simulated based on 2D numerical model proposed by [6] in which the propagation along the pipeline occurs after equilibrium depth scouring (S_{Max}) was reached. The results showed that the increasing of embedment depth, scouring propagation rate goes slower. The longitudinal scouring rate is determined by sediment property, bed shear stress (τ_∞) and the Shileds parameter (θcw) induced by wave-current interaction on flat bed, the embedment depth (e), and α - β coefficient. According to [6] shear stress amplification (α) decreases by the increasing embedment. The value of α is obtained by put it into the graph so that the result is e=-0.05D for α=2, e=0.15D for α=1.72, and e=-0.25D for α=1.58. So it can be concluded for the greater value of embedment condition will cause the rate of propagation scouring takes longer. The results obtained have been in accordance with research that has been done by [6,17].

Table 5. Longitudinal scour propagation for Zone 17 and Zone 18 with the pipeline diameter(D) = 32 inch in variation of the embedment value (e=-0.05 D, e=-0.15 and e=-0.25)

No	d (m)	θ_{cw}	e = -0.05D		e = -0.15D		e = -0.25D	
			θ_{co}	D=32inch c (m/h)	θco	D=32inch c (m/h)	θco	D=32inch c (m/h)
1	7	0.28	0.562	4.45	0.484	1.01	0.444	0.48
2	8.66	0.42	0.834	8.2	0.717	1.87	0.659	0.88
3	10.32	0.4	0.794	7.61	0.683	1.73	0.627	0.81
4	11.98	0.33	0.663	5.76	0.57	1.31	0.524	0.62
5	13.64	0.25	0.501	3.71	0.431	0.84	0.396	0.4
6	15.3	0.18	0.37	2.29	0.318	0.52	0.318	0.24

The slowest propagation rate exists at 15.30 m and it gradually increases to 8.66 m, while at water depth (d) = 7.00 m it goes back to decrease. This caused by the larger Shields parameter (θ_{cw}) indicating that the propagation rate is getting bigger. The results also show that greater embedment values gave smaller propagation rate values.

As the result it can be discovered that Shields parameter (θ_{cw}) becomes the determinant of longitudinal pipeline scour propagation characteristics. Shield parameters are representative of the shear stress and flow velocity. So the greater value of shield parameters, obtained greater stress and its velocity.

4. Conclusion

The numerical modelling of the longitudinal pipeline scour propagation induced by wave-current interaction for the South Sumatra-West Java Submarine Pipeline (SSWJ) has been done, It can be concluded that the Shields parameter (θ_{cw}) becomes the determinant of longitudinal pipeline scour propagation characteristic as representative of the shear stress and flow velocity. So the greater value of shield parameters, obtained greater stress and its velocity. The Shields parameters have an important role in determining propagation scouring where the greater the value of the Shields parameters obtain the faster of longitudinal scouring propagation. And the pipeline scour propagation decreases (decelerates) as the increases of the embedment ratio (e/D).

Acknowledgements
Authors would like to thank the LPPM-ITS and DRPM-RI for suporting. This research partially suported by grant of DRPM Ministry of Research, Technology and Higher Education of the Republic of Indonesia (No 135/SP2H/LT/DRPM/IV/ 2017) related with the research grants of PUPT, 2017.

References

[1] DNV-OS-F101. 2012. *Submarine Pipeline System*. Det Norske Veritas AS.
[2] Zhao, Z., Fernando, H. J. S. 2007. Numerical simulation of scour around pipelines using an Euler-Euler coupled two-phase model. *Environmental Fluid Mechanics*, 7(2), 121-142.
[3] Wang, R.-K., and Herbich, J. B. 1983. *Combined current and waveproduced scour around a single pile*. CEO Rep. No. 269, Texas Engrg. Experiment Station, Dept. Civ. Engrg., Texas University System.
[4] Eadie, R. W., and Herbich, J. B. 1986. Scour about a single, cylindrical pile due to combined random waves and current. *Proc., 20th Coast. Eng. Conf.*, 1858–1870.
[5] Sumer, B. M., Fredsøe, J., 2001. Scour around a pile in combined waves and current. *Journal of Hydraulic Engineering, ASCE*, Vol. 127, No. 5 403-411.
[6] Sumer, B. M., and Jorgen, Fredsoe. 2002. *The Mechanics Of Scour In The Marine Environment*. World Scientific : Denmark.
[7] Raaijmakers, T. and Rudolph, D. 2008. Time-dependent scour development under combined current and wave conditions – laboratory experiments with online monitoring technique. *Proceedings of the Fourth International Conference on Scour and Erosion*. Tokyo, 152-161.
[8] Myrhaug, D. and MC Ong. 2009. Burial and Scour of Short Cylinders under Combined Random Wave and Currents including Effect of Second Order Wave Asymmetry. *Elsevier: Journal of Coastal Engineering*. 56 (1), 73-81.
[9] Petersen, Thor Ugelvig, et al. 2012. Time Scale of Scour around a Pile in Combined Waves and Current. *Proc., 6th International Conf. on Scour and Erosion*. Paris. ICSE6-043.
[10] Suntoyo, Tanaka,H., and Sana, A. 2008. Characteristics of turbulent boundary layers over a rough bed under saw-tooth waves and its application to sediment transport. *Coastal Engineering*, 55 (12), 1102-1112.
[11] Suntoyo, and Tanaka, H. 2009. Effect of bed roughness on turbulent boundary layer and net sediment transport under asymmetric waves. *Coastal Engineering*, 56 (9), 960-969.
[12] Suntoyo, Fattah, A.H., Fahmi, M.Y., Rachman, T. and Tanaka, H. 2016. Bottom shear stress and bed load sediment transport due to irregular wave motion. *ARPN Journal of Engineering and Applied Sciences*, 11 (2), 825-829.
[13] Wijaya, M.M., Suntoyo, and Damerianne, H.A., 2016. Bottom shear stress and bed load sediment transport formula for modeling the morphological change in the canal water intake. *ARPN Journal of Engineering and Applied Sciences*, 11 (4), 2723-2728.
[14] Tanaka, H., Suntoyo and Nagasawa, T. 2003. Sediment intrusion into Gamo lagoon by wave overtopping. *Proceedings of the 28th International Conference*, pp. 823-835.
[15] Camenen, B and Larson, M. 2004. A General Formula For Non-Cohesive Bed Load Sediment Transport. *Estuarine, Coastal And Shelf Science 63 (2005) 249-260* DOI : 10.1016/j.ecss.2004.10.019.
[16] PT. Perusahaan Gas Negara (Persero) Tbk. 2005. Design Basis Manual Pipeline Offshore. South Sumatera – West Java Gas Pipeline Project Phase II Labuhan Maringgai – Muara Bekasi Offshore Pipeline, pp. 22.
[17] Suntoyo, Rahayu, N.D., Wisudawan, A. and Ikhwani, H. 2017. Experimental Study of Pipeline Scouring on Seabed and In-Trench Conditions under Regular Wave Motion. International Journal of Civil Engineering and Technology, 8(10), 659-666.

Mooring Design Selection of Aquaculture Cage for Indonesian Ocean

Y Mulyadi[1,a], N Syahroni[1], K Sambodho[1], M Zikra[1], Wahyudi[1] and H B P Adia[1]

[1]Ocean Engineering Department, Faculty of Marine Technology, Sepuluh Nopember Institute of Technology, Indonesia

Email : yeyesmulyadi@gmail.com

Abstract. Fish production is important for the economy in fishing community and for ensuring food security. Climate change will lead a threat to fish productivity. Therefore, a solution offered is to cultivate certain fish, especially those with high economic value by using offshore aquaculture technology. A Sea Station cage is one of the offshore aquaculture cage model that has been used in some locations. As a floating structure, the Sea Station cage need a mooring system to maintain its position. This paper presents the selection analysis of the mooring system designs of the Sea Station cage model that it is suitable with Indonesia Ocean. There are 3 mooring configurations that are linear array, rectangular array, and 4 points mooring type. The nylon mooring rope type has been selected to be used on the 3 mooring configurations and the rope has a diameter of 104 mm with a breaking force of 2.3 MN. Based on results from comparing the 3 mooring configurations, the best mooring configuration is linear array with the tension on the rope of 217 KN and has the safety factor of 0.2 based on DNVGL OS-E301

1. Introduction

Fish production from aquaculture technique Indonesia is 2% of total fish supply worldwide. Based on climate zones, Indonesia occupies the first position of the largest potential of marine products from aquaculture.Vietnam as a fellow Southeast Asian country with a smaller potential of marine products, has twice the production of larger than Indonesia [1]. Therefore, the development of offshore aquaculture is very important to be implemented in Indonesia.

One form that has been developed in the United States is the Sea Station cage. The Sea Station system has also been developed by Ocean Spar Technologies[2]. The structure of this building consists of a spar and a wire connected with a steel cable at 6, 8.10 or 12 sides. This structure can be placed half-afloat in the waters with a quiet environment. In the ferocious sea conditions this structure is drowned entirely about 15 meters below sea level. This structure has been used more than 50 units in the world. The Sea Station cage has a motion response due to the influence of the environmental load. The design of mooring system for offshore aquaculture is as important part of the main system because environment condition in open sea quite strong. So that this research aims to analysis and selection of the mooring system for the Sea Station cage. There are 3 mooring configurations that is analyzed in this research: a) linear array, b) rectangular array, and c) 4 points mooring type

2. Basic Theory

2.1 Aquaculture

Aquaculture is the cultivation of marine biota in a system that is controlled by humans. Ecosystems in aquaculture are made in such a way as to resemble the native habitat of marine biota cultivated. Aquaculture consists of several methods one of which is the open-net method. Offshore aquaculture offers many environmental benefits, and relative to other forms of animal protein production, the offshore aquaculture is an attractive option for expanding production [8]. The aquaculture structures must be integrated to accommodate a variety of needs such as automated food systems, protective nets from predators, and a dividing system for dead fish [3]. Besides, selection of the aquaculture cage location must be quite far from the shipping lanes and other structure in ocean. For example, the aquaculture cage cannot be installed in the Madura Strait is one of the busiest shipping lanes in Indonesia and many subsea pipeline has been installed in the Madura Strait [11]

Open-net method or open floating net is indeed still controversial because marine biota are cultivated directly interact with the environment. In some cases residues from aquaculture can contaminate the sea especially in closed waters but applications for high seas have a great potential for food security from marine resources.

2.2 Sea Station Cage

Sea Station cage is a offshore aquaculture structure that adopts the Single Point Anchor Reservoir (SPAR) offshore building [2]. The center of this structure is small SPAR and is connected to a ream using a steel cable on its side. This structure includes a simple and economical form. The structure of this model has advantages of relatively cheap construction costs. This form proves to be capable of being either submerged, or semi submerged. The source condition is usually applied to deep waters to reduce malignant ocean wave energy. Illustration of the structure model can be seen in Figure 1.

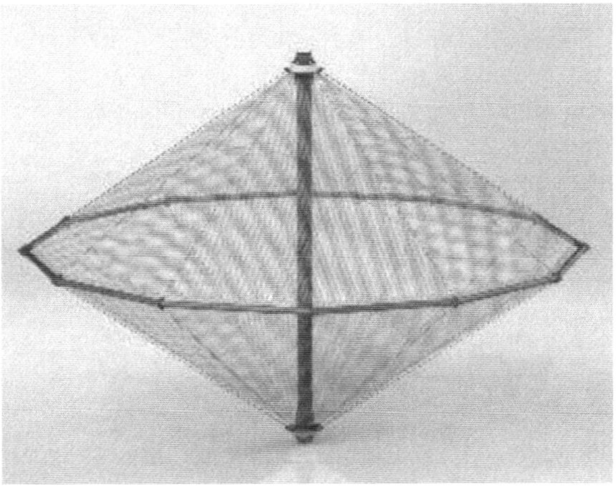

Figure 1. Sea Station Cage [2]

2.3 Aquaculture Mooring Configuration

With the development of aquaculture technology into the deep sea, the practice of mooring system commonly used on aquaculture technology is also growing. The mooring system configuration should be stable and reliable for a variety of offshore aquaculture structures. This mooring system also facilitates the installation of feeder systems and harvesting systems [4]. Analysis of the aquaculture is certainly not independent of the mooring system. Offshore aquaculture structures must take into account the effects of ocean waves and currents. Based on Finite Element Analysis (FEA) analysis, the nonlinear component of current and wave has an effect of up to 13% [9].

Here are some commonly mooring configurations that are the 3 mooring configurations: a) linear array, b) rectangular array, and c) 4 points mooring type. The mooring configuration model as illustrated in Figures 2, 3 and 4.

a) Rectangular Array Type

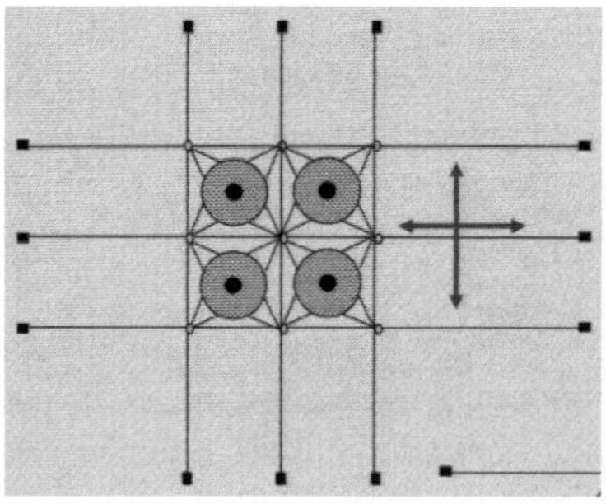

Figure 2. Rectangular array mooring configuration

b) Linear Array Type

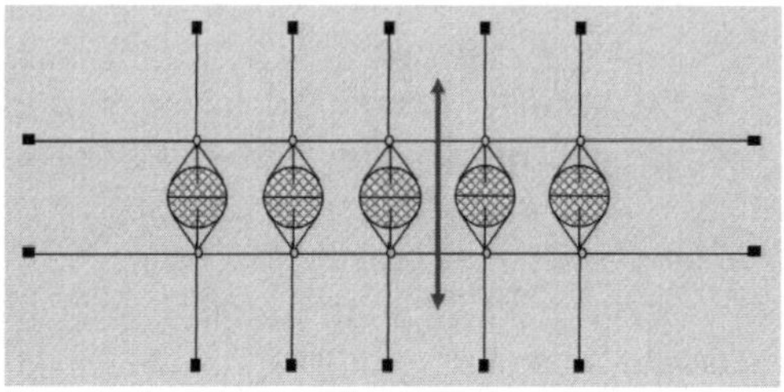

Figure 3. Linear array mooring configuration

The characteristics of this type of mooring configuration are as follows:
a) High efficiency oh multi-cages installation
b) Tolerance towards current from different direction
c) Local failure will have major effect on mooring grid system integrity
d) High risk of disease spreading in fish cage and fish residue concentration

c) Independent 4 Points Mooring Type

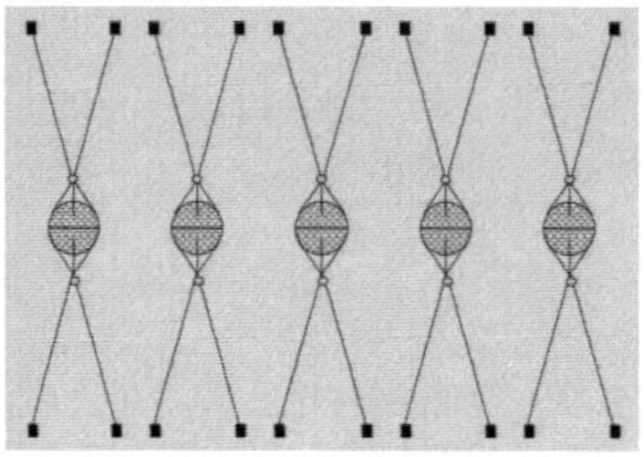

Figure 4. Independent 4 Points Mooring Type

Main characteristics of this mooring configuration present as follows:
a) High efficiency when two-directional current occur towards the mooring system
b) Minor integrity effect when failure occur in single cage
c) One of the most expensive mooring configuration in installation

2.4 JONSWAP Wave Spectrum
The wave spectrum used in this final project is the JONSWAP spectrum. The JONSWAP spectrum equation is a modification of the Pierson-Morkowitz spectrum equation that is adapted to existing

marine conditions. The JONSWAP spectrum equation according to Djatmiko (2012) [6] can be written according to Equation (1).

$$S(\omega) = \alpha g^2 \omega^{-5} \exp\left[-1.25\left(\frac{\omega}{\omega_0}\right)^{-4}\right] \gamma^{\exp\left[\frac{(\omega-\omega_0)^2}{2\tau\omega_0^2}\right]} \quad (1)$$

where:

α = $0.0076\ (X_0)^{-0.22}$

X_o = gX/U_w^2

X = fetch length (m)

α = fetch , 0.0081 if the fetch length is unknown

γ = peakedness parameter, the value vary from 1.0 up to 7.0.

τ = shape parameter

τ = 0.07 for $\omega \le \omega_0$

τ = 0.09 for $\omega \ge \omega_0$

The following is a spectrum JONSWAP derivative equation that uses significant wave height input and period. The equation can be seen in Equation (2) [5].

$$S(\omega) = 155\left(\frac{X_y^2}{T^4\omega^4}\right)\exp\left(\frac{-944}{T^4\omega^4}\right)\gamma^{\alpha} \quad (2)$$

2.5 Response Amplitude Operator (RAO)

Response-Amplitude Operator (RAO) is a response function that occurs due to waves in the frequency range of the structure. The result of RAO is used to estimate the motion response of structure due to wave load [6]. The formulation of RAO for translation motion i.e. surge, sway, and heave present as equation (3):

$$RAO = \frac{\zeta_{k0}}{\zeta_0}\left(m/m\right) \quad (3)$$

The RAO for rotation motions is the ratio between the amplitude of the rotational motion and the wave inclination. The formulation of RAO for rotational motion i.e. roll, pitch, and yaw present as equation (4) [6]:

$$RAO = \frac{\zeta_{k0}}{k_w\zeta_0} = \frac{\zeta_{k0}}{(\omega^2/g)\zeta_0}\left(rad/rad\right) \quad (4)$$

Where:

g : acceleration of gravity

ω : wave frequencies

ζ_{k0} : amplitude of structure motion

ζ_0 : amplitude of wave motion

2.6 Irregular Wave Motion Response

A response spectrum of floating structure is the result of interaction between RAO with wave spectra. A response spectrum is simply a plot of the peak or steady-state response of a series of oscillators of varying natural frequency. The resulting plot can then be used to pick off the response of any linear system, given its natural frequency of oscillation. The formulation of response spectrum of floating structure in a irregular wave presents as equation (5)[6]

$$S_{\zeta r}(\omega) = RAO^2 \cdot S_{\zeta}(\omega) \quad (5)$$

Where: $S_{\zeta r}(\omega)$: Spectra of the response and $S_{\zeta}(\omega)$: Wave spectrum

IOP Conf. Series: Earth and Environmental Science **135** (2018) 012022 doi:10.1088/1755-1315/135/1/012022

2.7 Mooring Tension

The environmental loads acting on the structure cause tension to mooring line. There is relationship between the mooring offset and the mooring tension. The mooring tension can be estimated using numerical modeling. Regarding with recommendations of DNV GL OS-E301, then the characteristic Strength, S_C, is estimated from the minimum breaking strength Smbs of new components, as equation (6):

$$S_C = 0.95 \text{ Smbs} \tag{6}$$

The mooring lines are considered to be intact in the analysis of the Ultimate Limit State (ULS) [7]. The design equation for the ULS is given by equation (7) [7]

$$S_c - T_{C-mean}\gamma_{mean} - T_{C-dyn}\gamma_{dyn} \geq 0 \tag{7}$$

The design equation of utilization factor, U, presents as equation (8) based on the recommendations of DNV GL OS-E301 [7]. The utilization factor, U, of the mooring system is the comparison between the actual mooring line tension with the mooring line capacity.

$$U = \frac{T_{c-mean} \cdot \gamma_{mean} + T_{c-dyn} \cdot \gamma_{dyn}}{S_c} \quad \text{where } U \leq 1 \tag{8}$$

Where

T_{C-mean} : Mean tension of the time series
T_{C-dyn} : TMPM - T_{C-mean}
T_{MPM} : Most probable max of the time series
S_C : Characteristic Strength of mooring line
γ_{mean} : Partial safety factor on mean tension
γ_{dyn} : Partial safety factor on dynamic tension

The partial of the safety factor for mean and dynamic tension can be seen in Table 1 [7].

Table 1. The partial of safety factor for ULS criteria [7]

Consequence Class	*Type of analysis of wave frequency tension*	*Partial Safety factor on mean tension* γ_{mean}	*Partial Safety factor on dynamic tension* γ_{dyn}
1	Dynamic	1.10	1.50
2	Dynamic	1.40	2.10
1	Quasi-static	1.70	
2	Quasi-static	2.50	

3. Data Collection

3.1 Environmental data

In this research, the environmental data uses Timor Sea data. The wave characteristics used in the analysis such as Table 2.

Table 2. Environmental data of Timor Sea [1]

Data of wave (100 years)		
Wave height	Hs =	4,8 m
	Tp =	11,5 s
	Tz =	8,3 s
Wave velocity (10 years)		
Current		1,1 m/s

3.2 Structure data

The structure data of Sea Station Cage is presented as in Table 3. As important as the structural weight, there is also support system weight that contributes to the cumulative weight as shown in Table 4.

Table 3. Structural dimension and weight of Sea Station

Dimension of Sea Station Structure					
Components	Length (m)	Inner diameter (m)	Outer diameter (m)	Number	Weight (ton)
SPAR	11.06	0.98	1.00	1	3.45
Rim Section	6.53	0.39	0.40	8	2.45
Sling	9.32	-	0.02	16	2.38
Flange	0.05	1.00	1.50	2	0.36
				Total of weight	8.65

Table 4. Support system of Sea Station cage

Items	Weight (ton)
Fish feeding	2.00
Live load (workers)	0.20
Net of cage	0.30

3.3 Mooring Line Sizing

The mooring line uses the type of the nylon mooring rope. The nylon rope characteristics used in the analysis such as Table 5[10].

Table 5. Properties of mooring line [10]

Diameter (mm)	Mass in air (kg/m)	MBL in wet (kN)
104	6.7	2310

4. Analysis

4.1 Structural Modeling

Modeling of structure is carried out using numerical modeling software. The modeling of the net is performed by simplified model with considering the same weight as the actual net weight. The modeling of structure and net is presented as Figure 5.

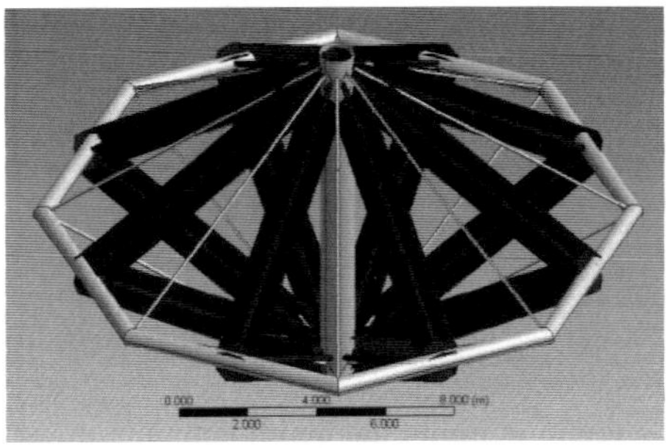

Figure 5. Structural modeling

4.2 Mooring Configuration Modeling

The mooring line configuration analysis is performed using numerical modeling software. This research uses some commonly mooring configurations that are the 3 mooring configurations: a) linear array, b) rectangular array, and c) 4 points mooring type. The rectangular array configuration is modeled as shown in Figure 6. The linear array configuration is modeled as shown in Figure 7. The model with 4 points mooring configuration is done as in Figure 8.

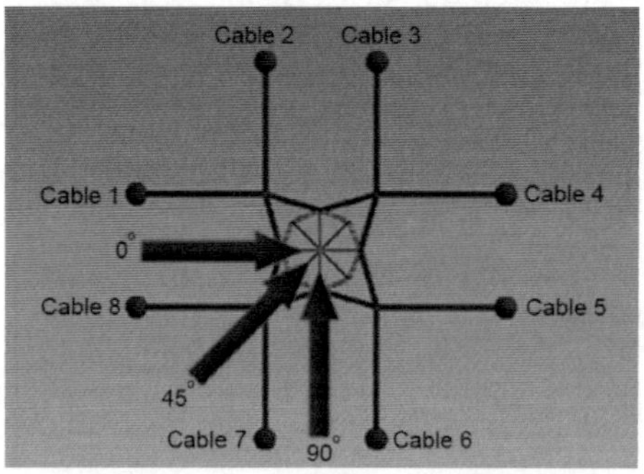

Figure 6. Rectangular array mooring configuration modeling

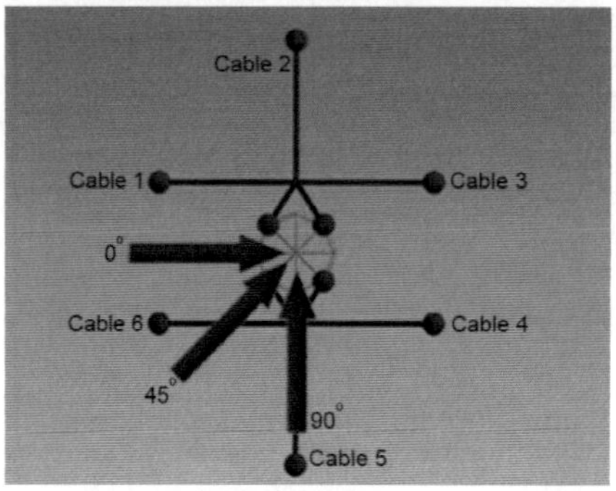

Figure7. Linear array mooring configuration modeling

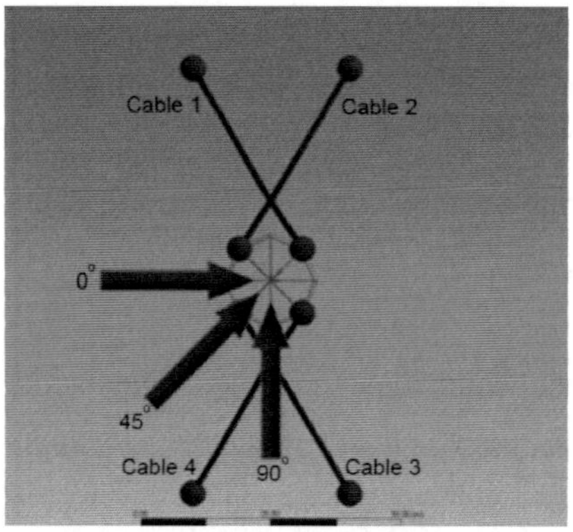

Figure 8. The 4 points mooring configuration modeling

4.3 Hydrodynamic Response

The response spectrum of a floating structure is the result of interaction between RAO with wave spectra. The response spectrum of the sea station cage moored is estimated by multiplication of the RAO quadrated with irregular wave spectrum. Here, the wave spectrum uses JONSWAP Spectrum. The analysis of the response spectrum of the sea station cage is carried out by numerical modeling.

The result of the response spectrum of the rectangular array mooring configuration in translation motion for wave direction 0 degree is presented as Figure 9.

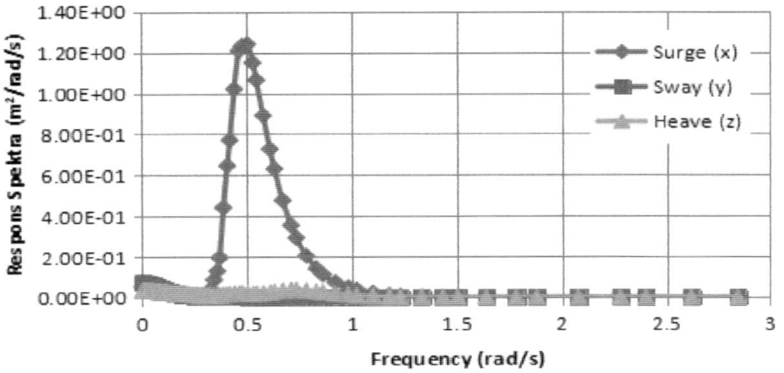

Figure 9. Response spectrum of the rectangular array configuration for wave direction 0 direction (translation motion)

The result of the response spectrum of the linear array mooring configuration in translation motion for wave direction 0 degree is presented as Figure 10.

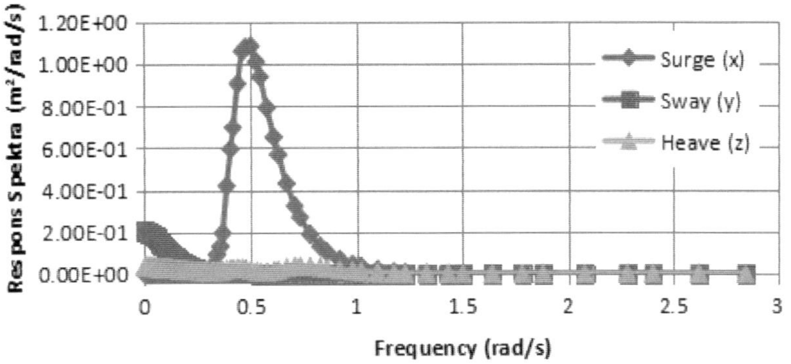

Figure 10. Response spectrum of the linear array configuration for wave direction 0 direction (translation motion)

The result of the response spectrum of the 4 points mooring configuration in translation motion for wave direction 0 degree is presented as Figure 11.

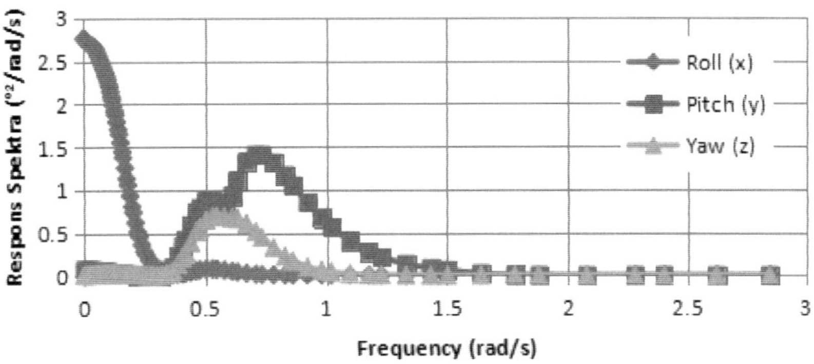

Figure 11. Response spectrum of the 4 points mooring configuration for wave direction 0 direction (translation motion)

The result of the response spectrum of the rectangular array mooring configuration in rotation motion for wave direction 0 degree is presented as Figure 12.

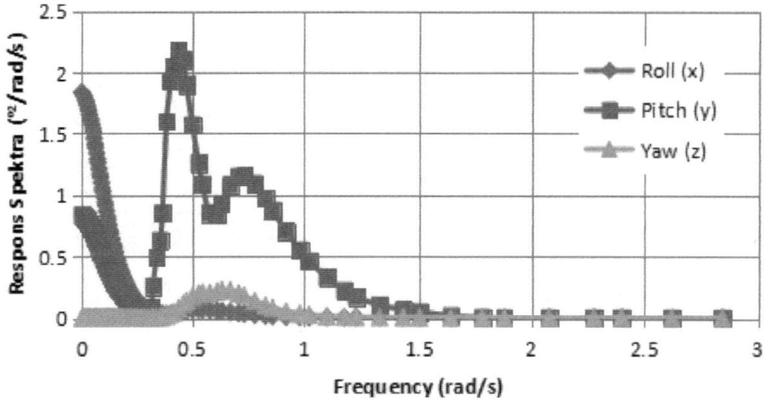

Figure 12. Response spectrum of the rectangular array configuration for wave direction 0 direction (rotation motion)

The result of the response spectrum of the linear array mooring configuration in rotation motion for wave direction 0 degree is presented as Figure 13.

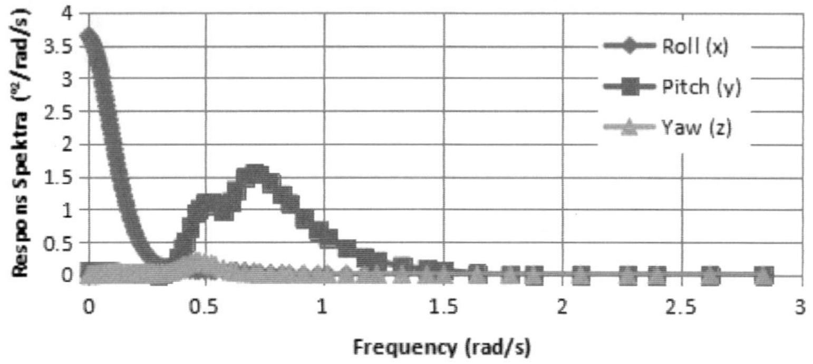

Figure 13. Response spectrum of the linear array configuration for wave direction 0 direction (rotation motion)

The result of the response spectrum of the 4 points mooring configuration in rotation motion for wave direction 0 degree is presented as Figure 14.

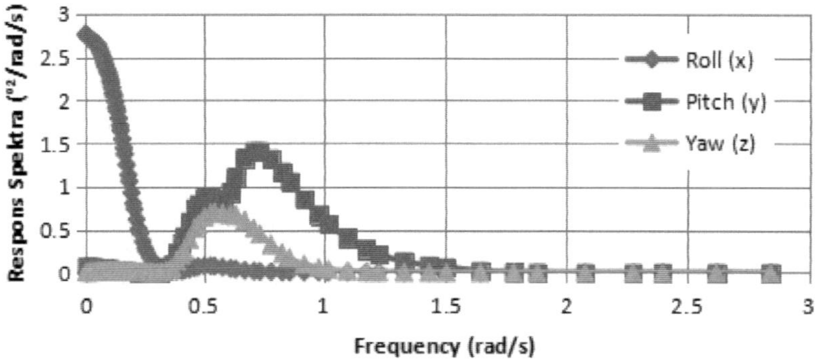

Figure 14. Response spectrum of the 4 points mooring configuration for wave direction 0 direction (rotation motion)

4.4 Tension and Offset Analysis

The analysis of the tension and maximum offset of the mooring line is carried out by numerical modeling. The results of the tension and maximum offset for the rectangular array mooring configuration in translation motion is presented as Table 6.

Table 6. The results of tension and offset for the rectangular array mooring configuration

Heading	0	45	90
	Tension (N)	Tension (N)	Tension (N)
Cable 1	94.167	168.121	8.973
Cable 2	6.837	159.781	70.165
Cable 3	9.987	9.916	7.072
Cable 4	75.045	29.587	40.186
Cable 5	8.658	24.765	13.894
Cable 6	61.167	290.606	69.414
Cable 7	14.091	33.832	100.238
Cable 8	73.750	7.601	7.660
Offset (m)	6.978	3.826	2.243

The results of the tension and maximum offset for the linear array configuration in translation motion is presented as Table 7.

Table 7. The results of the tension and maximum offset for the linear array configuration

Heading	0	45	90
	Tension (N)	Tension (N)	Tension (N)
Cable 1	217.001	29.612	62.699
Cable 2	154.686	979.319	139.698
Cable 3	49.237	105.329	40.872
Cable 4	34.909	36.821	59.403
Cable 5	137.502	105.328	96.706
Cable 6	61.579	36.820	57.677
Offset (m)	8.28	7.051	6.347

The results of the tension and maximum offset for the 4 points mooring configuration in translation motion is presented as Table 8.

Table 8. The results of the tension and maximum offset for the 4 points mooring configuration

Heading	0	45	90
	Tension (N)	Tension (N)	Tension (N)
Cable 1	14.254	32.721	1.078.514
Cable 2	14.048	581.579	615.658
Cable 3	16.466	20.094	243.680
Cable 4	16.108	20.834	363.111
Offset (m)	9.37	4.17	2.73

4.5 The utilization factor

The utilization factor, U, is calculated to check whether the mooring line meets with the limit criteria. The utilization factor, U, of the mooring system is the comparison between the actual mooring line tension with the mooring line capacity. The results of the utilization factor for the rectangular array mooring configuration is presented as Table 9.

Table 9. The results of the utilization factor for the rectangular array mooring configuration

Components	U	Remarks
Cable 1	0.15	pass
Cable 2	0.13	pass
Cable 3	0.06	pass
Cable 4	0.06	pass
Cable 5	0.02	pass
Cable 6	0.24	pass
Cable 7	0.08	pass
Cable 8	0.06	pass

The results of the utilization factor for the linear array mooring configuration is presented as Table 10.

Table 10. The results of the utilization factor for the linear array mooring configuration

Components	U	Remarks
Cable 1	0.18	pass
Cable 2	0.84	pass
Cable 3	0.08	pass
Cable 4	0.04	pass
Cable 5	0.10	pass
Cable 6	0.04	pass

The results of the utilization factor for the 4 points mooring mooring configuration is presented as Table 11.

Table 11. The results of the utilization factor for the 4 points mooring mooring configuration

Components	U	Remarks
Cable 1	0.945	pass
Cable 2	0.477	pass
Cable 3	0.211	pass
Cable 4	0.316	pass

5. Conclusion

Regarding to the results of the mooring design analysis for the sea station cage that have been performed above, here presents some conclusions as following:

1. The nylon mooring rope type has been selected to be used on the 3 mooring configurations: a) linear array, b) rectangular array, and c) 4 points mooring type. The rope has a diameter of 104 mm with a breaking force of 2.3 MN.
2. The largest tension is occurred on the configuration of the 4 points mooring type which the tension on mooring line is 1078 KN. The mooring configurations of the rectangular array and the linear array type have tensions of 290 KN and 979 KN, respectively.
3. The maximum offset is occurred on the configuration of the 4 points mooring type with offset of 9.3 m, whereas the linear array type has an offset of 8.3 m and the rectangular array type has an offset 6.9 m.
4. Based on results from comparing the 3 mooring configurations, the best mooring configuration is the rectangular array type with the tension on the rope of 290 KN, maximum offset of 6.9 m and the utilization factor of 0.24 based on DNVGL OS-E301.

.

References

[1] FAO. 2005. *Review of The State of World Marine Fishery Resources*. FAO Fisheries Technical Paper 457. Rome, FAO.

[2] D.C.B. Scott and J.F. Muir. 2000. *Offshore Cage Sytem-A Practical Overview*. Mediterranean offshore mariculture. Zaragoza, CIHEAM, p. 79-89 (Options Méditerranéennes : Série B. Etudes et Recherches; n. 30)

[3] Benetti, D. D. 2010. *Site Selection Criteria for Open Ocean Aquaculture*. The Marine Technology Society Journal. Sustainable U.S. Marine Aquaculture Expansion in the 21st Century.

[4] De Cew, J. et al. 2012. *Assessment of a Mooring System for Offshore Aquaculture*. World Aqua Culture Society. Baton Rouge, LA, USA.

[5] Djatmiko. E.B. and Murdijanto. 2003. *Seakeeping: Perilaku Bangunan Apung di Atas Gelombang*. Surabaya: Penerbit ITS Press. ITS Surabaya. Indonesia.

[6] Djatmiko, E. B., 2012. *Perilaku dan Operabilitas Bangunan Laut diatas Gelombang Acak*. Surabaya: ITS Press.

[7] DNV GL OS E301. 2015. *Offshore Standard : Position Mooring*. Norway.

[8] James Morris, et al. 2015. *Offshore Aquaculture in the Southern California Bight.* Aquaculture workshop report, April 28-29. California. USA

[9] Fredriksson, D. W. et al. 2005. *Moored Fish Cage Dynamics in Waves and Currents*. IEEE Journal Ocean Engineering. vol. 30, no. 1. pp. 28–36

[10] Bridon oil and gas. *Catalog of fibre rope solutions for the oil & gas industry.* United Kingdom: http://www.bridon.com/x/downloads/oilandgas/Oil%20and%20Gas%20brochure.pdf.

[11] Mulyadi Y, Kobayashi E, Wakabayashi N, Pitana T, Wahyudi. 2014. *Development of Ship Sinking Model over Subsea Pipeline for Madura Strait using AIS Data.* WMU Journal of Maritime Affairs, vol.13, pp.43-59.

IOP Conf. Series: Earth and Environmental Science **135** (2018) 012023 doi:10.1088/1755-1315/135/1/012023

The Usage of Geographical Information System in the Selection of Floating Cages Location for Aquaculture at Prigi Bay, Trenggalek Regency, East Java

H D Armono[1, a], B G Mahaputra[1, b], and M Zikra[1, c]

[1]Ocean Engineering Department, Faculty of Marine Technology, Sepuluh Nopember Institute of Technology, Indonesia

Email: armono@oe.its.ac.id

Abstract. Floating cages is one of the methods of fish farming (aqua culture) that can be developed at rivers, lakes or seas. To determine a proper location for floating cages, there are some requirements that need to be fulfilled to maintain sustainibility of floating cages. Those requirements are the quality of the environment. This paper will discuss the selection of best location for aquaculture activities using Weighted Overlay method in the Geographical Information System, based on the the concentration of chlorophyll-a, sea surface temperature presented by Aqua MODIS Level 1b satellite images. The satellite data will be associated with the measured field data on March and October 2016. The study take place on Prigi Bay, at Trenggalek Regency, East Java. Based on spatial analysis in the Geographical Information System, the Prigi bay generally suitable for aquaculture activities using floating net cages. The result of Weighted Overlay combinations in both periods showed a mean score of 2.18 of 3 where 8.33 km^2 (23.13% of the water area) considered as "very suitable" and 27.67 km^2 (76.87% of water area) considered " suitable".

1. Introduction

Aquaculture or fish farming with floating cages could provide an alternative to reduce negative effects of overfishing. The aquaculture activities using floating cages can be developed at sea, river or lake. A sufficient and good water quality provides a great option for foating cages aquacultures. The aquaculture activities also influenced by the conditions or oceanographic and meteorological parameters such as wind, ocean depths, sea brightness, current speed, salinity, acidity and dissolved oxygen.

In the current technology developments of remote sensing, the distribution of chlorophyll-a and sea surface temperature can be detected using satellite imagery. The image data can be processed to determine the area of fish farming. Such data must be processed first, and the available parameters must be evaluated to choose a location that has the potential for fish farming.

This paper will discuss the processing of satellite image data containing the distribution of chlorophyll-a concentration and sea surface temperature. By utilizing remote sensing technology, and overlaying the processed images containing those parameters, an information about the spatial and temporal information for the best option of Aquaculture wil be presented in form of thematic map. Prigi Bay, at Trenggalek Regency, East Java was selected as the area of study since the location was considered clean and calm.

2. Remote Sensing and Geographical Information System

Remote Sensing is a science or art to obtain data or information from an object at the surface the earth using equipments remotely from the object [1]. Remote sensing is also consider as variation technique developed to obtained and analyse information about earth [2]. In the remore sensing system, a natural or artificial energy is needed. The energy can be in the form of electromagnetic spectrum. Such as cosmic spretra, gamma ray, ultravolet, sunray, infra red, radio wave, and micro wave. The natural energy used for remote sensing was sun, as the sun transmitted energy to the earth. Energy transmitted from the object in the earth surface can be recorded by the sensor remotely installed in the satellite from the object, using infrared thermal spectrum. The information in the form of electromagnetics energy reflected, emitted, and transmitted from the objects in the earth surface to the sensors attached in the satellite. Some of energy transmitted to earth disipated by atmosphere and some portion reach the earth surface and hit the object. The object will absorp some of the energy and transmitted to the sensor attached in the satellite. The electromagnetics from satelille captured by earth station and send the data to Data Processing Centre for further analysis.

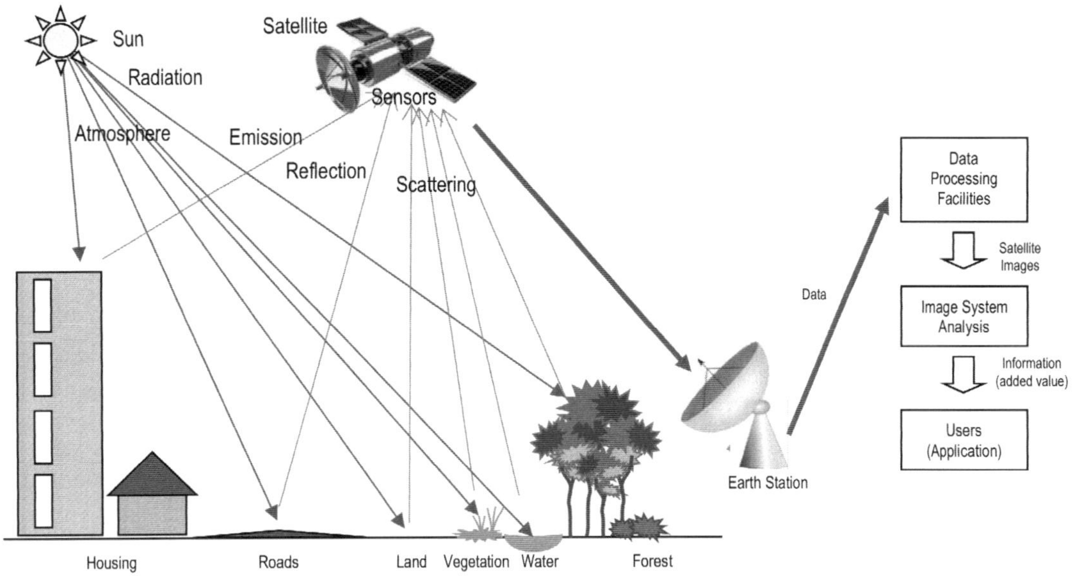

Figure 1. Remote Sensing System [3]

One of the object in the earth covered by remote sensing is the sea surface. The energy trasmitted form the sea surface, may contain the concentration of chlorphyll-a. The chlorophyll-a concentration is a significant parameter for determining areas with potential fishery resources. The temperature contours that shows a dense temperature gradient over the surrounding area and high chlorophyll-a concentrations indicate the areas with high fish potential [4]. Therefore, by examinizing the distribution profile of chlorophyll-a concentration obtained from satellites and other related parameters for aquaculture, one can select the suitability of the area for floating cages placement for fish aquaculture. The selection process is integrated in the geographic information system.

Geographic Information System (GIS) is a set of tools used for collecting, storing, transforming, and presenting the spatial data of a real surface phenomenon of the earth for specific purposes [5]. The technology is growing rapidly due to the development of information technology or computer technology [6] which capable of handling databases, displaying an image (graph) and consider as the alternatives in presenting a map. GIS graphic data can be presented in two data models, namely raster data model and vector data model (spatial). The raster data model represents data expressed by grid or cell, whereas the vector data model presents graphical data (dots, lines, polygons) in vector format structures or in coordinates system (x, y). In the vector data, one can made a comparison between the line and area information in the form of data that have magnitudes, directions, and relations.

In the field of fisheries and marine science, GIS can be a valuable input for fishermen or fisheries entrepreneurs to predict the best locations for fishing. The interaction of attribute data with spatial data is very useful at fishing sites, where regular reports can be made to inform the potential fisheries around the ship landing site. Such information can be utilized in the selection process to find suitable floating cage location.

IOP Conf. Series: Earth and Environmental Science **135** (2018) 012023 doi:10.1088/1755-1315/135/1/012023

There are two factors that determine the suitability of the location of fishery cultivation, namely environmental factors and water quality [6]. Environmental factors are determined by the depth, brightness, current and wave while water quality is determined from temperature, salinity, acidity and oxygen. The rate of water fertility is also influential by considering the distribution of a-chlorophyll concentration in the waters. The use of satellite image data of chlorophyll-a concentration and sea surface temperature can be used to predict the location of floating cage using overlay analysis of parameters that have been classified into a raster map [7].

3. **Analysis and Discussion**

3.1 Area of Study

Prigi Bay in Trenggalek Regency, about 190km from Surabaya in the southern part of East Java Province, located at coordinates 8°19'39"S and 111° 43'43" E as shown in **Figure 2**. The Ministry of Fisheries and Ocean have a fishing port in this place and equipped with a fish auction facility.

Figure 2. Area of Study

3.2 Data Collection

The data in this study is collected from various sources:

1. Distribution of chlorophyll-a concentration data in Prigi Bay obtained from the LADSWEB (ladsweb.modaps.eosdis.nasa.gov/search/) on March 2016 and October 2016. Data selected position limit to 8°19'39" S - 111°43'43" E.
2. Wind data obtained from the Agency of Meteorology, Climate and Geophysics (BMKG).
3. Oceanographic data such as depth of the sea, the sea brightness, current speed, salinity, dissolved oxygen and acidity obtained from Prigi Fishing Port and the Oceanographical Research Center – Indonesian Institute of Science (LIPI).
4. The map obtained from The National Agency of Survey and Mapping (ww.bakosurtanal.go.id/).

3.3 Digital Image Processing

The satellite images were processed to obtain distribution of chlorophyll-a concentration in the Prigi Bay waters. The processing stages include:

1. **Geometric correction**. Prior to image data processing, projection or map coordinate system must be adjusted to the area of interest or with previous spatial data. Geometric correction or rectification is the stage process in adjusting the image data with the appropriate coordinate

IOP Conf. Series: Earth and Environmental Science **135** (2018) 012023 doi:10.1088/1755-1315/135/1/012023

system. The references coordinate used in geometric correction were either a base map or previous image data that has been corrected. The geometric correction on a digitized topographical map of Prigi Bay performed using available Ground Control Point (GCP)..

2. **Radiometric correction.** To clearly read and interpreted the information contained in the image data, the radiometric correction is performed. In this stage, the earth surface images are separated with clouds images using cloud masking processes. Satellite imagery that has been corrected geometrically were defined on the mask feature definition. Next, the images were overlaid with the desired canals, such as sea surface temperature or the concentration of chlorophyll-a.
3. The application of the algorithm associated with the prefered output using Band Math features on the cloud masked satellite images. After applying a predetermined algorithm, the processed files are stored and cropped by GIS software, subsequently.
4. The final stage is image cropping. This stage is performed to restrain the observed areas into the area of interest and reduce the size of the images. This stage will ensure that more focused, detailed and optimized data processing was achieved. The cropping image processes can be carried out following the desired polygon shapes such as following on the district, sub-district or village shape.

3.4 Processing of chlorophyll-a concentration and SST in MODIS Satellite Imagery

The image data processed in this study was the MODIS level 1b freely obtained from NASA Ocean Color Web. The file has the format of Hierarchical Data Format (*.hdf) which describe the captured time and spatial resolution of 1 km. The radiometric and geometric correction, as well as cloud masking will performed on MODIS data using satellite image processing software.

After image correction, the algorithm to determine the content of chlorophyll and sea surface temperature were applied. For chlorophyll concentration estimation, the algorithms OC4 [8] as shown below is applied:.

X = log10 (Rrs (λ1) / Rrs (λ2) .... (1)

$Chl\text{-}a$ = 10 [$a_0 + a_1 + (a_2 . X_2) + (a_3 . X_3) + (a_4 . X_4)$ ....(2)

where:

Rrs (λ1) = RRS channel a blue (488 nm)

Rrs (λ2) = RRS in the green channel (531 nm)

a_0 = 0.3272

a_1 = -2.9940

a_2 = 2.7218

a_3 = -1.2259

a_4 = -0.5683

The algorithm for sea surface temperature (SST) is expressed as follows [9].

SST Modis = C1+ C2 * T31 + C3*T31-32*Tb+C4*(sec(θ)-1) *T31-2.....................................(3)

where: Tb = Channel 20

C1 = constant radiation that is worth 1.1911x108 W m-2 sr-1 μm4

C2 = constants of radiation that is worth 1.4388x104 K μm

Θ = sSatellite zenith angle

The constants C1, C2, C3 and C4 available in **Table 1**

Table 1. The coefficient of C in B31 and 32 [9]

Coefficient	ΔT ≤ 0.7	ΔT> 0.7
C1	1.228552	1.69521
C2	0.9576555	0.9558419
C3	1.1182196	0.0873754
C4	1.774631	1.199584

3.5. Classification for Aquaculture Suitability

The evaluated parameters used for classification of aquaculture suitability in this study were based on environmental factors turbidity, depth, current speed, and wave as presented in **Table 2**, and water quality such as temperature, salinity, pH, and dissolved oxygen as presented in **Table 3**. The weight values for environmental and water quality factors were presented in **Table 4**.

Table 2. The Weight values based on Environmental Factors [7]

Environment	Turbidity	Depth	Current	Wave	Weight
Turbidity	1	2	4	7	0:53
Depth	½	1	2	3	0:26
Current	¼	1/2	1	2	0:14
Wave	1/7	1/3	1/2	1	0:07
Consistency ratio	0.0029				

Table 3. The Weight values based on Water Quality. [7]

Water quality	Temperature	Salinity	PH	Oxygen	Weight
Temperature	1	2	7	4	0:52
Salinity	½	1	4	2	0:27
pH	1/7	1/4	1	½	0:07
Oxygen	¼	1/2	2	1	0:14
Consistency ratio	0.0008				

Table 4. The Weight values based on Environmental Factors and Water Quality [7]

Land Eligibility Criteria	Environment	Water quality	Weight
Environment	1	3/2	0.6
Water quality	2/3	1	0.4
Consistency ratio	0.0000		

Furthermore, scoring method is used in determination of the criteria and the division suitability class (3 for very suitable, 2 for suitable, and 1 for not suitable) for floating cage aquaculture activities using physical and chemical parameters by reducing some parameters such as sheltering location, substrate condition and distance from pollution. The weighting is performed by Pair Wise Comparison analysis as presented in matrix form as presented in **Table 5** [7] [10] [11]. Parameters that have strong influence to the aquaculture activities have higher weighting factors than those with less influence. However, due to limited space, from various suitability parameter for Aquaculture presented in **Table 5** below, only distribution of chrolophyll-a and sea surface temperature on Prigi Bay is discussed in this paper.

Table 5. Matrix of Parameter Suitability for Floating Cage Aquaculture

No	**Parameters**	**Weight (%)**	**Unit**	**Very Suitable Score 3**	**Suitable Score 2**	**Not Suitable Score 1**
1	Turbidity	31,8	NTU	<11	11–40	40>
2	Temperature	20,8	°C	27–29	26–27 and 29–32	<26 and >32
3	Depth	15,6	m	15–25	5–15 and 25–40	<5 and >40
4	Salinity	10,8	‰	30–35	20–30	<20 and >35
5	Current Speeed	8,4	cm/s	15–35	10–15 and 35–100	<10 and >100
6	Dissolved Oxygen	5,6	mg/l	>6	4–6	<4
7	Wave	4,2	cm	<20	20-40	>40
8	pH	2,8	-	7,0–8,5	4–7 and 8,5–9	<4 and >9

3.6. Determination of Class Classification

The total score of each class is defined by multiplying the suitability level score with the parameter weight as presented in **Table 5** using the following expression. [12].

Y = Σ (ai x Xn)..(4)

where:

Y = Total Score
ai = Weighting factors
Xn = Suitability level score

Furthermore, the interval calss for suitability level were estimated using Equal Interval method [12] to divide the range of attribute values into sub-range with the same size. Mathematically it can be expressed as follows:

$I = (\Sigma\ (ai \times Xn)_{max} - \Sigma(ai \times Xn)_{min})/k$ ....(5)

Where:

I = Interval of suitability classes
k = Number of the desired suitability classes

As presented in **Table 5** above, there are three suitability classes: "very suitable", "suitable" and "not suitable". Based on equation 5, if the value of Nmax is 3 and the Nmin is 1, and the desired suitability classes is 3, it is found that the class interval was 0.667, and the score of suitability were:

2.333 to 3 = Very Suitable (S1)
1.668 to 2.332 = Suitable (S2)
1 to 1.667 = Not Suitable (S3)

The level of suitability is defined as [13]:

- Very Suitable (S1)

This grade indicates that the reviewed location is potentially suitable for aquaculture activities. At this location there are minor to no limiting factor and will not affect the productivity of aquaculture significantly.

- Suitable (S2)

This grade indicates that the aquaculture activities can be implemented in this location, but less suitable because it has a considerable limiting factor that affecting the productivity. Therefore, it required additional treatment and modification for aquaculture activities.

- Not Suitable (S3)

This grade indicates that the location is unsuitable for aquaculture due to limiting factors that affect the productivity of aquaculture activities.

3.7. Sampling Data Interpolation

The interpolation method called Inverse Distance Weighted (IDW) was employed to considers the spatial distance as "weights". The distance meant in this study here is the distance of the sample points to a block of data to be estimated. So, the closer the distance between sample points and a block to be estimated, the greater the "weight", and vice versa. The IDW method assumes that the estimated value at the point which is not recorded is a function of the distance and the average value point located nearby. The interpolation depends on how strong a data point affects the area around it. It is also the number of points in the vicinity which are used to calculate the average value, and the pixel size / raster desired. IDW interpolation method provides more accurate results than Kriging method [14]. This is because the value of data generated in IDW, are close to the minimum and maximum value of the sample data. While the method of interpolation Kriging sometimes results in a low range and have less influence on the sampling variation.

IOP Conf. Series: Earth and Environmental Science **135** (2018) 012023 doi:10.1088/1755-1315/135/1/012023

4. RESULTS

4.1. Distribution of Chlorophyll-a concentration on Prigi Bay

The minimum value of concentration in March 2016 was 0.4 mg/m^3 while the maximum value was 2 mg / m^3, with the average concentration was 0.8 mg/m^3. Meanwhile, the concentration on October 2016 had a minimum value of 0.5 mg/m^3 and a maximum value of 2.1 mg / m^3, with the average of 0.9 mg /m^3.

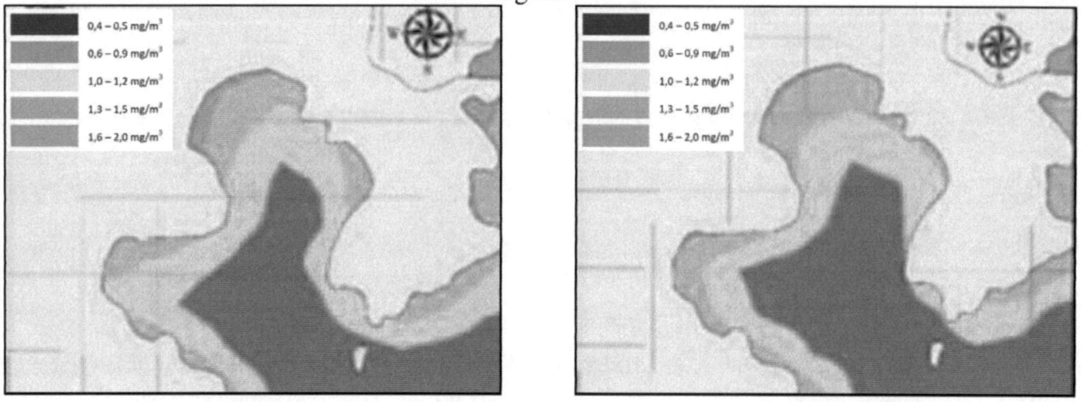

a. March 2016 b. October 2016

Figure 3 shows the map of chlorophyll-a distribution on March and October 2016.

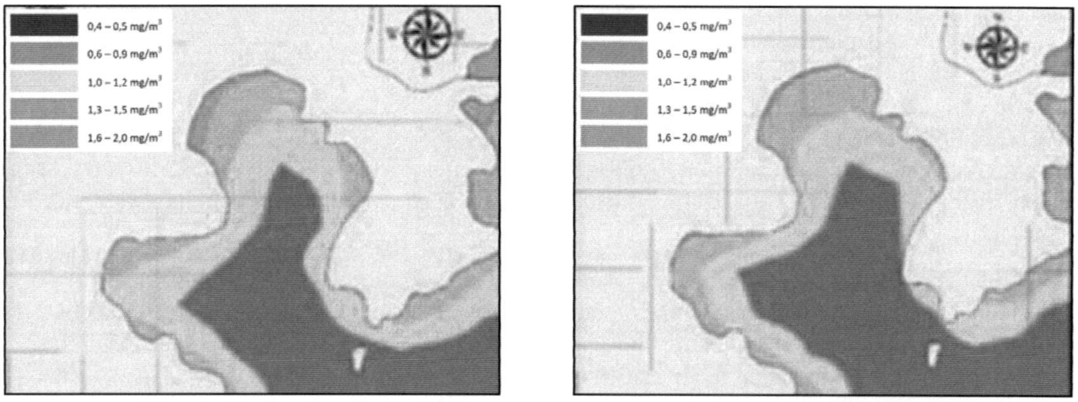

c. March 2016 d. October 2016

Figure 3. Distribution Map concentration of chlorophyll-a in Prigi Bay (5 ranges)

The concentration of chlorophyll-a in March 2016 did not have a significant difference compared to the distribution of chlorophyll-a concentration in October 2016 as the average differences is only 0.1

mg / m^3.

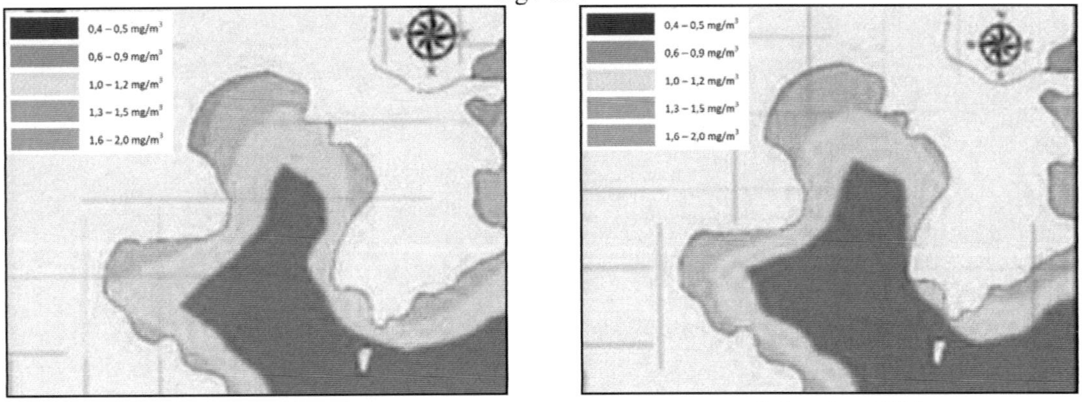

e. March 2016 f. October 2016

Figure 3 were re analysed to classify the water quality based on chlorophyll-a concentration and sedimen suspension [17] resulting map as shown in **Figure 4** below. Again, there are no significant difference of concentration between March and October 2016.

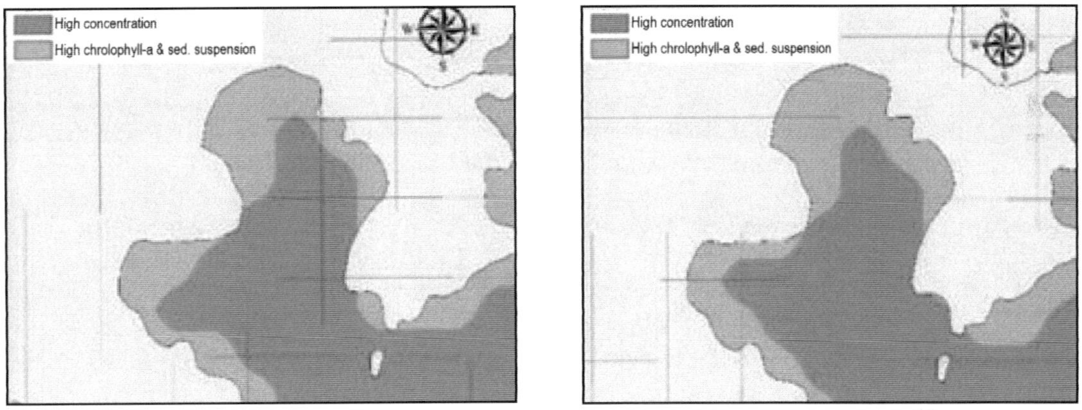

a. March 2016 b. October 2016

Figure 4. Distribution of Chlorophyll-a concentrations in Prigi Bay

4.2 Distribution of Sea Surface Temperature Prigi Bay

The sea-surface temperatures in March and October 2016 also did not have a significant differences, although in the month of October, the sea surface temperatures tend to be lower compared to March. Sea surface temperature in March has the lowest temperature and the highest temperature of 30,25°C and 30,96°C, while the average temperature was 30,54°C. Meanwhile on October the sea surface temperatur has 29,47°C as lowest temperature and 30,43°C as highest temperature with average value of 30,1°C. The temperature distribution are shown in **Figure 5** below.

IOP Conf. Series: Earth and Environmental Science **135** (2018) 012023 doi:10.1088/1755-1315/135/1/012023

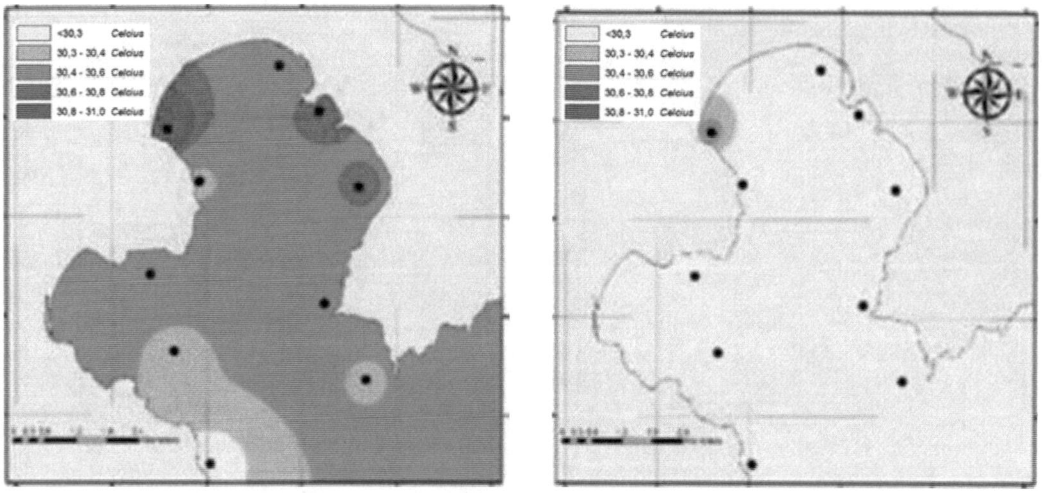

a. March 2016 b. October 2016

Figure 5. Distribution Map Sea Surface Temperature in Prigi Bay from Field Data.

The sea surface temperatures are very feasible for aquaculture in the range of temperature 28-32°C [7]. Therefore, based on field data and spatial analysis shown in

Figure 5 above, the sea surface temperatures in the Prigi bay are very suitable for aquaculture activities. Although the processed satellite images had average temperatures slightly lower than the data sample points, but the temperature was still feasible for aquaculture activities.

4.3 Suitability Location using Weighted Overlay Method

The spatial analysis using various parameters presented in **Table 5** above can be reclassified based on three interval suitability classes as discussed previously; i. Very Suitable (S1), ii.Suitable (S2) and iii. Not Suitable (S3). A suitability score can be evaluated using equation 4, and the result of aquaculture suitability location is shown in **Figure 6**.

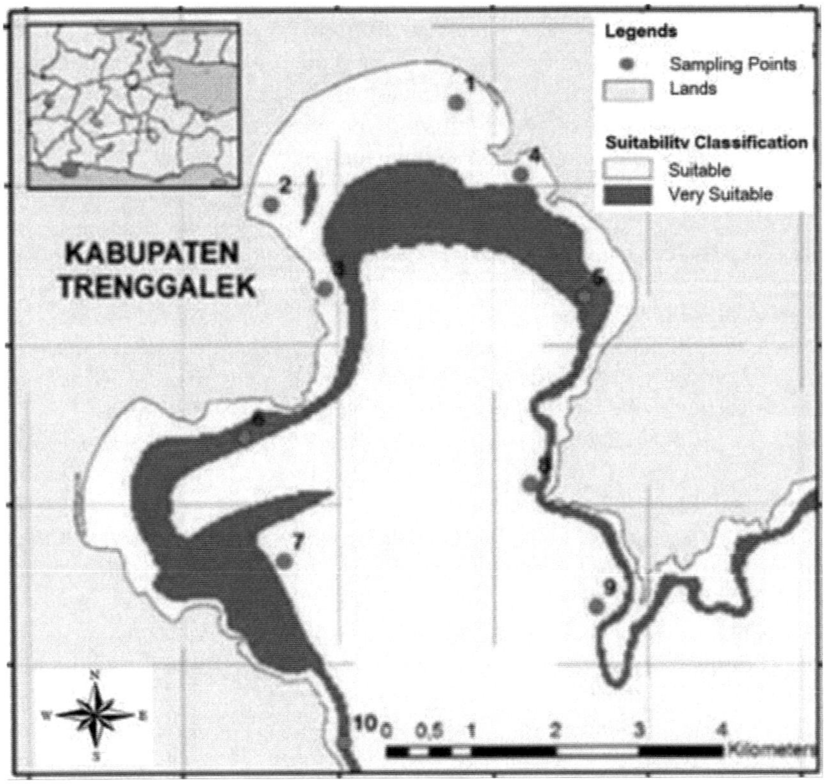

Figure 6. Map of Suitability Location in Prigi bay based on March and October 2016

The green color in **Figure 6** represents the very suitable location and yellow colors indicate suitable location for floating cages aquaculture. In general, the waters of the Prigi Bay quite suitable for floating cage aquaculture activities.

By weighted overlay methods and combining March and October 2016 analysis results, the average suitablity score was 2.18 of 3 and categorized as "suitable". Further analysis shows that "very suitable" area in Prigi Bay was 8.33 km^2 (23.13% of the water area) while "suitable" area was 27.67 km^2 (76.87% of the total water). The most suitable location for Aquacuture activities are stations 5, 6 and 10 in **Figure 6**.

5. Conclusion

From the results of the analysis conducted in this study could be concluded as follows:

1. The waters of the Prigi Bay in March and October 2016 has high chlorophyll-a concentration or considers rich phytoplankton. The average distribution of chlorophyll-a concentration reaches from 0.8 to 0.9 mg / m^3.
2. The spatial analysis performed at Prigi Bay shows that the area generally suitable for Aquaculture activities. A "very suitable" area was found of 8.33 km^2 (23.13% of the water area) while "suitable" was 27.67 km^2 (76.87% of the total water).

References

[1] Lillesand.T.M. and Kiefer, R.W., 1979. *Remote Sensing and Image Interpretation.* John. Willey and Sons. New York
[2] Lindgren, D. 1985. *Land Use Planning and Remote Sensing.* Nijhoff Publishers. Dordrecht
[3] Sutanto. 1994. *Penginderaan Jauh Jilid II.* Gajah Mada University Press. Yogyakarta.

[4] Hasyim, B., and Priyanti, N.S. 1999. *Analisis Distribusi Suhu Permukaan Laut and Kaitannya dengan Lokasi Penangkapan Ikan*. Prosiding Seminar Validasi Data Inderaja untuk Biandg Perikanan. Baand Pengkajian and Penerapan Teknologi. Jakarta. ISBN;979-95760-1-6.

[5] Burrough, P.A. 1986. *Principles of Ge. Oographic Information Systems for Land Resource Assessment.* Oxford Science Publications, New York.

[6] Beveridge, M.C.M.,2008. *Cage Aquaculture*, 3rd Ed. ISBN: 978-1-4051-0842-3, 380 pages, Wiley-Blackwell

[7] Radiarta, I.N., Saputra, A.J., and Prihadi, T.H. 2006. *Pemetaan Kelayakan Lahan Budi Daya Ikan Laut Di Kecataman Moro, Kepulauan Riau: Dengan Pendekatan Sistem Informasi Geografis.* Jurnal Riset Akuakultur Volume 1 Nomor 2

[8] Prinina, E., Jaelani, L.M. and Tarigan, S. 2016. *Validasi Algoritma Estimasi konsentrasi Klorofil-a and Padatan Tersuspensi Menggunakan Citra Terra and Aqua Modis dengan Data In situ.* JURNAL TEKNIK ITS Vol. 5, No. 2, (2016) ISSN: 2337-3539. Surabaya.

[9] Brown, O.B.; Minnett, P.J. *MODIS Infrared Sea Surface Temperature Algorithm—Algorithm Theoretical Basis (ATBD) 25 Version 2.0*, Document; under Contract NAS5-31361; University of Miami: Miami, FL, USA, 1999

[10] Sari, K.Y, 2011. *Analisis Spasial Citra Satelit Landsat Untuk Penentuan Lokasi Budi Daya Keramba Jaring Apung Ikan Kerapu di Perairan Pulau Semujur Kabupaten Bangka Tengah.* Institut Pertanian Bogor, Bogor

[11] Adipu, Y., Lumenta, C., Kaligis, E., and Sinjal, H.J. 2013. *Kesesuaian Lahan Budidaya Laut Di Perairan Bolaang Mongondow Selatan, Sulawesi Utara.* Universitas Sam Ratulangi, Manado, Sulawesi Utara.

[12] Ariyanti, R.W., Sya'rani, L., and Arini, E. 2007. *Analisis Kesesuaian Perairan Pulau Karimunjawa and Pulau Kemujan sebagai Lahan Budidaya Rumput Laut Menggunakan Sistem Informasi Geografis*. Universitas Diponegoro, Semarang

[13] Food and Agriculture Organization. 1976. *A Framework for Land Evaluation*. FAO Soils Bulletin. (32): 51-54

[14] Pramono, G.H., 2008. *Akurasi Metode IDW and Kriging untuk Interpolasi Sebaran Sedimen Tersuspensi*. Forum Geografi, Vol. 22, No. 1 Bakosurtanal

[15] Arsjad, M., Siswantoro, Y., and D. Sari. 2004. *Inventarisasi Sumberdaya Alam dan Lingkungan Hidup. Sebaran Chlorophyll-a di Perairan Indonesia*. Pusat Survei Sumberdaya Alam Laut. Badan Koordinasi Survei dan Pemetaan Nasional. Cibinong. Bogor.

Hydrodynamic and Sediment Transport Modelling of Suralaya Coastal Area, Cilegon, Indonesia

A. H. Fattah[1a], Suntoyo[1], H. A. Damerianne[1] and Wahyudi[1]

[1]Ocean Engineering Department, Faculty of Marine Technology, Institut Teknologi Sepuluh Nopember (ITS), Surabaya, Indonesia (60111)

[a]Email: harisfattah@gmail.com

Abstract. The coastal zone of Suralaya is located in the district Pulomerak, Cilegon City, Province Banten. This region is a part of the Sunda Strait region that is very important area to support the ongoing activities such as, industries, power plant, ports, and tourism. However, those various activities will certainly give effect to the surrounding environment. To determine the environmental conditions of Suralaya Coast, it is necessary to study the hydrodynamics analysis and sediment transport modelling including the analysis of currents patterns. Tidal elevation observation was conducted for 15 days used to validate the water elevation simulation results, in which a good agreement between the observed data and the model result was obtained with the error value of 1.6%. The dominant current direction is from northeast in west season, while in the east season predominant current direction is from northwest with a speed average current 12,44 cm/s. The dominant wave direction is from the west. The average temperature is at 27°C and the bottom sediment dominant form is fine sand.

Keywords: Suralaya Coast, Modelling, Hydrodynamic, Sediment Transport

1. Introduction

Coastal area of Suralaya is located in the Suralaya village, Pulomerak district, Cilegon, Banten Province. As part of the Sunda Strait area, it is very important as a support area for various activities of industry, power plant, ferry port, and tourism, which led to very strategic location as the gateway of the voyage between the island of Java and Sumatra.

However, the various activities will certainly have an impact on the coastal environment and surrounding areas. To know the condition of coastal area, hence needed a comprehensive understanding related to coast hydrodynamics process and sediment transport. Therefore, it is necessary to conduct a research study on hydrodynamic and sediment transport modeling analysis of coastal area of Suralaya. Thus, hydrodynamic and sediment transport modelling in coastal area has been done by some reserachers [1,2,3]. The bottom shear stress and sediment transport model induced by non-linear wave shown the difference characteristics that can increase the accuracy of the costal morphological change [4, 5]. Moreover, the wave over topping also could lead to sediment intrusion into Lagoon [6].

Hydrodynamic and sediment transport modelling lead us possible to know and estimate hydrodynamic behavior and quantify erosion, deposition and transport sediment for given period over large or tight area [7]. It has been developed rapidly in recent decades since 20th century until now, either the experimental or the model simulation, and commonly used to predict the hydrodynamic and the morphodynamic response of low crested structures [8]. The environmental data used in this modelling were obtained by direct measurement, namely bathymetriy, tidal, currents velocity and bottom sediment sample data, while for met-ocean data such as wind, wave and tidal data for long time was provided by BMKG and Dishidros. And then the obtained data are used to model validation.

In the present paper, aims to investigate the hydrodynamic behavior, the current velocity and sediment transport pattern and bed level change occurring in the Suralaya Coastal Area. Thus, the outcomes of the study can be used as the basis information to manage the Suralaya and surrounding coastal areas.

2. Study Area for Numerical Modelling

The study area of this research is located in coastal area of Suralaya, Cilegon, Banten Province with total area about 7.05 km^2 shown in Figure 1, it has a land boundary involved the coastal area of Suralaya Power Plant as well as three open sea boundaries. The Suralaya coastel water is a part of the waters of the Sunda Strait with a fairly steep coastal condition with a slope range of 0.5. The bottom sediment of the beach is fine sand.

Figure 1. Study Area of Suralaya Coastal Water

3. Numerical Modelling of Hydrodynamics and Sediment Transport

The most important to get the high accuracy of numerical model results is the set-up of modeling supported by the high quality of fields survey data. The better the data quality and set-up, then the results obtained will also be satisfactory [9]. The model setup requires some important data ie bathymetry data and field data obtained directly through the hydro-oceanography survey in December 2015. The data consist of bathymetry, tidal, current, water quality, and basic sediment data. The data is very important because it is used as input in the modeling setup. The data required in hydrodynamic and sediment transport modeling include bathymetry, tidal time series, wind data, bottom sediment sampling data. But for wind data use the wind data obtained from BMKG.

3.1. Model Description

The numerical model used in this study is Mike 21 which has been developed by DHI. Mike 21 is a numerical 2D modeling for modeling simulation of hydraulic and environment phenomena in lakes, bays, coastal waters and seas, estuaries, river where stratification can be neglected [9]. It consists of several models module that can be applied to 2D free surface flows modeling. Mike 21 can solve models with complex and multidisciplinary problems such as hydraulic, coastal engineering, and environmental science [9].

In this study, there are 3 modeling modules that will be used for the simulation, namely the Hydrodynamic (HD), the Nearshore Spectral Wind-Wave (SW), and the Non-Cohesive Sediment Transport (ST).

3.2. The Hydrodynamic Model

Hydrodynamic (HD) module is the basic module in the Mike 21 flow model [9]. It provide hydrodynamic basis modelling for computations performed. This module simulates water level variations and its flows to time in response to the force acting on it and the output results used as input in other modules.

The basic principle in this module solves continuity equation and vertically integral momentum conservation mass for 2HD (two horizontal dimension), so it can explain the interaction of fluid variation to depth by [10]. The equations used in this module are as follows:

Continuity equation:

$$\frac{\partial \zeta}{\partial t} + \frac{\partial p}{\partial x} + \frac{\partial q}{\partial y} = \frac{\partial d}{\partial t} \quad (1)$$

Momentum of mass in the X-directon:

$$\frac{\partial p}{\partial t} + \frac{\partial}{\partial t}\left(\frac{p^2}{h}\right) + \frac{\partial}{\partial y}\left(\frac{pq}{h}\right) + gh\frac{\partial \zeta}{\partial x} + \frac{gp\sqrt{p^2+q^2}}{C^2.h^2} - \frac{1}{\rho_w}\left[\frac{\partial}{\partial x}(h\tau_{xx}) + \frac{\partial}{\partial y}(h\tau_{xy})\right] - \Omega p - fVV_x + \frac{h}{\rho_w}\frac{\partial}{\partial_x}(p_a) = 0 \quad (2)$$

Momentum of mass in the Y-directon:

$$\frac{\partial q}{\partial t} + \frac{\partial}{\partial y}\left(\frac{q^2}{h}\right) + \frac{\partial}{\partial x}\left(\frac{pq}{h}\right) + gh\frac{\partial \zeta}{\partial y} + \frac{gq\sqrt{p^2+q^2}}{C^2.h^2} - \frac{1}{\rho_w}\left[\frac{\partial}{\partial y}(h\tau_{yy}) + \frac{\partial}{\partial x}(h\tau_{xy})\right] - \Omega p - fVV_y + \frac{h}{\rho_w}\frac{\partial}{\partial_y}(p_a) = 0 \quad (3)$$

where,

$h(x,y,t)$	: *water depth* $(=\zeta-d,m)$
$d(x,y,t)$	: *time varying water depth (m)*
$\zeta(x,y,t)$	: *surface elevation (m)*
$p,q(x,y,t)$	: *flux densities in x- and y-directions* $(m^3/s/m)=(uh,vh)$; (u,v)=*depth averaged velocities in x- and y-directions*
$C(x,y)$	: *Chezy resistance* $(m^{1/2}/s)$
g	: *acceleration due to gravity* (m/s^2)
$f(V)$	: *wind friction factor*
$V, V_x, V_y(x,y,t)$	: *wind speed and components in x- and y- direction (m/s)*
$\Omega(x,y)$	: *Coriolis parameter, latitude dependent* (s^{-1})
$p_a(x,y,t)$	: *atmospheric pressure (kg/m/s2)*
ρ_w	: *density of water (kg/m3)*
x,y	: *space coordinates (m)*
t	: *time (s)*
$\tau_{xx}, \tau_{xx}, \tau_{xx}$	: *component of effective shear tress*

For effective shear stress calculation in x- and y-direction can be calculated using following calculations by [11]:

$$\tau_{bx} = \rho c_f U\sqrt{U^2+V^2}\left[1+\left(\frac{\partial z_b}{\partial x}\right)^2+\left(\frac{\partial z_b}{\partial y}\right)^2\right]^{1/2} \quad (4)$$

$$\tau_{by} = \rho c_f V\sqrt{U^2+V^2}\left[1+\left(\frac{\partial z_b}{\partial x}\right)^2+\left(\frac{\partial z_b}{\partial y}\right)^2\right]^{1/2} \quad (5)$$

where *cf* is friction coefficient.

For heat dispersion analysis, heat dispersion modeling with Advection/Dispersion equation was used. This equation as given by [12] is as follows:

$$\frac{\partial T}{\partial t}+\frac{\partial uT}{\partial x}+\frac{\partial vT}{\partial y}+\frac{\partial wT}{\partial z}=F_T+\frac{\partial}{\partial z}\left(D_v\frac{\partial T}{\partial z}\right)+\hat{H}+T_s \quad (6)$$

Where, D_V is vertical eddy viscosity coefficient, $\hat{H}$ is atmospheric heat transfer, T_S is temperature of outfall and F_T is horizontal diffusion.

3.3. Nearshore Spectral Wind-Wave Model

This module used to simulate the propagation, growth, decay and transformation of wind-generated waves and swell both in offshore and coastal areas. Basic principle of this module is derived from the conservation equation of spectral wave action density and solved using the Eulerian finite difference technique [13]. And the output of this module are integral wave parameters and spectral output data such as significant wave height, mean wave period, mean wave direction, directional standard deviation, radiation stresses and wave energy dissipation.

Wave radiation stress is included in the momentum equation by the mean of motion equations over depth and time (wave period):

$$\text{x-momentum: } \frac{\partial S_{xx}}{\partial x}+\frac{\partial S_{xy}}{\partial y} \quad (7)$$

$$\text{y-momentum: } \frac{\partial S_{yy}}{\partial y}+\frac{\partial S_{xy}}{\partial x} \quad (8)$$

where S_{xx}, S_{xy}, and S_{yy} are components of radiation stress.

3.4. Sediment Transport Model

The basic principle of this module is for calculating of non-cohesive sediment transport capacity as a function of local wave, current, and sediment conditions. Continuity equation describing rates of bed level change and solved by finite difference method. Optimum input data such as tidal, wind, wave, and current will influence optimum precision in the simulation [11].

In this module, Bijker's transport model is used to calculate sediment transport [14]. The total load sediment transport calculated as the sum of bed load transport and suspended load transport as following equation:

$$q_t = q_b + q_s = q_{bt}(1+1.83Q) \quad (9)$$

Q in Equation (9) is a dimensionless factor defined as

$$Q=\left[I_1 ln\left(\frac{33h}{r}\right)+I_2\right] \quad (10)$$

where h is water depth, r is bed roughness, I_1 and I_2 are Einstein's integrals, which should be evaluated on the basis of dimensionless reference level, A = r/h, and z^*, defined as following equation:

$$z^* = \frac{w}{\kappa U_{f,wc}} \quad (11)$$

With w is settling velocity of the suspended sediment, κ is von Karman's constant and $U_{f,wc}$ is the shear velocity under combined waves and current. The influence of the waves on the suspended-load transport is therefore taken into account through the shear velocity. The roughness, r, can be related to Chezy number, C, by:

$$C = 18log\left(\frac{12h}{r}\right) \quad (12)$$

The shear velocity in combined waves and current is found as:

$$U_{f,wc} = U_{f,c}\sqrt{1+\frac{1}{2}\left(\xi\frac{\hat{u}_b}{V}\right)} = \frac{\sqrt{gV}}{C}\sqrt{1+\frac{1}{2}\left(\xi\frac{\hat{u}_b}{V}\right)^2} \quad (13)$$

Where $U_{f,c}$ is current-related shear velocity, V is the depth-averaged current velocity, $\hat{u}_b$ is the amplitude of the wave-induced oscillatory velocity at the bottom, and ζ is a dimensionless factor that can be expressed in terms of wave friction factor f_w and chezy's number C.

$$\zeta = C\sqrt{\frac{f_w}{2g}} \quad (14)$$

The wave friction factor f_w is calculated as

$$f_w = exp\left[-5.977 + 5.213\left(\frac{a_b}{r}\right)^{-0.194}\right] \text{ If } 1.47 < \frac{a_b}{r} < 3000 \quad (14a)$$

$$f_w = 0.32 \text{ If } \frac{a_b}{r} \leq 1.47 \quad (14b)$$

Where a_b is the amplitude of the wave motion at the bottom.

$$a_b = \frac{\hat{u}_b T}{2\pi} \quad (15)$$

a_b and $\hat{u}_b$ are evaluated using linear wave theory. And the bed load and suspended load transport are calculated according to:

$$q_b = Bd_{50}U_{f,c}exp\left(-\frac{0.27\Delta d_{50}g}{\mu U_{f,wc}^2}\right) \quad (16)$$

B is a dimensionless bed total load transport coefficient, Δ is the relative density of sediments and μ is the ripple factor. Δ and μ are defined as:

$$\Delta = s - 1 = \frac{\rho_s}{\rho} - 1 \quad (17)$$

$$\left(\frac{C}{C'}\right)^{3/2} = 1/4 \quad (18)$$

ρ_s is density of the sediment, ρ is the density of water, and C' is the Chezy number related to the geometric characteristics of the bed material. It is calculated as:

$$C' = 18log\left(\frac{12h}{d_{90}}\right) \quad (19)$$

d_{90} is the sediment size for which 90% in weight of the bed material is finer. For uniform bed material d_{90} become identical to the specified value of d_{50}.

4. Results and Discussion

The result of the model simulation in Suralaya coastal area are presented. The first part of discussion is model validation. In the second part, results concerning the hydrodynamic simulation and in the third part, results concerning sediment transport rates are described.

4.1. *Model Validation*

To obtain accurate results, the validation is required by comparison of measured fields data and numerical model results. So the data validity for the next simulation is obtained. Model validation of field data can be performed on tidal, current, and temperature modeling results by Root Mean Square (RMSE) and percentage errors:

$$RMSE = \sqrt{\frac{1}{N}\left[\sum_{i=1}^{N}\left[\hat{X}_i - X_i\right]^2\right]} \quad (20)$$

$$Error = \frac{1}{N}\left[\sum_{i=1}^{N}\left|\frac{\hat{X}_i - X_i}{TP}\right|\right] * 100\ \% \quad (21)$$

Where, $RMSE$ is root mean square from errors, $\hat{X}_i$ is model result, X_i is field data and N is amount of data, TP is range of tide observation data.

Figure 2 showed the comparison between the measured tidal data and the model results, it is found that the average error value is 1.6%. So it can be concluded that the result of tidal parameter modeling approaching actual condition in the field.

IOP Conf. Series: Earth and Environmental Science **135** (2018) 012024 doi:10.1088/1755-1315/135/1/012024

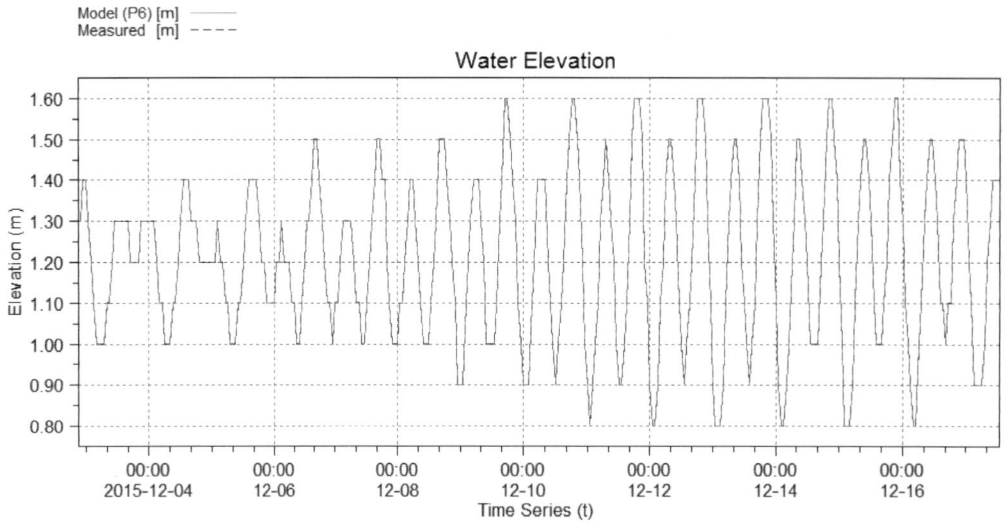

Figure 2. Comparison of measured and model water level at location 5, from 2th-17th December 2015. Measured data indicated with blue dash line and model data indicated with red line.

4.2. Hydrodynamic Simulation

The current velocity obtained by observing in the fields study area (for the five observation points) as given in Figure 1 is compared with the current velocity of simulation results shown in Table 1 and Figure 3. It can be seen that point 6 has the highest maximum current velocity about 0.224 m/s and 0.235 m/s for observed data and simulation results, respectively. And the comparison between the water level obtained from observing and that of obtained from simulation results showed a good agreement with the error is only 1.6%. The hydrodynamics model results simulated for West and East Season condition showed that the dominant current direction is from the Northeast and the Northwest for West Season and East Season, respectively within an average flow velocity of 12.44 cm/s as shown in Figure 3.

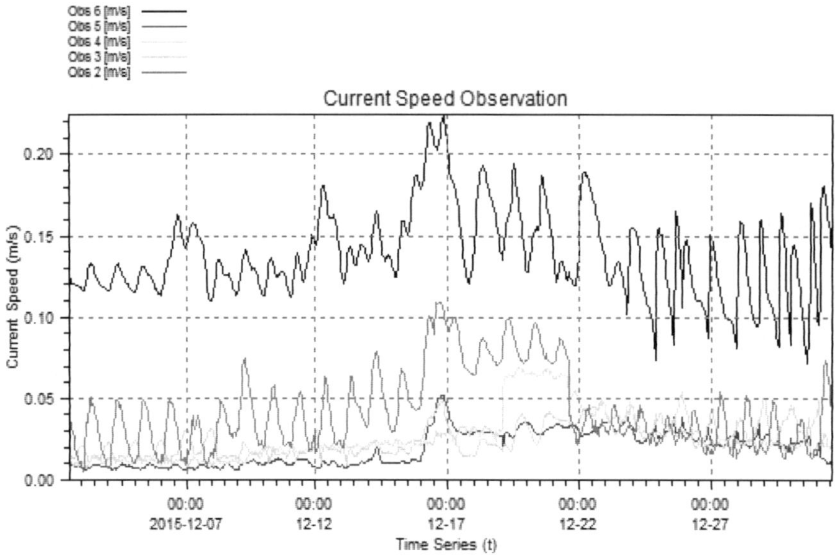

Figure 3. Current velocity on five different observation points

Table 1. Comparison current speed values between simulation data and field measured data

Observation point	**Model** (m/s)			**Measured** (m/s)		
	Max	**Min**	**Mean**	**Max**	**Min**	**Mean**
6	0.224	0.011	0.138	0.235	0.037	0.124
5	0.052	0.006	0.019	0.378	0.040	0.187
4	0.054	0.004	0.025	0.511	0,090	0.273
3	0.069	0.003	0.025	0.418	0.030	0.194
2	0.110	0.003	0.042	0.647	0.056	0.313

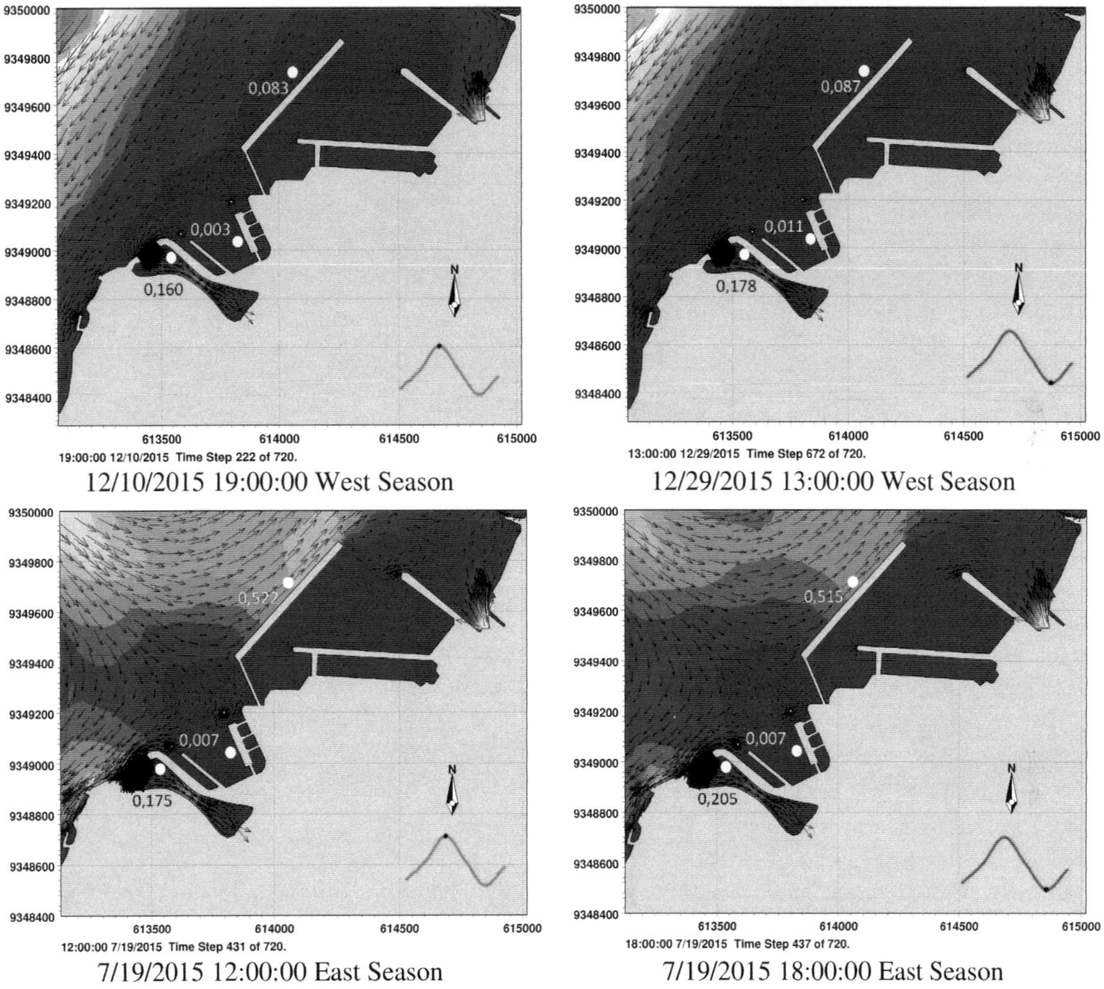

Figure 4. Current velocity pattern of Suralaya coastal area in two seasons (West and East Season) with two tidal condition (high and low tide)

The changing of seasons, followed by changing direction of wind movement. The season change give effect to the changes in the direction of waves and currents in the sea as shown in Figure 4. The current pattern change parallel to the shore in the West Season, while in the East Season, the flow pattern is perpendicular and then reach the land the current pattern becomes parallel to the coast.

4.3. Sediment Transport Simulation

Sediment transport modeling is an important step to know the pattern of sediment movement in a region to find out how many sediment transport rates and it also can be used to identify the area with tendency of erosion or sedimentation. Figure 5, shows that areas of light blue tend to erode, while dark blue areas tend to be deposition.

Based on numerical calculations on modeling, sediment transport rates in Suralaya coastal area in general in this study dominantly move to the southwest. This is congruence to the direction of simulated current movement in hydrodynamic modeling. The numerical model results show that the sediment transport rate is 51.4 m3 / day (in the point 5) and 32.32 m3 / day (in the point 6).

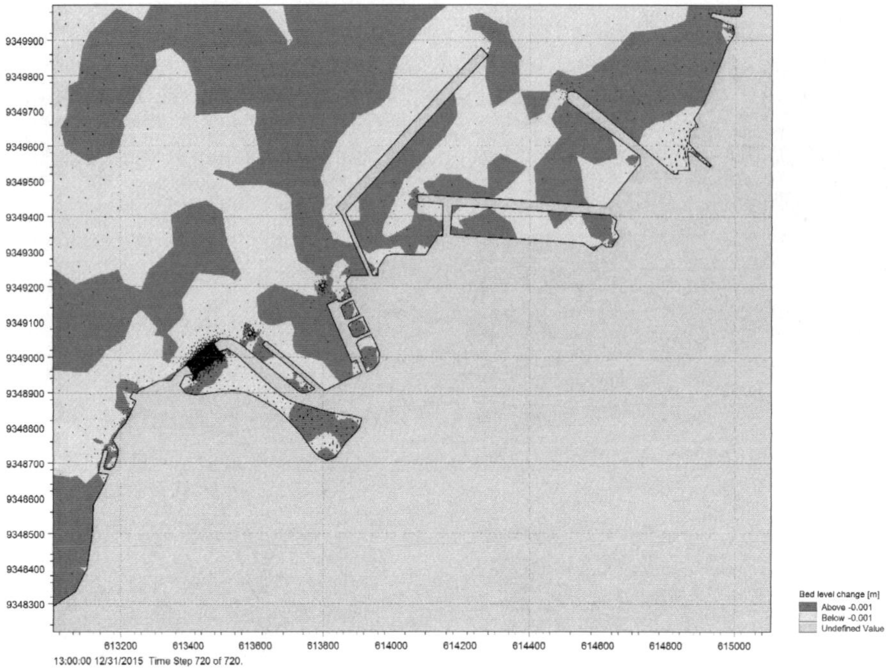

Figure 5. Bed Level Change in Last time step

5. Conclusions

Based on hydrodynamic and sediment transport modelling for Suralaya coastal area, it at can be concluded as follows:

- First, Hydrodynamic numerical model (Mike21 HD) gives simulation results of current velocity along with its movement pattern and water surface elevation which is very satisfying with existing conditions in the field. This has been proved by the validation between the water surface elevation models with the measurement with very satisfactory results.
- The simulation results of the current velocity have shown that point 6 has the highest current velocity between the other points. Maximum Speed Point 6 to be 0.224 ms^{-1} and minimum speed of 0.011 ms^{-1}. Second, water surface elevation has also got very good results, it can be seen from the error error is only 1.6%. The highest water level of simulation was 1,603 m and the lowest was 0.772 m.
- Third, the sediment transport module (MIKE21 ST) result shown that the direction of sediment transport movement in this study is dominant to the southwest. The numerical model calculations show that the sediment transport rate is 51.4 m3 / day (Point 5) and 32.32 m3 / day (Point 6).

Acknowledgements

The first author is grateful for the supported by Higher-Education, Ministry of Education and Culture RI, LPPM-ITS, Institut Teknologi Sepuluh Nopember (ITS) Surabaya, Indonesia. This research was partially supported by Grant of PMDSU-LPPM-ITS (No.135/SP2H/LT/DPRM/IV/2017). And authors also thank to the Centre of Research and Development of Marine and Coastal Resources, the Ministry of Marine Affairs and Fisheries Republic of Indonesia for providing the facilities of DHI's Mike 21/3 model.

References

[1] Liang, B., Li, H., and Lee, D. 2007. Numerical study of three-dimensional suspended sediment transport in waves and currents. *Ocean Engineering*, *34*, 1569–1583.

[2] Suntoyo, Fattah, A.H., Fahmi, M.Y., Rachman, T. and Tanaka, H. 2016. Bottom shear stress and bed load sediment transport due to irregular wave motion. *ARPN Journal of Engineering and Applied Sciences*, 11 (2), 825-829.

[3] Wijaya, M.M., Suntoyo, and Damerianne, H.A., 2016. Bottom shear stress and bed load sediment transport formula for modeling the morphological change in the canal water intake. *ARPN Journal of Engineering and Applied Sciences*, 11 (4), 2723-2728.

[4] Suntoyo, Tanaka,H., and Sana, A. 2008. Characteristics of turbulent boundary layers over a rough bed under saw-tooth waves and its application to sediment transport. *Coastal Engineering*, 55 (12), 1102-1112.

[5] Suntoyo, and Tanaka, H. 2009. Effect of bed roughness on turbulent boundary layer and net sediment transport under asymmetric waves. *Coastal Engineering*, 56 (9), 960-969.

[6] Tanaka, H.. Suntoyo and Nagasawa, T. 2003. Sediment intrusion into Gamo lagoon by wave overtopping. *Proceedings of the 28th International Conference*, pp. 823-835.

[7] Lumborg, U., & Windelin, A. (2003). Hydrography and cohesive sediment modelling: Application to the Romo Dyb tidal area. *Journal of Marine Systems*, *38*, 287–303.doi:10.1016/S0924-7963 (02)00247-6

[8] Martinelli, L., Zanuttigh, B., & Lamberti, A. 2006. Hydrodynamic and morphodynamic respone of isolated and multiple low crested structures: experiments and simulations. An International Journal for Coastal. *Harbour and Offshore Engineers*, *53*, 363–379.

[9] Danish Hydraulic Institute. 2007. *MIKE21 HD-Flow model hydrodynamic module user guide.* Horsholm, Denmark: Danish Hydraulic Institute.

[10] Abbott, M.B., H.M. Petersen and O. Skovgard. 1978. On the Numerical Modelling of Short Waves in Shallow Water. Journal of Hydraulic Research, 16(3).

[11] Mellor, G. L. 1998. User's Guide for a 3D, Primitive Equation, Numerical Ocean Model, Atmospheric and Ocean Sciences Program, Princeton University, NJ.

[12] Ramming, H.G., Z. Kowalik. 1980. Numerical modelling of marine hydrodynamics: Applications to dynamics physical processes. Elsevier publishing for Oceanography Series 26, Amsterdam (NL).

[13] Danish Hydraulic Institute. 2007. *MIKE21 NSW-Nearshore spectral wave module user guide.* Horsholm, Denmark: Danish Hydraulic Institute.

[14] Danish Hydraulic Institute. 2007. *MIKE21 ST Non-cohesive sediment transport module user guide.* Horsholm, Denmark: Danish Hydraulic Institute.

[15] Bijker, E.W., 1971. Longshore transport computation. J. Waterways Harbors Division 97, WW4, 687--701.

Experimental Study on Interference Effects of Two Tandem Cylinders Wrapped Around by Triple Helical Rods with Gap on Induced Drag

R.W. Prastianto[1,a], K. H. Dwipayana[1,b], N. Syahroni[1,c] and B. Pumbarino[1,d]

[1]Department of Ocean Engineering, Institut Teknologi Sepuluh Nopember, Kampus ITS Sukolilo Surabaya (60111), Indonesia

[a]rudiwp@oe.its.ac.id, [b]khdwipayana@gmail.com, [c]nur.syahroni@gmail.com, [d]pumbarino@gmail.com

Abstract. This paper examines the results of laboratory experiments to investigate the effect of interference of two tandem cylinders covered by triple helical rods with gap to the induced drag force. Two identical rigid models are horizontally positioned with roll support on both ends of each cylinder. Uniform air flow in subcritical regime that correspond to Reynolds number (Re) of $1.6 \times 10^4 \sim 6.5 \times 10^4$ perpendicularly flowed to the models in the wind tunnel with three variations of the distance between the cylinders which are 1.75D, 3D and 5D. At Re $= 4.2 \times 10^4$ the results show that the maximum shielding effects occur in the rear cylinder at the distance of 1.75D so the drag coefficient (C_D) is reduced to 93.6% compared to single cylinder case. This shielding effect will weaken with increasing the distance between the cylinder. In contrast, the fluid flow interference effect on the front cylinder increases due to increasing of spacing between the two cylinders and still occurred at that spacing of 5D until C_D reduction reached 10% of the single cylinder case.

Keywords: Triple helical rods with gap, Drag coefficient, Tandem configuration, Interference effects.

1. Introduction

Conductors, raisers and jacket legs of offshore platforms are cylindrical structures that generally receive the load of current during their operation. The current load on a certain Reynolds numbers will evoke the phenomenon of Vortex Induced Vibration (VIV). Such structures are generally installed in an array that allows interference of their vortex flow caused by the current.

The classification of fluid flow interference is given by Zdravkovich [1] where tandem configuration is classified as wake interference. Pappaioannou, et al. [2] examined the interference of two bare cylinders through numerical study. The results show that the mean drag of the front cylinder is always greater than the rear cylinder in all cases of reduced velocity and distance variations. This caused by shielding effect.

Previously, experimental studies on the VIV of free-hanging cantilever cylinder already done by Prastianto, et al. The studies covered a single, two cylinders in tandem, side-by-side and staggered configurations and three cylinders in a group cases. In all the case studied drag and motion amplitude of the cylinders were evaluated as well as wake interference effects to the cylinders responses [3, 4, 5, 6].

Helical rods as a suppression device were reviewed by Zdravkovich [7] which is the one solution to reduce VIV. Then Sugiwanto, et al. [8] found the optimum configuration for the triple helical rods with gap through a numerical study, i.e. with a rod diameter of 0.0625D where D is the diameter of the cylinder. Referring to the Sugiwanto, et al. [8]; Prastianto, et al. [9] proceed it with laboratory experiments and the results indicated that the device is able to reduce the drag force up to 50.0% at Re $= 2.36 \times 10^4$ compared to the single cylinder case.

This paper presents the effect of interference of two rigid cylinders covered by triple helical rods with gap to the drag force at Reynolds number of 4.2×10^4 due to variation of the distance between them. The results are validated and evaluated to previous relevant study which conducted on multiple tandem bare cylinders [11].

2. Experimental Method

Cylinders of the model are made of aluminum tube with diameter (D) of 5.08cm and 48cm in length (l). While the rods with diameter (d) of 0.3cm are made of stainless steel wire. The rods are helically wrapped around on each cylinder in triple configurations with 0.0625D gap (g) as depicted in Figures 1 and 2.

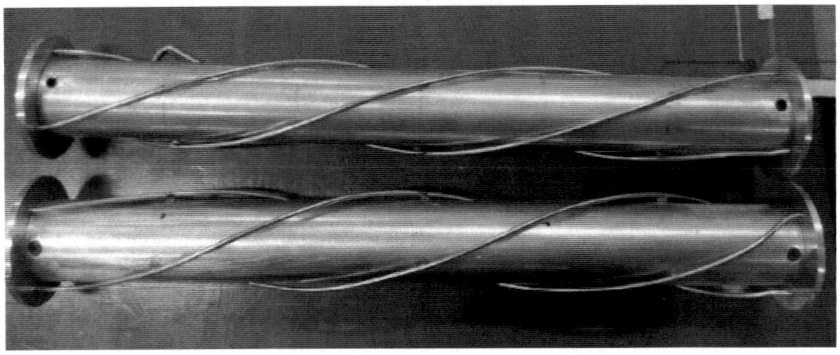

Figure 1. The model of cylinders wrapped around by triple helical rods with gap

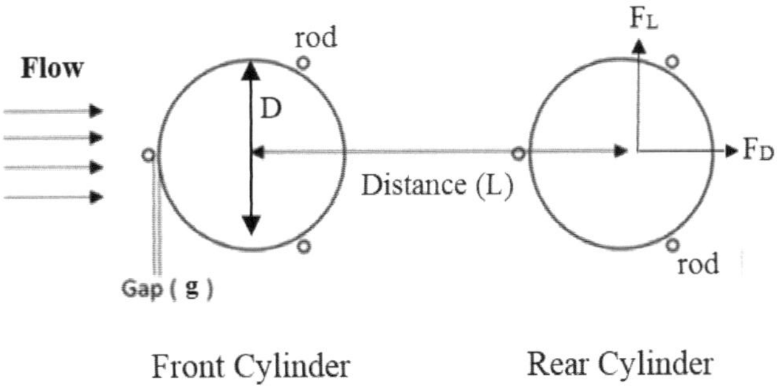

Figure 2. Two dimensional illustration of the experiment

Experiments were performed in Low Speed Wind Tunnel (LSWT) with 0.5m × 0.5m × 1.25m test section size with the model are horizontally installed in tandem arrangement. The tunnel belongs to a laboratory of BBTA3, Agency for the Assessment and Application of Technology (BPPT), South Tangerang, Indonesia. The test used three variations of center-to-center distance (L) between the cylinders which are 1.75D, 3D and 5D, respectively. Direction of in-coming air flow is perpendicular to the model at speed range of 5m/s ~ 20m/s that correspond to Re of 1.6×10^4 up to 6.5×10^4. Detail of the experiment is explained in Prastianto et al. [12]. Load cells for measuring drag are installed at the all ends of the model cylinders which is supported by special frame as roll supports (see Figures 2 and 3).

3. Results and Discussions

Validation of the experiment was done by comparing the drag coefficient (C_D) for the case of single bare cylinder at tested range of Re to the relevant reference data [10]. The result showed a very good agreement at Re range of 3.5×10^4 up to 4.8×10^4 as shown in Figure 4. Thus the model could be used for further intended cases at such Re range. The analysis in the present paper is focused for Re = 4.2×10^4. At relatively low Re, the measured drag were smaller than the reference data which is may caused by unstable flow speed generated in the test section by propeller system of the tunnel at very low rotation frequencies.

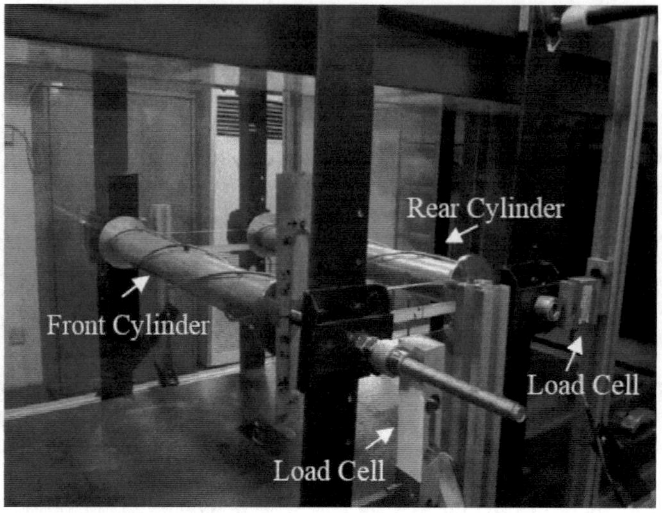

Figure 3. The model installation in the test section (e.g. for a case of L/D = 5).

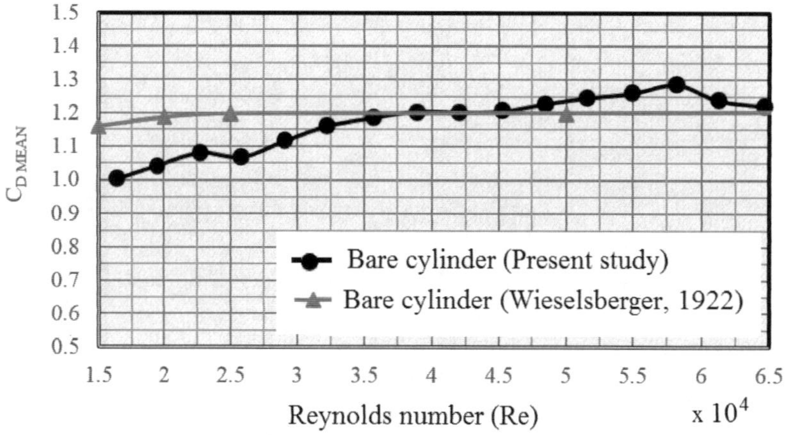

Figure 4. Validation of the experiment.

Figure 5 shows the experiment result of C_D of the two cylinders at Re = 4.2×10^4. For all three variations of the distance between the cylinders (L/D), C_D of the rear cylinder are smaller than that of the front cylinder which are 0.07, 0.23 and 0.55, respectively. The lowest C_D (at a value of 0.07) is due to strong shielding effect onto the rear cylinder caused by existence of the front cylinder that occurs at the L/D = 1.75. The rear cylinder was almost totally covered from the incoming flow that in turn caused a significant reduction on the C_D of about 93.6% compared to the C_D of single cylinder with helical rods. Such shielding effect will gradually weaken with increasing the distance between the cylinders that caused the C_D gradually increased and became 0.55 at L/D = 5.0.

Meanwhile, interference effect of the fluid flow around the cylinders also causes reduction on C_D of the front cylinder. C_D of the front cylinder at L/D = 1.75 is 1.06 that slightly decreased of about 2.5% from the case of single cylinder with helical rods. This reduction slightly increase due to increasing the distance between cylinders. At L/D = 5.0 the flow interference around the two cylinders still occurred which causes reduction on C_D of the front cylinder by about 10% compared to the C_D of the single cylinder with helical rods.

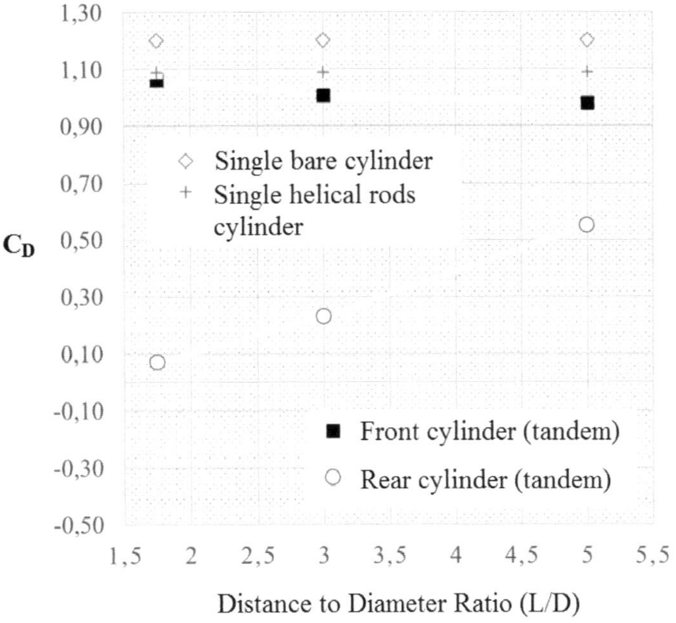

Figure 5. Drag coefficient (C_D) of the cylinders for present study.

The result of the present study has also been compared to the relevant existing data such as given by Igarashi on Kitagawa, et al [11]; Zdravkovich, et al on Kitagawa, et al [11] and Kitagawa, et al [11] (see Figure 6). All these previous studies were also investigating the characteristics of drag and flow around circular cylinders in tandem arrangement but for bare cylinders cases and were performed at subcritical Reynolds number (2.2×10^4, 3.1×10^4, and 3.55×10^4) that are comparable to the present study. Generally, the comparison showed a good agreement of C_D at almost all of the distance variation (L/D) tested. Slight differences on the C_D of the present study to the other data may caused by the existance of the additional triple helical rods on the cylinders.

4. Conclusions

In the tandem arrangement of two cylinders, the front cylinder causing a shielding effect to the rear cylinder for certain spacing between them and fluid flow interference from around the two cylinders resulting in reduction of the drag coefficient of the rear cylinder. The largest reduction occured for the distance between cylinders of 1.75D by about 93.4%. This shielding effect decreased (that means the drag coefficient increased) with increasing the spacing between the cylinders. At the same time, in contrast, drag coefficient of the front cylinder gradually decreased as the spacing between the cylinders increasing and reached a largest reduction of 10% at the spacing of 5D. At this spacing the flow interference around the two cylinders still occurred. In the present study, the increase of the C_D of the rear cylinder occured much more drastic than that the reduction of the C_D of the front cylinder as the spacing between the cylinders increased.

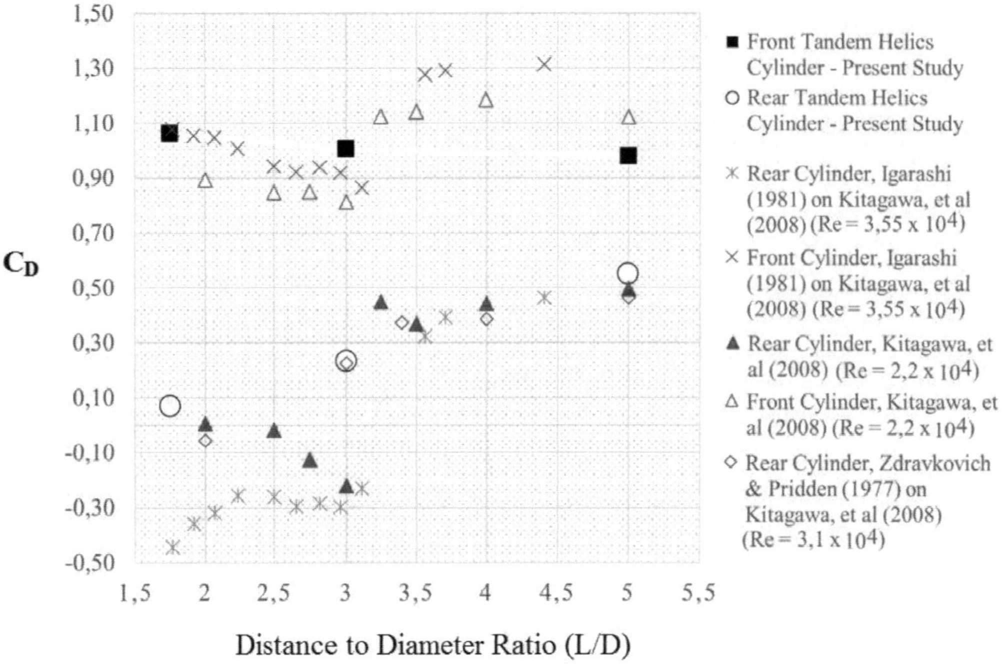

Figure 6. Comparation of the present study to the relevant existing data.

Acknowledgements

The authors wish to thank Directorate General of Higher Education (DIKTI) – Ministry of Research, Technology and Higher Education, Republic of Indonesia for the financial support through a research grant scheme of “Penelitian Unggulan Perguruan Tinggi (PUPT)” 2016. Also thanks to Institute for Research and Community Services (LPPM) – Institut Teknologi Sepuluh Nopember (ITS) Surabaya for the assistance and guidance. All contributions from National Laboratory for Aerodynamics, Aeroelastics and Aeroacoustics Technology, Agency for the Assessment and Application of Technology (BPPT), Indonesia, are highly apreciated.

References

[1] Zdravkovich, M. M. (1987). The effects of Interference between circular cylinders in cross flow. *Journal of Fluids and Structures,* 239-261.
[2] Pappaioannou, G. V., Yue, D. K., Triantafyllou, M. S., & Karniadakis, G. E. (2007). On the Effect of Spacing on the Vortex-Induced Vibrations of Two Tandem Cylinders. *Journal of Fluids and Structures*, 833-854.
[3] Prastianto, R. W., Otsuka, K., & Ikeda, Y. (2009). Vortex-induced Vibrations of a Flexible Free-hanging Circular Cantilever. ITB *J. Eng. Sci., Vol. 41, No. 2*, 111-125.
[4] Prastianto, R. W., Otsuka, K., & Ikeda, Y. (2008). Dynamics of Two Flexible Hanging-off Circular Cylinders in Staggered Configurations, *Proceedings of the Eighth (2008) ISOPE Pacific/Asia Offshore Mechanics Symposium*, Bangkok, Thailand, November 10-14, 2008.
[5] Prastianto, R. W., Otsuka, K., & Ikeda, Y. (2008). Experimental Study on Two Flexible Hanging-off Circular Cylinders undergoing Wake Interference, OCEANS ’08 MTS/IEEE Kobe-Techno_Ocean’08 – Voyage toward the Future OTO’08, Kobe, Japan, April 8-11, 2008.

[6] Prastianto, R. W., Otsuka, K., & Ikeda, Y. (2008). Hydrodynamic Forces on Multiple Hanging-off Circular Cylinders in Uniform Flows, 18th International Offshore and Polar Engineering Conference (ISOPE 2008), Vancouver-BC, Canada, July 6-11, 2008.
[7] Zdravkovich, M. M. (1981). Review and Classification of Various Aerodynamics and Hydrodynamics Means for Supressing Vortex Shedding. *Journal of Wind Engineering and Industrial Aerodynamics*, 145-189.
[8] Sugiwanto, A., Prastianto, R. W., Murdjito, Djatmiko, E. B. (2013). A numerical study on cylinders with passive control device of helical rods with gap for reducing vortex-induced vibration. *International Conference on Sustainable Infrastructure and Built Environment (hal 357- 368).* Bandung.
[9] Prastianto, R. W., Musthofa A. Z., Arianti, E., Handayanu, Murdjito, Suntoyo, and Fariduzzaman. (2014). Triple Helical Rods with Gap as Passive Control Device for Reducing Fluid Forces on A Cylinder. *The 9th International Conference on Marine Technology,* Institut Teknologi Sepuluh Nopember, Surabaya, Indonesia, MT-27, Oktober 24-25.
[10] Wieselsberger, C. (1922). New Data on the Laws of Fluid Resistance. T. N. No. 84, N. A. C. A.
[11] Kitagawa, T and Ohta, H. (2008). Numerical investigation on flow around circular cylinders in tandem arrangement at a subcritical Reynolds number. *Journal of Fluids and Structures 24,* 680-699.
[12] Prastianto, R. W., Syahroni, N., and Murdjito (2016). Kajian Evaluasi Kinerja Piranti Inovasi Baru Pereduksi Getaran akibat Vorteks pada Komponen Struktur Anjungan Lepas Pantai dalam Konfigurasi Jamak. Penelitian Unggulan Perguruan Tinggi (PUPT) – Tahun ke-2, LPPM - Institut Teknologi Sepuluh Nopember, Surabaya, Indonesia (in Indonesia Language).

IOP Conf. Series: Earth and Environmental Science **135** (2018) 012026 doi:10.1088/1755-1315/135/1/012026

Project Scheduling Based on Risk of Gas Transmission Pipe

Silvianita[1,a], A Nurbaity[1,b], Y Mulyadi[1,c], Suntoyo[1,d] and D M Chamelia[1,e]

[1]Ocean Engineering Department, Faculty of Marine Technology, Sepuluh Nopember Institute of Technology, Indonesia

Email : silvianita.pc@gmail.com

Abstract. The planning of a project has a time limit on which must be completed before or right at a predetermined time. Thus, in a project planning, it is necessary to have scheduling management that is useful for completing a project to achieve maximum results by considering the constraints that will exists. Scheduling management is undertaken to deal with uncertainties and negative impacts of time and cost in project completion. This paper explains about scheduling management in gas transmission pipeline project Gresik-Semarang to find out which scheduling plan is most effectively used in accordance with its risk value. Scheduling management in this paper is assissted by Microsoft Project software to find the critical path of existing project scheduling planning data. Critical path is the longest scheduling path with the fastest completion time. The result is found a critical path on project scheduling with completion time is 152 days. Furthermore, the calculation of risk is done by using *House of Risk* (HOR) method and it is found that the critical path has a share of 40.98 percent of all causes of the occurence of risk events that will be experienced.

1. Introduction

Project has a time limit on which the project should be completed before or right at the designated time. However, in reality on the ground, a project does not always run according to the time set since there are factors that hinder the course of project work. Thus, it is necessary to have good project planning so as not to disrupt the project activity.

The function of planning is to reduce uncertainty [1]. By doing good planning, like what to do, the timing, resources, etc., it will make the activity run more clear, sure, and efficient. Improving the efficiency of operations, one with good planning, will make the project run smoothly. Therefore, planning is an important factor in the execution of a project. Planning a project starts from the scheduling or time of project work.

Scheduling is the allocation of time available to carry out each job in order to complete a project until the maximum results are achieved taking into account the constraints. To avoid various obstacles in the process of project work, good scheduling management is required. Scheduling management is done to deal with uncertainty and negative impact on project completion time.

The gas transmission pipeline project is a major project that is expected to supply gas to the areas of East and Central Java. This project certainly requires an optimal scheduling and control planning in order to avoid any deviation in project work. The first objective of this paper is to identify the critical path of gas transmission pipeline project. The second objective is to identify the priority agency of risk agents using the House Of Risk (HOR) method and determines which critical path has the smallest value of the risk agent, thereby reducing the risk that occurs during the course of the project. The advantage of the House of Risk (HOR) method knows what factors can cause the occurrence of risk events. In addition, with this method can be determined priority risk agents and risk percentage, so that can be done appropriate risk handling.

2. Project Scheduling

2.1 Scheduling Management

Scheduling management consists of two words of management and scheduling. Management is a science of the art of organizational leadership that consists of planning, organizing, implementing and controlling of limited resources in order to achieve effective and efficient goals and objectives [2]. Management is an integrated process in which individuals as part of an organization are involved in maintaining, developing, controlling, and running programs that are all directed towards a set goal and are ongoing over time [3].

Scheduling is the coordination of time in production activities, so that the allocation of raw materials and auxiliary materials and equipment to the facility or part of processing within the factory can be made at a given time [4]. In scheduling planning there has been provided specific guidelines for completing a project activity more quickly and efficiently [5].
Efficient scheduling management criteria that achieve the greatest reduction in the time required to complete a project in question while maintaining the economic feasibility of the use of available resources [6].

Thus, scheduling management is planning, organizing, implementing, and controlling activities on production activities schedules and limited resources, so as to achieve goals and objectives more quickly and efficiently.

Scheduling management is used to control project activity every day. The aspects of scheduling management are determining project scheduling, measuring and reporting on project progress, comparing scheduling with actual project progress in the field, determining the consequences of comparisons of schedules with progress in the field at the end of project completion, planning for handling to address those outcomes, and renew project scheduling.
Management continues to seek new and better control techniques to address the complexity, mass data, and narrow time that characterize many of today's highly competitive industries and environments, and look for better methods to provide technical data and costs to customers [7].

There are various software to support the optimization of project scheduling management. One of the most popular software is Microsoft Project (Ms Project). Ms Project developed since 1984 with Ms-DOS base. Later, the software was developed on a Windows basis and known as Microsoft Project [8]. Meanwhile, Ms Project is a software project management program developed and sold by Microsoft designed to assist project managers in developing plans, assigning resources for tasks, tracking progress, managing budgets, and analyzing workloads [9].

2.2 Risk

The definition of risk is the probability of an event that will cause a loss or failure or potential failure [10]. Risk is a decision-making condition in which the probability of a certain alternative will be known to lead to a desired goal or outcome [11]. The general formula used to perform the calculation of risk values is [12] :

$$\text{Risk} = \text{Consequence x Likelihood} \quad (1)$$

A project is an effort that mobilizes available resources, organized to achieve certain important goals, objectives, and expectations and must be completed within a limited time in accordance with the agreement.

2.3 *House of Risk* (HOR)

Risk Analysis to be used in this research is by using House of Risk (HOR) method. This method has the advantage of knowing which risk agents have an effect in the course of a project, so that risk management can be done accordingly.
House Of Risk Method (HOR) to identify, analyze, measure and mitigate potential risks in a project.

The application of House Of Risk (HOR) consists of two stages as follows [13] :

a. House Of Risk (HOR) 1, is used to identify potential risk events and risk agents. Results of House Of Risk (HOR) 1 are grouping of risk agents into priority risk agents in accordance with Aggregate Risk Potential (ARP) and Pareto Diagrams.
b. House Of Risk (HOR) 2, is used for the design of mitigation strategies undertaken for the handling of priority risk agents. Results of House Of Risk (HOR) 1 will be used as initial data on House Of Risk (HOR) 2.
   And also House Of Risk (HOR) 2 gives priority proactive step effective financial ability and other resources. The steps of House Of Risk (HOR) 2, among others :
   - Determine some of the top ranking risk agents to be the cause of the risks that will be prioritized for handling.
   - Identify relevant Proactive Action (PA) steps to prevent risk agents.
   - Determine the level of relationship between each PA and the risk agent.
   - Calculates the total effectiveness of each proactive action.
   - Assess difficulty level (Dk) in implementing PA.
   - Calculates the total effectiveness ratio with difficulty level.
   - Priority ranking on proactive action is most effective at reducing risk occurrence according to company capability.

In this study, the risk analysis will only be done until the House Of Risk (HOR) stage. Steps in House Of Risk (HOR) 1 are as follows :

a. Identify the occurrence of risk events (Risk Event, Ei).
   These risks are all possible events of project implementation.
b. Assess the level of impact or severity (Severity, Si).
   This severity value states how much of a disruption caused by an event risk to the implementation of the project.
   The severity value consists of 1, 2, 3, and 4. Where 1) shows that the impact is very small (negligible value), 2) illustrates that impacts have little effect on the project (no injuries, low financial loss), 3) means impacts in the medium category on the sustainability of the project (first aid treatment, medium finacial loss), and 4) indicate that the impacts are included in the medical treatment requires, high financial loss category.
c. Identify the risk-causing agent (Risk Agent, Aj), ie any factor that can lead to occurrence of identified risk events.
d. Determining the probability of occurrence (Occurance, Oj) of a risk agent.
   This occupation represents the probability of occurrence of a risk agent, resulting in the emergence of one or more risk effects. The occurance value of 1, 2, and 3. 1) means only occurring once a month, 2) means occurring once a week, and 3) means occurring once a day.
e. Measure the correlation (Correlation, Rij) between a risk event and the risk-causing agent.
   If a risk agent causes an occurrence of a risk event, then there can be a correlation. Correlation value consists of 0, 1, 3, 9, where 0) indicates no correlation relationship, 1) describes the correlation relationship is small, 3) represents medium correlation, and 9) describes the high correlation.
f. Calculates the value of the risk priority index or Aggregate Risk Potential (ARPj).
   This priority index will be used as a consideration for determining the priority of risk handling in data entry on House Of Risk (HOR) 2 and determining which agents or causes of risk are most influential in the sustainability of a project. Calculation of ARP value is as follows:

IOP Conf. Series: Earth and Environmental Science **135** (2018) 012026 doi:10.1088/1755-1315/135/1/012026

$$ARPj = Oj \ \Sigma Si.Rij \qquad (2)$$

Where,
ARPj: Aggregate Risk Potential (Risk Priority Index)
Oj : Occurance (Opportunity Emergence Risk Agent)
Si : Severity (Impact of Risk Events)
Rij : Correlation Between Risk Agent and Risk Event

3. Analysis and Discussion

3.1 Project Scheduling Data

Project scheduling data was obtained from PT. X who contributed as sub-contractor to the ongoing project. This scheduling data includes data on the name of the activity, the duration of the activity (in the day), the start date and its completion of an activity, and the sequence of interdependencies between the project activities.

3.2 Critical Path Determination

After knowing and writing the relationship between activities into Microsoft Project, the next step is to find a critical path. The critical path is the path with the longest path with the fastest completion time. To obtain a critical path, forward calculations are required to obtain ES and EF values as well as backward calculations to obtain LS and LF values. Afterwards, the total float values obtained from the calculation of LF - ES or LS - ES values. The critical path can be determined by looking at activities that have a total value of float = 0 (LF - EF = 0 or LS - ES = 0).
In accordance with the results obtained from the Microsoft Project program, two critical paths are obtained from the scheduling of this project:

a. Critical path with duration of 194 days.
   It consists of activities numbered 1-2-3-7-10-30-32-43-44-45-46-47-48.
b. A critical path with duration of 152 days.
   It consists of activities numbered 1-2-3-7-11-12-13-19-32-43-44-45-46-48

It is known that the critical path is the longest path with the completion of the fastest time, so it can be concluded that the critical path used in this paper is a critical path with a duration of 152 days consisting of Mobility and Survey activities - ROW Clearing - Load and Unload Pipe - Stringing - Line Up Welding - NDT - FJC - Holiday Test - Lowering - Backfill - Pigging - Hydrotest - Aerial Marker, Warning Sign, Marker Sign - Re-Instatement.

3.3 Risk Analysis with House of Risk (HOR) Method

Risk Analysis to be used in this paper is to use House of Risk (HOR) method. Reference [13] uses the HOR method to identify, analyze, measure and mitigate potential risks in a project.
Application of HOR consists of two stages, namely :

a. HOR 1 is used to identify potential risk events and risk agents. The result of HOR 1 is grouping of risk agents into priority risk agents in accordance with the results of Aggregate Risk Potential (ARP) calculations.
b. HOR 2 is used for the design of mitigation strategies undertaken for the handling of priority risk agents. The result of this HOR 1 will be used as initial data on HOR 2.

In this paper, risk analysis will only be done until stage HOR 1. This is because there is no data about risk mitigation. The steps in HOR 1 are as follows:

a. Identify the occurrence of risk events (Risk Event, Ej). These risks are all possible events of project implementation.
b. Assess the level of impact or severity (Severity, Sj). This severity value states how much of a disruption caused by an event risk to the implementation of the project. The severity value consists of 1, 2, 3, and 4. Where 1 indicates that the impact is very small (negligible value), 2 illustrates that the impacts have little effect on the project (no injuries, low financial loss), 3 means impacts in the moderate category on the sustainability of the project (first aid treatment, medium finacial loss), and 4 indicates that the impacts are in the category of medical treatment requires, high financial loss.
c. Identify the risk-causing agent (Risk Agent, Aj), ie any factor that can lead to occurrence of identified risk events.
d. Determining the probability of occurrence (Occurance, Oj) of a risk agent. This Occurance states the probability of occurrence of an occurrence of a risk agent, resulting in the emergence of one or more risk effects. The occurance value of 1, 2, and 3. 1 means that it only happens once a month, 2 means it occurs once a week, and 3 means occurs once a day.
e. Measure the correlation (Correlation, Rij) between a risk event and the risk-causing agent. If a risk agent causes a risk to occur, then there can be a correlation. The correlation value consists of 0, 1, 3, 9, where 0 indicates no correlation relationship, 1 represents a small correlation relationship, 3 represents a moderate correlation, and 9 represents a high correlation.
f. Calculates the value of the risk priority index or Aggregate Risk Potential (ARPj). This priority index will be used as a consideration for determining priority risk handling in data entry in HOR 2 and determining which agents or causes of risk are most influential in the sustainability of a project. Calculation of ARP value is as follows:

$$ARPj = Oj\ \Sigma Sj.Rij$$

Where Oj is the occurence value of the risk agent, Si is the severity value of the risk event, and Rij is the correlation value between the risk events and the risk agent.

The risk analysis in this paper uses the HOR phase 1 method that includes the stage of identification and measurement of risk. Risk identification includes the risk events and its severity, risk agents and chance occurrence opportunities, as well as correlations between risk events and risk agents (correlations).

1. Identify Risk Events and Value of Severity

Risk events are the risks that will occur in the project and the impact on the sustainability (severity) of the project itself. This severity value can be seen in Table 1.

Table 1. Severity Value

Severity Value	Description
1	the impact is very small (negligible value)
2	the impacts have little effect on the project (no injuries, low financial loss)
3	the impacts in the moderate category on the sustainability of the project (first aid treatment, medium finacial loss)
4	the impacts are in the category of medical treatment requires, high financial loss.

The results of identification of risk events and severity values are as follows :
It is known that there are 10 risk events that have scores of two that indicate that the impact of the risk event has little effect on the sustainability of the project activity, 12 scores of three that indicating that the risk event has a moderate impact on the sustainability of the project activity, and Two four-point scores indicating that such risk events have a serious impact on the course of the project activity.

2. Identify Risk Agent and Occurence Value
Risk agents are what factors are responsible for the occurrence of risk events. The value of occurrence is the probability of occurrence of risk agents. The value of this occurence can be seen in Table 2.

Table 2. Occurence Value

Occurance Value	Description
1	Once a month
2	Once a week
3	Once a day

The results of identification of risk agents and occurrence values are consist of 33 risk agents that potentially trigger the occurrence of risk events. Based on the value of the probability scale, the risk agent is divided into 20 risk agents with occurence value of one which indicates that the possibility of the risk agent occurs only once in a month. 11 risk agents that have a two occurence value that suggests that the possibility of a risk agent appears only once in a week of operation and there are two risk agents that have a three occurence value indicating that the agent occurs once in a single day of operation.

3. Calculation of Aggregate Risk Potential Value (ARP)
Prior to calculating the value of Aggregate Risk Potential (ARP), first create a House of Risk table (HOR) phase 1. In this HOR table there is the value of occurence and risk agents, severity and risk events, and correlation values between risk agents and risk events . These values will then be used in calculating the ARP value. Through the results of risk analysis, obtained ARP value which is the result of HOR output phase 1.
Aggregate Risk Potential (ARP) calculation aims to determine the priority level in the process of handling a risk agent. The risk agent will then be sorted by the highest to lowest ARP values. The ARP calculation process involves several elements, ie the level of impact of a risk event (Si), the chance level of the occurrence of the risk agent (Oj), and the level of connectedness between the risk agent and the risk event (Rij). The calculation of ARP is obtained by using the following calculation :

$$ARPj = Oj\ \Sigma Si.Rij$$

Where,
ARPj: Aggregate Risk Potential (Risk Priority Index)
Oj : Occurance (Opportunity Emergence Risk Agent)
Si : Severity (Impact of Risk Events)
Rij : Correlation Between Risk Agent and Risk Event.

In Table 3 there is an ARP calculation result and a priority risk ranking agent. These ratings are arranged based on ARP values from the largest to the smallest. The greater the value of ARP, the smaller the ratings and risk agents are also increasingly prioritized. The ARP calculation results and the 10 ranking priority risk agents can be seen in Table 3.

Table 3. ARP Calculation Results and Risk Priority Ranking

Risk Agent Code	Ranking	ARP
A7	1	172
A25	2	146
A4	3	117
A1	4	102
A11	5	99
A29	6	94
A12	7	88
A14	8	86
A10	9	78
A5	10	71

Based on Table 3 the highest ARP value is found in risk agents with code A7 with ARP value is 172 and ranking 1. This indicates that the risk agent has the highest priority in handling it compared to the others. This is because, the higher the ARP value of a risk agent, it will be directly proportional to the level of impact that will be generated in the process of completion of a project.

4. Grouping of Priority Risk Agent With Pareto Calculation

In handling risk not all risk agents get a handling. This is caused by several factors, such as the cost of the handling process and the level of impact is considered too small. Thus, not all risk agents are handled by the project owner, unless the risk agent is considered a priority.

The Pareto diagram is a diagram developed by an Italian economist named Vilfredo Pareto in the XIX Century [14]. Determination of the category of risk agents is done by using Pareto law or known as 80 : 20 law. The application of Pareto law at risk is that 80% of project owner losses are attributable to a crucial 20% risk. By focusing 20% of crucial risks, the impact of project owner risk by 80% can be overcome.

The process of constructing a pareto diagram consists of six steps, namely [15], [16] :

a. Determine the method or meaning of data classification, eg based on problem, cause, type of mismatch, and so on.
b. Determine the units used to sequence these characteristics, such as dollars, frequency, units, and so forth.
c. Collect data according to predetermined time interval.
d. Summarizes the data and ranks the categories of data from the largest to the smallest.
e. Calculates the cumulative frequency or cumulative percentage used.

Draw a bar chart showing the relative importance of each issue. Identify some things that are important to get attention.

In Table 4 we can see seven risk agents that included in the priority category in accordance with the Pareto calculations.

Table 4. Grouping of Priority Risk Category

Risk Agent	Ranking	ARP	ARP Cumulative	% ARP	% ARP Cumulative	Category
A7	1	172	172	8.62	8.62	Priority
A25	2	146	318	7.31	15.93	
A4	3	117	435	5.86	21.79	
A1	4	102	537	5.11	26.90	
A11	5	99	636	4.96	31.86	
A29	6	94	730	4.71	36.57	
A12	7	88	818	4.41	40.98	

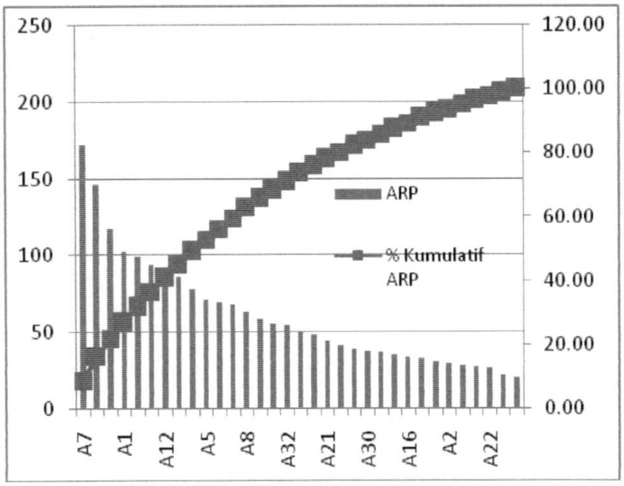

Figure 1. Pareto Diagram of Risk Agent

In Figure 1 it can be seen that getting to the right, the ARP value of a risk agent is getting smaller and the % cumulative ARP is getting bigger. In Figure 1 it is also known which risks are included in the priority category, ie with an allusion between the ARP value and its % cumulative ARP. It is known in Figure 1 that the priority risk agent ends at a risk agent with code A12. Subsequently, the risk agent is incorporated into the grouping of non-priority risk agents.

4. Conclusion

Based on the results of the analysis conducted, can be drawn some conclusions as follows :

1. The critical path on project scheduling consist of activities Mobility and Survey - ROW Clearing - Load and Unload Pipes - Stringing - Line Up Welding - NDT - FJC - Holiday Test - Lowering - Backfill - Pigging - Hydrotest - Aerial Marker, Warning Sign, Marker Sign - Re-Instatement with a duration of completion of 152 days.
2. There are seven category fall into the priority agency of 33 identified risk agents. The seven risk agents are limited to human resources, miscommunication between project executor and owner, seasonal change factor, sudden material damage, non compliance with employee safety standards, transport / transportation equipment is damaged, employee accidents is occur. The category of priority risk agents indicates that these seven risk agents account for 40.98 percent of all causes of the risk occurrence experienced in accordance with the results obtained from the Pareto calculations.

5. References

[1] Santosa, B. 2003. *Manajemen Proyek, Edisi Pertama.* Surabaya : Guna Widya

[2] Husen, A. 2009. *Manajemen Proyek : Perencanaan, Penjadwalan, dan Pengendalian Proyek, Edisi Revisi.* Yogyakarta : Andi

[3] Dipohusodo, I. 1995. *Manajemen Proyek & Konstruksi Jilid I.* Yogyakarta : Kanisius

[4] Soeharto, I. 1995. *Manajemen Proyek.* Jakarta : Erlangga.

[5] Clough, R. H. dan Scars, G. A. 1991. *Construction Project Management.* Canada : John Willey & Sons Inc.

[6] Taha, H. A. 1996. *Riset Operasi Edisi Kelima.* Jakarta : Penerbit Binarupa Aksara

[7] Kerzner, H. 2000. *Project Management : A System Approach to Planning, Schedulling, and Controlling Seventh Edition.* New Jersey : John and Wiley Inc.

[8] Harsanto, B. 2011. "Manajemen Proyek Menggunakan Ms. Project 2010". Makalah Pada

Pelatihan Manajemen Proyek 25-29 Juli 2011. Jakarta.
[9] Erizal. 2007. "Manajemen Konstruksi : Pengenalan Ms. Project". Diktat Departemen Teknik Sipil dan Lingkungan. Institut Pertanian Bogor.
[10] Muhlbauer, W. K. 2004. *Pipeline Risk Management Manual 3rd Edition*. Canada : Gulf Professional Publishing.
[11] Scott, B. dan Bringham, E. F. 2000. *Essentials of Managerial Finance Twelfth Edition.* Orlando : Harcourt Inc.
[12] AS/NZS 4360. 2004. *Australian/New Zealand Standard Risk Management*. Standard Association of Australia, New South Wales.
[13] Pujawan, I.N. dan Geraldin, L.H. 2009. "Supply Chain Risk Management (SCRM) : A Case Study on the Automotive and Electronic Industries in Brazil". *Supply Chain Management : An International Journal,* Vol. 14, No. 4, p 247-252.
[14] Nasution, M. N. 2004. *Manajemen Mutu Terpadu Total Quality Management.* Bogor : Graha Indonesia.
[15] Mitra, A. 1993. *Fundamentals of Quality Control and Improvement.* Singapore : Mac Millan Publishing Co.
[16] Besterfield, D. H. 1998. *Quality Control*. New Jersey : Prentice-Hall Inc.

Numerical Analysis of Residual Stress and Distortion Use Finite Element Method on Inner Bottom Construction of Geomarin IV Survey Ship with Welding Sequence Variations

N Syahroni [1,a], A B W Hartono [1], and M Murtedjo[1]

[1]Ocean Engineering Department, Faculty of Marine Technology, Sepuluh Nopember Institute of Technology, Indonesia

[a]Email : nur.syahroni@gmail.com

Abstract. In the ship fabrication industry, welding is the most critical stage. If the quality of welding on ship fabrication is not good, then it will affect the strength and overall appearance of the structure. One of the factors that affect the quality of welding is residual stress and distortion. In this research welding simulation is performed on the inner bottom construction of Geomarin IV Ship Survey using shell element and has variation to welding sequence. In this study, welding simulations produced peak temperatures at 2490 K at variation 4. While the lowest peak temperature was produced by variation 2 with a temperature of 2339 K. After welding simulation, it continued simulating residual stresses and distortion. The smallest maximum tensile residual stress found in the inner bottom construction is 375.23 MPa, and the maximum tensile pressure is -20.18 MPa. The residual stress is obtained from variation 3. The distortion occurring in the inner bottom construction for X=720 mm is 4.2 mm and for X=-720 mm, the distortion is 4.92 mm. The distortion is obtained from the variation 3. Near the welding area, distortion value reaches its minimum point. This is because the stiffeners in the form of frames serves as anchoring.

Keywords: Finite Element Method, Inner Bottom, Frames, Residual Stress, Distortion

1. Introduction

The fabrication stage of the survey vessel basically has the same principle with other commercial vessels, as in the assembly stage to the build stage. However, in this study, one part of the survey ship block, the inner bottom, receives special attention, especially the welding process. Welding is the process of combining two or more metals based on the principle of diffusion process that has the advantage of being able to make lightweight construction, can withstand high strength, easy in process, and economical [1].

In the construction of a ship, welding has a percentage of about 35% of the total activity. During the welding process, heat transfer occurs along the weld area allowing uneven distribution of heat leading to thermal development. In areas that have cooled down, a stretch barrier arises, which is capable of causing a shape change or is often referred to as residual stress [2]. During heating and cooling cycles, there will be a heat strain between the weld metal and the base metal in the area close to the weld bead, the heat strain is capable of triggering the buckling, bending, and rotation called distortion [3].

The presence of residual stress and distortion can reduce load capacity which contributes to the occurrence of brittle fracture and surface shape change [4]. Distortion can reduce bending strength of the structure to affect reliability [5].

In order to know the magnitude of residual stress and distortion, various experiments and simulations have been performed even with simple objects. In this Study will be implemented simulation to know the amount of residual stress and distortion in more complex construction. The study was performed using finite element method with the help of ANSYS Workbench 16.0 software to obtain residual stress and distortion stress on the construction of Geomarin IV inner bottom vessel that varied against the welding sequences.

2. Research Methodology

The systematic work of this research has been described as follows:

2.1 Data Collection

Datas collected to support the research needs include:

- Block 1.6 drawing of Geomarin IV survey ship.
- Welding Procedure Specification for Submerged Arc Welding (SAW).

2.2 Geometry Modelling

The modeling in this research is 2D and 3D geometry of inner bottom construction along with frames located in block 1.6. The modeling was carried out using AutoCAD 2016 software support. Block 1.6 has 18 frames with solid floor type, open floor, and watertight floor. However, in this study, the number of frames that will be analyzed only 3, due to the ability of supporting computers that do not meet, and due to the size of the construction is large enough. Three frames that are used as the object of analysis are frames 74, 75, and 76. Frame 74 is solid floor, frame 75 is watertight floor, and frame 76 is open floor. The geometry modeling results are listed in Figures 1 to 3.

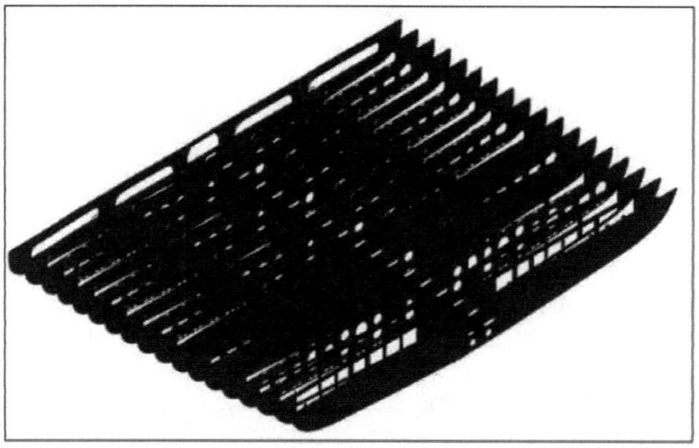

Figure 1. Frames Modelling

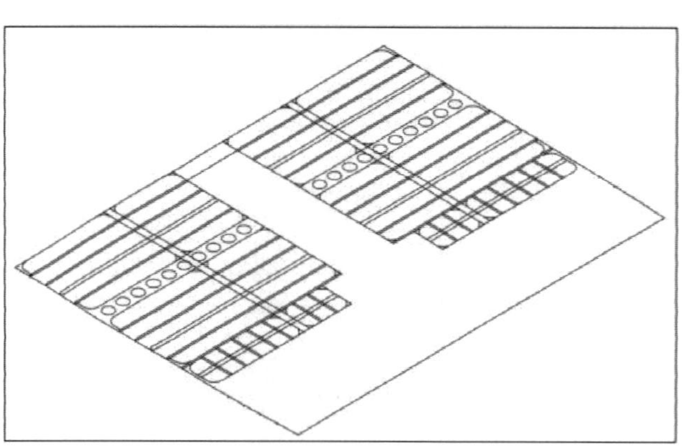

Figure 2. Inner Bottom Modelling

2.3 Geometry Model Input

In this study, the geometry model is modeled using shell elements. This is because the shell element is faster in helping to obtain the desired solution [6]. Geometries that have been modeled in AutoCAD 2016 must be saved with an .iges extension to be readable by ANSYS Workbench 16.0. Geometry modeling that must be entered is a 2D (area) geometry because the element used is shell. Figure 4. shows the geometry modeling that has been inserted into ANSYS.

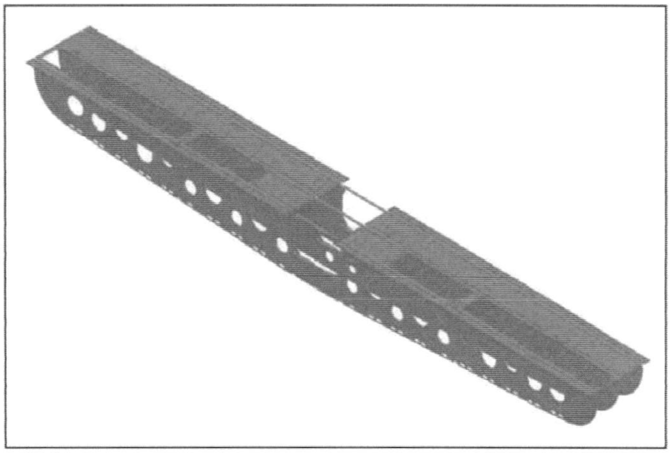

Figure 3. Frames for Research Objects

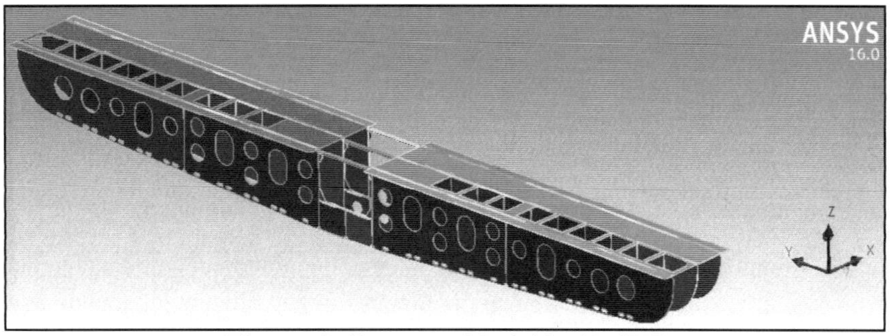

Figure 4. Geometry Model in ANSYS

2.4 Transient Thermal Analysis

Thermal analysis is a basic analysis in the simulation of the welding process to obtain heat distribution. In this study, the type of welding used is Submerged Arc Welding (SAW) with a 2-sided fillet welded T-joint connection type. The welding data used is 575 A electric current, 34 V voltage, and welding speed of 9.31 mm/s. Electrode used is EM13K type. The analyzed construction is composed of ASTM A131 Grade AH36 steel. Based on these parameters, the heat flux used is 179.9 x 106 W m^2.

2.5 Transient Structural Analysis

After transient thermal analysis, the next analysis is transient structural analysis. The purpose of this analysis is to determine residual stress and distortion that occurs in the object of research. At this stage, construction is given boundary condition (fixed boundary condition) in the form of fixed supports (pedestal flops). The pinch is placed on the ends of the research object along with the center that is not covered by the inner bottom. AH36 steel has a yield stress value of 350 MPa.

IOP Conf. Series: Earth and Environmental Science **135** (2018) 012027 doi:10.1088/1755-1315/135/1/012027

3. Analysis and Discussion

3.1 Partition of Load Steps

The welding simulation process is numerically performed by dividing the welding area into several hot loading areas. Figures 5 to 7 show the length of the welding area in each frame, and Table 1 shows the details of the welding area division.

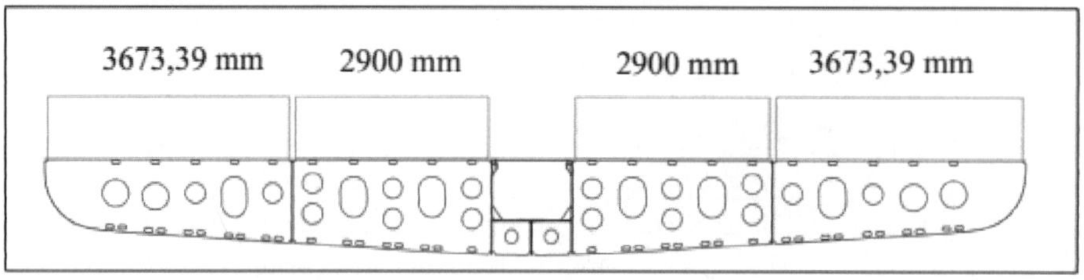

Figure 5. Welding Area Length of Frame 74

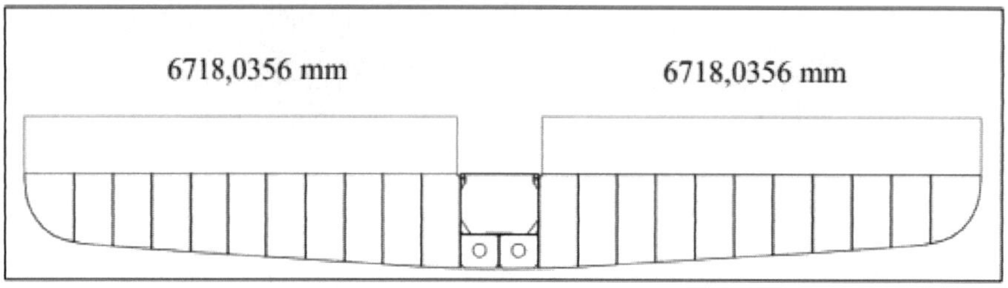

Figure 6. Welding Area Length of Frame 75

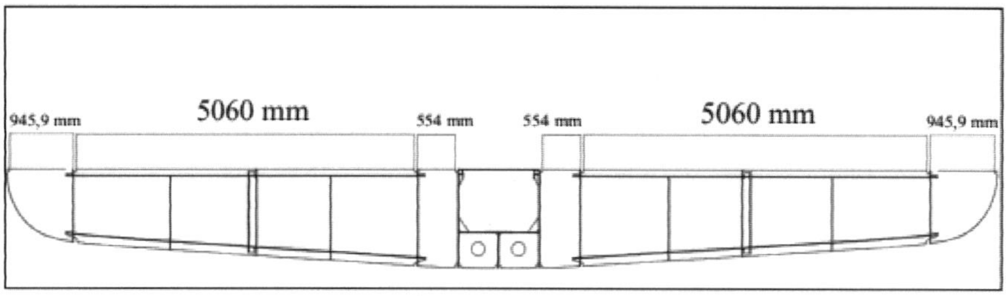

Figure 7. Welding Area Length of Frame 76

Table 1. Partition of Load Steps

Frame	Welding Area Length (mm)	Partition of Load Steps (mm)	Size of 1 Element (mm)
74	3673.39	60	61.22
	2900	45	64.4
75	6718.04	105	63.98
76	945.9	15	63.06
	5060	80	63.25
	554	10	55.4

According those partition, one frame has 210 loadsteps, and all of frames which researched have 630 loadsteps.

3.2 Heat Flux Load Calculation

The formulation which used in the determination of heat flux value is:

$$q_{el} = q_l \frac{A_l}{A_f} \quad (1)$$

where,
q_{el} : Heat flux of element (W/mm^2)
q_l : Heat flux of electrode (W/mm2)
A_l : Area width under heat load (mm^2)
A_f : Loading area width from welding process (mm^2)

3.3 Meshing Sensitivity

Meshing sensitivity has a function to know the number of elements change in numerical analysis. The smaller the divisor element, the smaller the completion value of the analysis. But it affects the duration of completion of the analysis, as it gets longer in giving results. Elements used in this study is a shell element, because it has a running time that tends to be faster than solid elements. Figure 8 illustrates meshing sensitivity in this study.

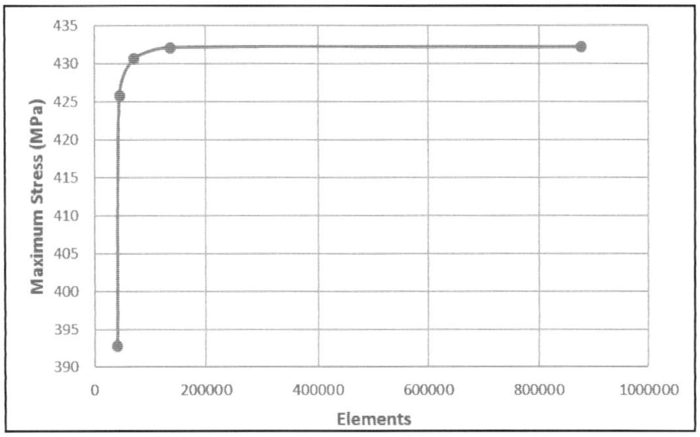

Figure 8. Meshing Sensitivity Graphic

Based on the meshing sensitivity that has been done, the type of elements used in this study is 0.2 mm quadrilateral and 0.5 mm triangle. Meshing results can be seen in Figure 9.

Figure 9. Meshing with 0.2 mm Quadrilateral and 0.5 mm Triangle

3.4. Heat Distribution

The heat distribution validation was performed using research result owned by Chen [7], Wibowo [5], and Wibisono [6]. Analysis on ANSYS with transient thermal analysis type is required to obtain heat distribution.

At this stage, variations in the welding sequence are performed to improve efficiency during the fabrication process. Figure 10 show variations in this research.

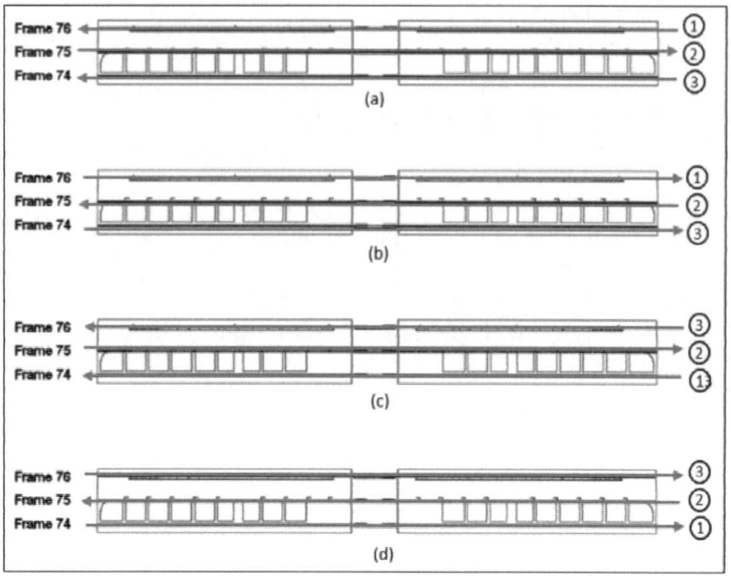

Figure 10. Welding Sequence Variations

Table 2. Explanation of Welding Sequence Variations

Variation	*Frame 76*	*Frame 75*	*Frame 74*
1	*Starboard → Portside*	*Portside → Starboard*	*Starboard → Portside*
2	*Portside → Starboard*	*Starboard → Portside*	*Portside → Starboard*
	Frame 74	*Frame 75*	*Frame 76*
3	*Starboard → Portside*	*Portside → Starboard*	*Starboard → Portside*
4	*Portside → Starboard*	*Starboard → Portside*	*Portside → Starboard*

After the analysis, heat distribution graphs have shown in Figure 11 as sample. Based on the results, peak temperature each welding sequence variations have shown in Table 3.

Table 3. Peak Temperature Each Variations

Variation	Peak Temp. (K)
1	2339,1
2	2339
3	2483,4
4	2490

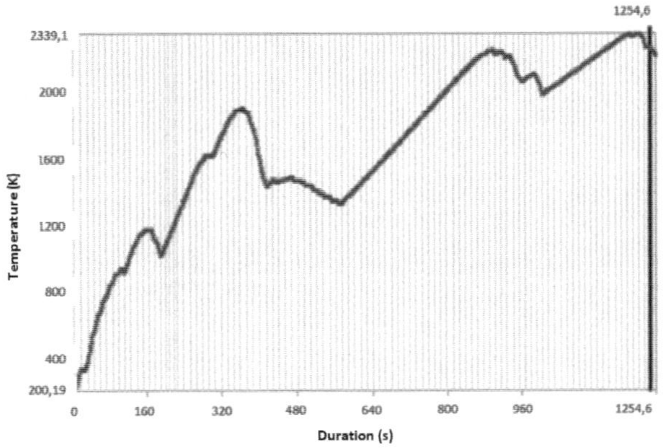

Figure 11. Heat Distribution Graph Variation 1

Figure 12 below shows final condition of heat distribution variation 1 as sample.

Figure 12. Heat Distribution Variation 1

Peak temperature in each variations have passed 1800 K of material's melting point

3.5 Residual Stress

The residual stress analyzed in this research is longitudinal residual stress. The illustration of the longitudinal residual stress illustrated by a pathline has been shown in Figure 13.

Figure 13. Residual Stress Analysis' Pathline

IOP Conf. Series: Earth and Environmental Science **135** (2018) 012027 doi:10.1088/1755-1315/135/1/012027

Using 4 variations of the predetermined welding sequence, the residual stress analysis results have shown in the graphs in Figure 14 to 17 below.

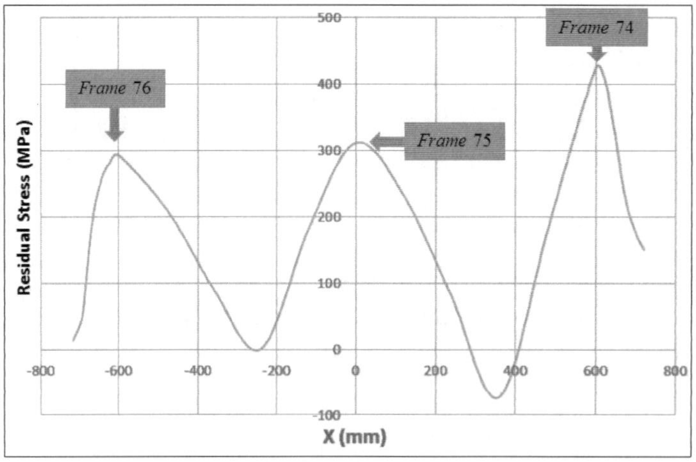

Figure 14. Graph of Residual Stress Variation 1

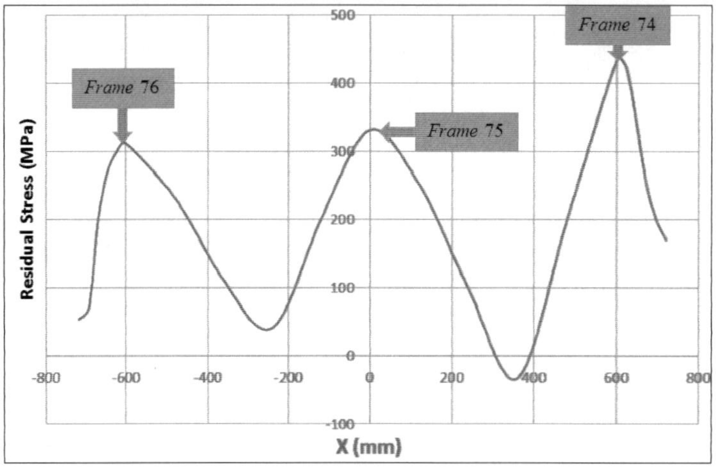

Figure 15. Graph of Residual Stress Variation 2

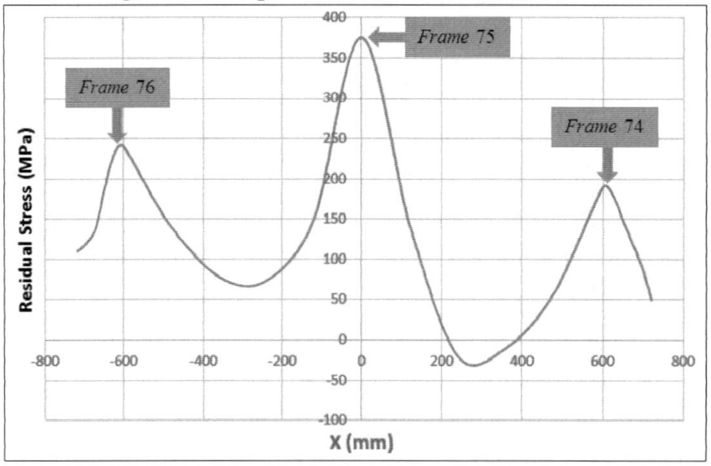

Figure 16. Graph of Residual Stress Variation 3

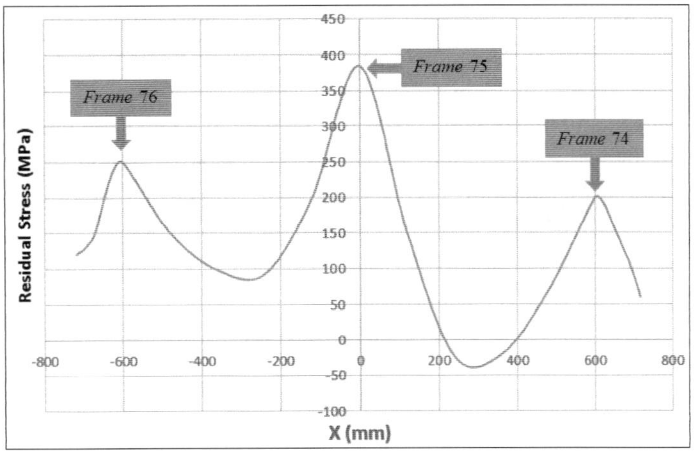

Figure 17. Graph of Residual Stress Variation 4

Based on analysis, 3rd variation is a welding sequence variation which has minimum tensile residual stress and minimum compressive residual stress. Maximum tensile residual stress in 3rd variation is 375.23 MPa, and maximum compressive residual stress is -20.18 MPa. Table 4 gives the summary of residual stresses in all variation.

Table 4. Summary of Residual Stress All Variations

Variation	**Max. Tensile Residual Stress (MPa)**	**Max. Compressive Residual Stress (MPa)**
1	424,81	-72,97
2	434,73	-33,76
3	375,23	-20,18
4	384,85	-25,22

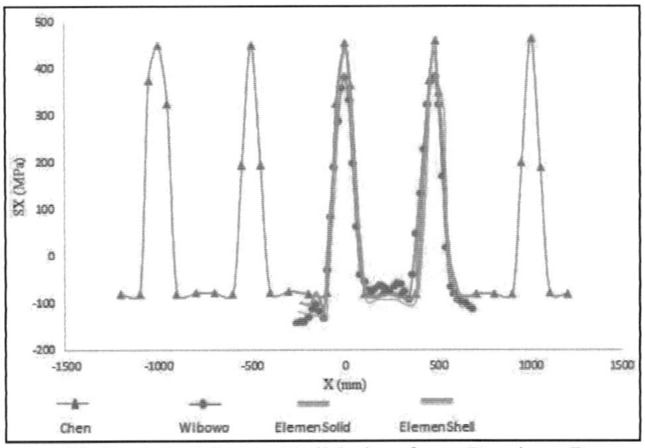

Figure 18. Residual Stress Validation from Previous Research

From the graph shown in Figure 18 it is clear that the graph of the longitudinal residual stress distribution of this study has the same tendency traits. This property has a maximum residual stress value in the region close to the welding area, and the minimum value is in areas far from the welding area. The illustration of the residual stress yield of all variations in the welding sequence has been shown in Figure 19 to 22.

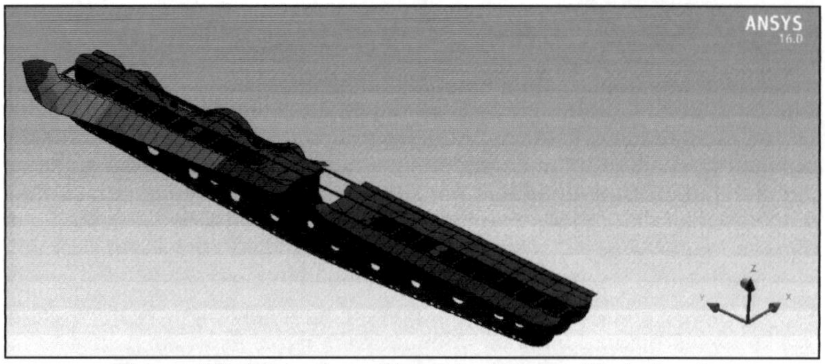

Figure 19. Illusration of Residual Stress Variation 1

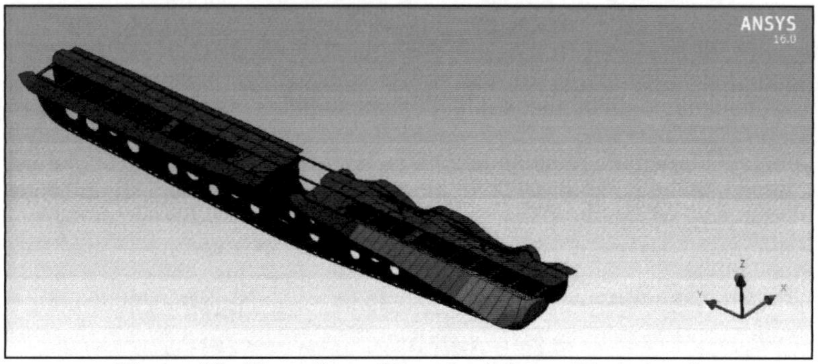

Figure 20. Illusration of Residual Stress Variation 2

Figure 21. Illusration of Residual Stress Variation 3

Figure 22. Illusration of Residual Stress Variation 4

3.6 Distortion

The results of distortion in this study were conducted on the Z axis in millimeters. In the pathline 1 as in Figure 13, the length of the distortion occurrence begins at the end of the frame 76 to frame 74, or in the direction of the X axis. Figures 23 to 26 have shown graphic distortion after weld simulation in all variations of the welding sequence.

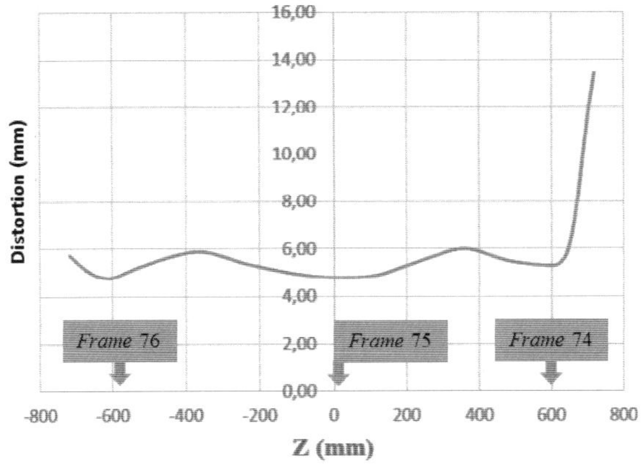

Figure 23. Graph of Distortion Variation 1

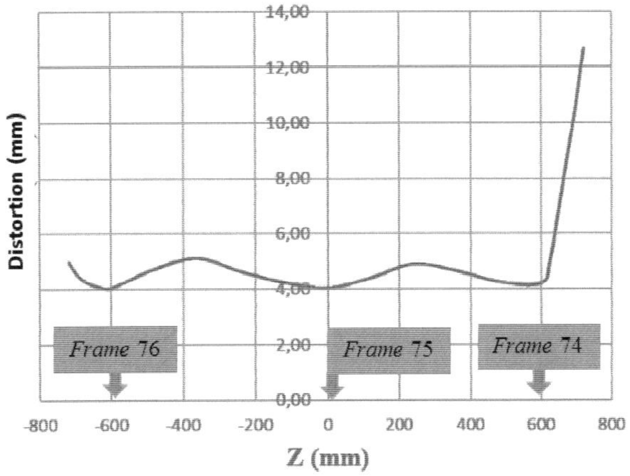

Figure 24. Graph of Distortion Variation 2

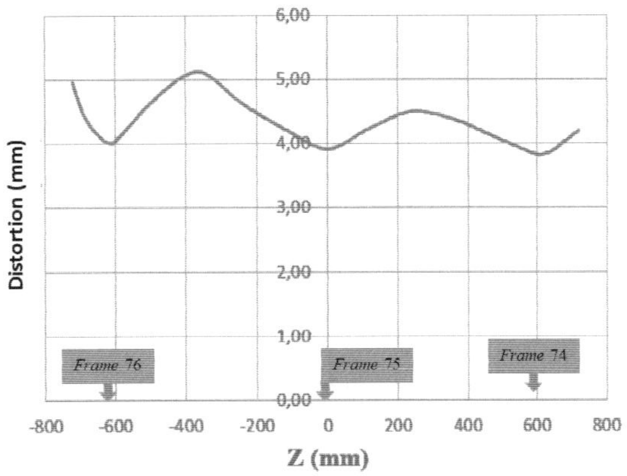

Figure 25. Graph of Distortion Variation 3

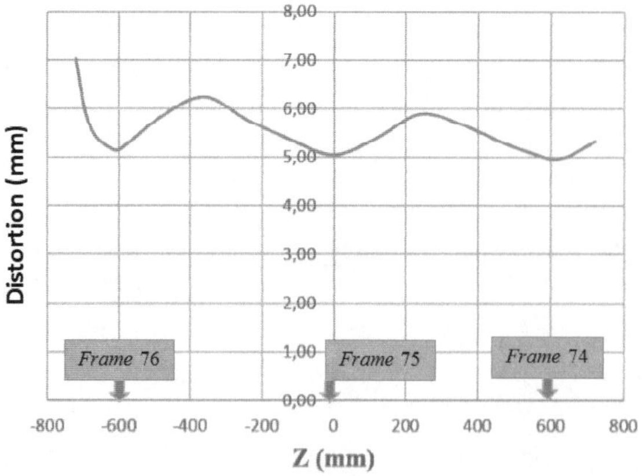

Figure 26. Graph of Distrotion Variation 4

From the graph shown in Figure 23 to 26, it is seen that the maximum distortion value occurs at the free end of the plate (free edge). This is because of the stiffener that serves as a barrier so that no distortion occurs in the area, in this study stiffener is frames. The results of this study, have also been validated by previous studies that show similar things, namely maximum distortion occurs in areas far from welding areas.

Based on the analysis that has been done, it is known that variation 3 has the minimum distortion value, while variation 1 has the greatest distortion value. Table 5. shows the distortion results of all the variations in the welding sequence.

Table 5. Summary of Distortion All Variations

Variation	UZ			
	The Inner Bottom Edge Near Frame 76	Location (X = -720 mm)	The Inner Bottom Edge Near Frame 74	Location (X = 720 mm)
1	5,74	*Portside Frame 76*	13,45	*Portside Frame 74*
2	4,97	*Starboard Frame 76*	12,68	*Starboard Frame 76*
3	4,92	*Portside Frame 76*	4,2	*Portside Frame 74*
4	7,03	*Starboard Frame 76*	5,32	*Starboard Frame 76*

Illustration of distortion in all variations have shown in Figure 27 to 30.

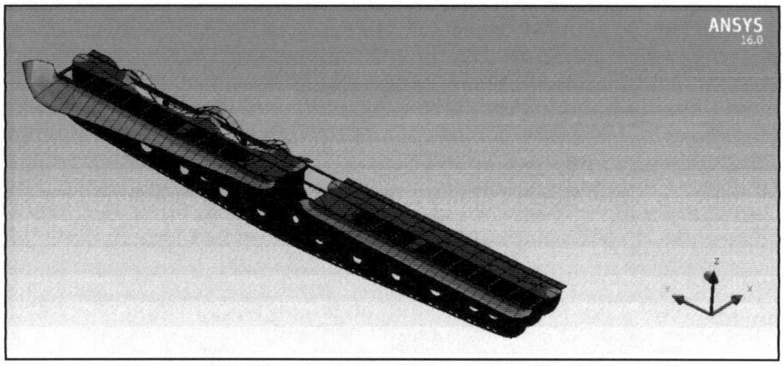

Figure 27. Illustration of Distortion Variation 1

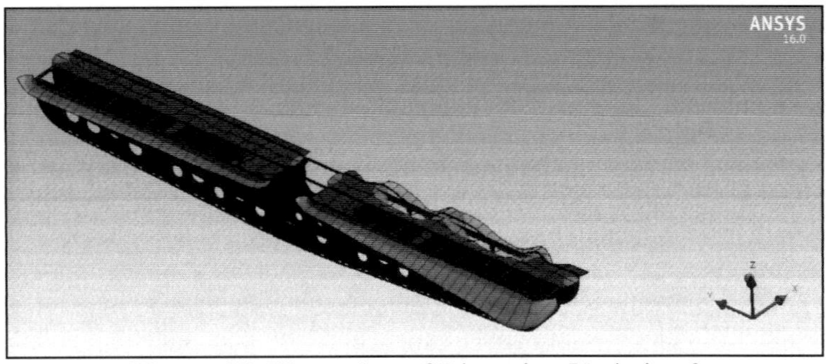

Figure 28. Illustration of Distortion Variation 2

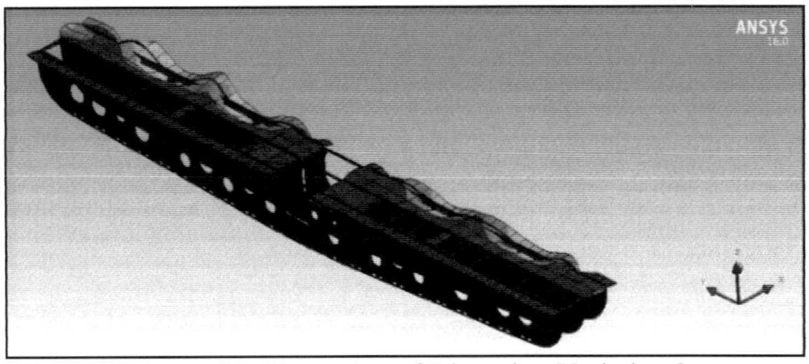

Figure 29. Illustration of Distortion Variation 3

Figure 30. Illuastration of Distortion Variation 4

4. Conclusion

The following is the conclusion that can be summarized from this research.

a. Heat distribution with the highest peak temperature occurs during welding simulation with the fourth variation with a temperature of 2490 K at the 808.7 second or about 13.5 minutes. While the lowest peak temperature occurs during the welding simulation with a second variation of 2339 K at seconds to the 1243.7 or about 20.7 minutes. This peak temperature has exceeded the material melting point limit of 1800 K.

b. In each variation of the welding sequence, the maximum residual stress with the smallest value is generated by the third welding variation sequence. The maximum tensile residual stress of the third variation is 375.23 MPa, and the maximum compressive stress is -20.18 MPa. Thus, the variation of the welding sequence that can be used to improve the efficiency that is started from the starboard frame 74 to portside frame 76.

c. The distortion with the smallest value that arises after the running of the structural analysis is obtained from the third welding sequence variation as well. The distortion results are reviewed at

X = 720 mm and X = -720 mm. The result of distortion due to variations in the third welding sequence has a large 4.92 mm at X = -720 mm, and 4.2 at X = 720 mm. Thus, the variation of the welding sequence that can be used to improve the efficiency that is started from the starboard frame 74 to portside frame 76.

d. Based on thermal transient analysis and structural transient analysis, the third welding variation variation has advantages when compared to other variations. Because, the third variation has a welding sequence that leads to inner bottom construction without holes, so that the residual stress distribution and distortion can be spread evenly. As for the welding variations approaching the perforated inner bottom construction, the residual stress distribution and distortion are less evenly distributed, and are concentrated only on certain areas with high residual stress and deformation values.

References

[1] Djamiko, R. D. 2008. Modul Teori Pengelasan Logam. Yogyakarta: Jurusan Pendidikan Teknik Mesin, Universitas Negeri Yogyakarta.

[2] Magga, Ramang. 2009. "Pengaruh Pembentukan Tegangan Sisa pada Hasil Pengelasan", JIMT. Vol. 6: 54-62.

[3] Putra, Y.A., S. 2007. Analisa Tegangan Sisa dan Distorsi pada Pengelasan Fillet T-Joint dengan Metode Elemen Hingga. Tugas Akhir. Surabaya: Jurusan Teknik Material dan Metalurgi, Institut Teknologi Sepuluh Nopember.

[4] Zaroog, O. S. 2014. "Current and Challenge of Residual Stress Measurement Technique", International Journal of Science and Research. Vol. 3: 210-216.

[5] Wibowo, B. A. 2016. Analisa Numerik Tegangan Sisa dan Distorsi pada Struktur Plat Berpenegar Jamak dengan Variasi Urutan Pengelasan. Tugas Akhir. Surabaya: Jurusan Teknik Kelautan, Institut Teknologi Sepuluh Nopember.

[6] Wibisono, B. O. B. 2017. Analisa Tegangan Sisa pada Panel Plat dengan Metode Elemen Hingga Menggunakan Jenis Elemen Solid dan Elemen Shell. Tugas Akhir. Surabaya: Jurusan Teknik Kelautan, Institut Teknologi Sepuluh Nopember.

[7] Shen, C. 2013. Low Distortion Welding for Shipbuilding Industry. Thesis Master. Wollongong: School of Mechanical, Materials, & Mechatronic Engineering, University of Wollongong.